INTRODUCTORY QUANTUM CHEMISTRY

HARPER'S CHEMISTRY SERIES
Under the Editorship of **Stuart Alan Rice**

INTRODUCTORY QUANTUM CHEMISTRY

S. R. LA PAGLIA
Georgetown University

HARPER & ROW, PUBLISHERS
New York, Evanston, and London

To my mother

INTRODUCTORY QUANTUM CHEMISTRY
Copyright © 1971 by **S. R. La Paglia**
Printed in the United States of America. All rights reserved. No part of this book may be used or reproduced in any manner whatsoever without written permission except in the case of brief quotations embodied in critical articles and reviews. For information address Harper & Row, Publishers, Inc., 49 East 33rd Street, New York, N.Y. 10016.
Library of Congress Catalog Card Number: 78-129478

CONTENTS

A Table of Fundamental Physical Constants — ix
A Conversion Table for Energy Units — ix
The Greek Alphabet — x
Common Abbreviations — x

PREFACE — xi

1 THE CHEMICAL BOND — 1

2 DEVELOPMENT OF THE WAVE THEORY OF MATTER — 11
    2.1 Conservation Laws and Classical Mechanics — 13
    2.2 Atomic Physics — 18
    2.3 Matter Waves — 32
    2.4 The Schrödinger Equation — 36
    2.5 Eigenvalues, Eigenfunctions, and Operators — 40
    2.6 The Variational Principle — 46
    2.7 The Perturbation Approach — 49

3 THE PRINCIPLES OF QUANTUM MECHANICS WITH APPLICATIONS — 53
    3.1 Probability and the Wavefunction — 55
    3.2 Particle in a One-Dimensional Box — 56
    3.3 Eigenvalues and Operators — 59
    3.4 Average Values — 62
    3.5 Electron Spin, Spin-Orbitals, and the Pauli Principle — 64
    3.6 Many Particles in a Box — 67
    3.7 Free-Electron Molecular Orbital Method for Conjugated Molecules — 71

	3.8	Localized Orbitals	73
	3.9	Particle in a Two-Dimensional Box: Degeneracy	75

4 ATOMS AND THE PERIODIC SYSTEM — 79

	4.1	The Hydrogen Atom	81
	4.2	The Helium Atom	95
	4.3	Many-Electron Atoms: Hydrogenic Model	99
	4.4	The Electronic Charge Density	101
	4.5	The Periodic System of the Elements	102
	4.6	Angular Momentum	107
	4.7	The Spectroscopic States of Atoms: Vector Coupling	109
	4.8	Energy Level Diagrams and Spectroscopic Transitions	117
	4.9	Screening, Slater Orbitals, Ionization Potentials, and Atomic Radius	119
	4.10	The Hartree-Fock Method	122
	4.11	Electron Correlation	127

5 THE DIATOMIC MOLECULE — 133

Molecular Methods — 135

	5.1	H_2^+	135
	5.2	The MO as Simple LCAO	137
	5.3	Observable Features of the Chemical Bond	143
	5.4	H_2 and Other Small Species	146
		The General LCAO–MO Method	150
	5.5	Overlapping Atomic Orbitals: Symmetry	150
	5.6	The Secular Equation	155
	5.7	The Hartree-Fock-Roothaan Procedure	156
	5.8	Angular Momentum	160
	5.9	Linear Transformations of the Wavefunction	161

The Diatomic Molecule — 162

	5.10	Building up the Homonuclear Diatomic Molecule	162
	5.11	Second-Row and Other Homonuclear Diatomic Molecules	167
	5.12	Homonuclear Diatomic Charge Distributions	169
	5.13	The Heteronuclear Diatomic Molecule	174
	5.14	Localized Orbitals	181
	5.15	Beyond the Hartree-Fock Limit	186
	5.16	Valence Bond Theory	189

6 MOLECULAR SPECTRA AND MOLECULAR SYMMETRY — 195

Molecular Spectroscopy — 197
- 6.1 Vibration–Rotation Spectra — 199
- 6.2 Electronic Spectra — 210

Molecular Symmetry — 215
- 6.3 Symmetry Elements of Molecules — 215
- 6.4 Group Theory — 219
- 6.5 Group Representations — 222
- 6.6 Group Theory in Quantum Mechanics — 226
- 6.7 Vibrational Spectra of Polyatomic Molecules — 231
- 6.8 Optical Spectra and Selection Rules — 235

Magnetic Resonance Spectra — 238
- 6.9 Introduction to Magnetic Resonance Spectroscopy — 238
- 6.10 Magnetic Resonance in Hydrogen and Helium — 244
- 6.11 The Chemical Shift and Spin–Spin Coupling in NMR — 247

7 POLYATOMIC MOLECULES — 257
- 7.1 The Molecular Schrödinger Equation — 259
- 7.2 Linear Molecules — 260
- 7.3 Trigonal Planar Molecules — 269
- 7.4 Tetrahedral Molecules — 272
- 7.5 Angular Triatomic and Trigonal Pyramidal Molecules — 276
- 7.6 Molecular Shape—The Bond Concept and Molecular Properties — 282
- 7.7 Formaldehyde Structure and Spectra—Population Analysis — 287

8 π-ELECTRON THEORY — 295
- 8.1 The π-Electron Approximation — 297
- 8.2 The Hückel Molecular Orbital Method — 301
- 8.3 Two Applications of Symmetry — 314
- 8.4 Extensions of the Hückel Method — 319

9 LIGAND FIELD THEORY — 323
- 9.1 Crystal Field Theory — 327
- 9.2 Parametrization of the Crystal Field — 331
- 9.3 Δ, The Crystal Field Parameter — 334
- 9.4 Weak Field Complexes — 337
- 9.5 Strong Field Complexes — 339
- 9.6 Electronic Spectra — 342
- 9.7 The Molecular Orbital Theory of the Octahedral Complex — 345

APPENDIXES   **351**

    A. Mathematical Tools   353
    A.1 Operators, Functions, and Coordinate Systems   355
    A.2 Vectors, Matrices, Determinants, and
        Simultaneous Equations   360
    B. Evaluation of Coulomb Integral J_{1s1s}
    C. The Hartree-Fock Equation for the Helium Atom and
        Other Closed-Shell Systems   371
    D. The Born-Oppenheimer Approximation   375
    E. The Hellmann-Feynman Theorem   379
    F. Electronegativities and Covalent Radii   383
    G. Molecular Symmetry   389
    H. The Probability of Absorption and Emission of Radiation   393

INDEX   **399**

A Table of Fundamental Physical Constants[a]

Electron mass	m_e	9.10908×10^{-28}	gm
Proton mass	m_p	1.67252×10^{-24}	gm
Electron charge	e	4.80298×10^{-10}	esu
Bohr radius	a_0	0.529167×10^{-8}	cm
Velocity of light in vacuum	c	2.997925×10^{10}	cm sec^{-1}
Planck's constant	h	6.62559×10^{-27}	erg sec
	$\hbar$	1.05449×10^{-27}	erg sec
Rydberg constant	R	1.097373×10^{5}	cm^{-1}
Boltzmann constant	k	1.3805×10^{-16}	erg deg^{-1}

[a] From the fundamental constants as given by E. R. Cohen and J. W. M. DuMond, *Rev. Mod. Phys.* **37**, 537 (1965); also see B. N. Taylor, W. H. Parker, and D. N. Langenberg, *Rev. Mod. Phys.* **41**, 375 (1969).

A Conversion Table for Energy Units[a]

	hartree	eV	cm^{-1}	kcal/mole	ergs
1 hartree =	1	27.210	2.1947×10^{5}	6.2750×10^{2}	4.3594×10^{-11}
1 eV =	3.6750×10^{-2}	1	8065.7	23.061	1.6021×10^{-12}
1 cm^{-1} =	4.5563×10^{-6}	1.2398×10^{-4}	1	2.8591×10^{-3}	1.9863×10^{-16}
1 kcal/mole =	1.5936×10^{-3}	4.3363×10^{-2}	3.4975×10^{2}	1	6.9472×10^{-14}
1 erg =	2.2938×10^{10}	6.2418×10^{11}	5.0344×10^{15}	1.4394×10^{13}	1

[a] From the physical constants given in the Table of Fundamental Physical Constants.

The Greek Alphabet

α	alpha	ν	nu
β	beta	ξ	xi
γ	gamma	o	omicron
δ	delta	π	pi
ϵ	epsilon	ρ	rho
ζ	zeta	σ	sigma
η	eta	τ	tau
θ	theta	υ	upsilon
ι	iota	ϕ	phi
κ	kappa	χ	chi
λ	lambda	ψ	psi
μ	mu	ω	omega

Common Abbreviations

MO	Molecular Orbital
LO	Localized Orbital
STO	Slater-Type Orbital
SCF	Self-Consistent Field
LCAO	Linear Combination of Atomic Orbitals
CI	Configuration Interaction
IP	Ionization Potential
VB	Valence Bond
BO	Bonding Orbital
BOA	Born-Oppenheimer Approximation
CFSE	Crystal Field Stabilization Energy

PREFACE

This text is designed to be a realistic and satisfying introduction to chemical bonding and the elementary methods of modern quantum chemistry. The most notable development in the teaching of chemistry in the last thirty years has been the increasing reliance on quantum theoretical explanations of chemical properties. Orbitals and symmetry are used very naturally in discussions of molecular structure and re-reaction mechanisms, interpretations which have increasing predictive power. Within the lifetime of today's student the quantum mechanical interpretation of chemical phenomena, and their computation via electronic devices which are ever more available and easier to use, will become so important and commonplace that it is difficult to define the degree of mastery and understanding the student will require. The present text tries to set a minimum level of comprehension and skill for the chemist, biologist, or physicist concerned with chemical bonding.

The treatment should be suitable for students with a variety of mathematical backgrounds. It is expected the student will have a year of university level calculus before embarking on the study of quantum phenomena. The author has found it fruitful to integrate the necessary additional mathematics into the logical development of the quantum theory. Thus, the last few sections of Chapter 2 constitute an exposition of the mathematical operations, notation, and vocabulary which are used in elementary quantum mechanics, short of matrix methods. Matrix methods are similarly integrated into Chapters 5, 6, and 8. At no time is the student presented with mathematics for its own sake; rather, the physical necessity behind the mathematics is emphasized. Finally, the entire mathematical presentation is gathered together in Appendix A for reference.

Chapter 1, which is a sketch of the development of the concept of chemical bonding, should provide the student with the rationale and motivation behind our study of quantum chemistry. It is difficult to know the level of preparation of the average student in classical physics and atomic physics, hence the inclusion of the most relevant topics from these fields in the first sections of Chapter 2. Then the Schrödinger equation and other eigenvalue equations are made plausible and the mathematical consequences studied under the general properties of Sturm-Liouville differential equations and their solutions. In this way the mathematical apparatus is established. Chapter 3 presents the basic principles of quantum mechanics in conjunction with the problem of the particle(s) in a box. The simplicity of this problem allows the instructor and the student to manipulate even many-particle wavefunctions in an initial study of eigenvalues, boundary conditions, average values, antisymmetrization and the Pauli Principle, orbital transformations, symmetry, degeneracy, electron spin and spin-orbitals, etc. A chemical application (the free-electron theory of conjugated molecules) serves to show the student the relation between the abstract theory, physical models, and real molecules.

The remainder of the text (except for Chapter 6) treats characteristic physical systems individually chapter by chapter. A strong effort has been made to keep the data as up to date as possibile, because of the rapid developments in quantum chemistry over the last few years. In Chapters 4, 5, and 7, which are concerned with atoms, diatomic molecules, and polyatomic molecules, respectively, the approach is to emphasize recent precise calculations and the chemically intuitive information obtainable from them, such as electronic density distributions, dipole moments, localized orbitals, and atomic populations. A constant theme is the complementarity of localized orbital and delocalized orbital representations of the electronic structure of molecules and the possibility of transformation between the two. The residual problem of electron correlation in these small systems is not ignored. In dealing with larger chemical systems in Chapters 8 and 9 (π-electron molecules and transition metal complexes) the successful semiempirical methods of quantum chemistry are developed in some detail.

Chapter 6 is concerned with the optical and magnetic spectra of molecules. After the treatment of diatomic spectroscopy, group theory is introduced. Although group theory is necessary for the discussion of selection rules and polyatomic vibrational spectra which follow, it is chiefly used in Chapter 7. The author realizes that group theory will probably be omitted from a one-semester course. As a result, it has been introduced rather late in the text and is not vital for a qualitative understanding of the later chapters.

Many examples and exercises have been inserted directly into the text of each chapter. These were designed to make the book suitable for self-study by a motivated student and as an aid to the instructor.

It is a pleasure to acknowledge the assistance afforded by my classes at Georgetown University in the preparation of this text, especially that of two students, Michael Marchetti and William Lamb, who made substantial contributions in reading an early draft. I also take this opportunity to thank the teachers of quantum chemistry who through their classes introduced me to their points of view: P.-O. Löwdin, R. S. Mulliken, R. Pauncz, C. A. Coulson, A. B. F. Duncan, F. Buff, and O. Sinanoğlu.

I also wish to acknowledge the generosity of the authors who have lent me the use of their figures.

S. R. LA PAGLIA

Washington, D.C.
October, 1970

I
THE CHEMICAL BOND

Historical Sketch of the Development of the Concepts of Chemical Bonding

"... We are perhaps not far removed
from the time when we shall be able
to submit the bulk of chemical phenomena
to calculation."

JOSEPH LOUIS GAY-LUSSAC
Mémoires de la Société d'Arcueil
2, 207 (1808)

G AY-LUSSAC'S hope of chemical calculation reflected the tremendous impact of Newtonian mechanics on eighteenth-century science and the hope engendered in the power of the rational mind. Being based on Newton's laws these hopes were premature, to say the least. The correct principles of chemical combination are manifestations of quantum mechanics (1926). It is only within the last 40 years that Gay-Lussac's hopes have *begun* to be realized.

Chemistry is based on the fact that *in spite of chemical transformations the elements are recoverable and are never transmuted into other elemental bodies by chemical change.* Dalton's atomic theory[1] (1808) explains this fundamental chemical fact and provides a formal basis for chemistry. Dalton's theory is contained in a few postulates.

1. Matter is composed of atoms which are indivisible.
2. Atoms of a given element are identical.
3. Atoms of different elements have different weights.
4. Atoms are indestructible and preserve their identity in all chemical transformations.

As evidence of modernity and of the power of these postulates, we mention that Dalton is known to have constructed molecular models from pins and balls, much as today's chemist does, and Wollaston[2] perceived that four identical atoms would tend to adopt a tetrahedral arrangement around a fifth for stable equilibrium. In effect, geometrical conjecture is a natural outgrowth of an atomic theory of chemistry.

Gay-Lussac's law of combining volumes (1808) was opposed by Dalton because it suggests that equal volumes of gases under the same conditions contain equal numbers of particles. Dalton could not reconcile this with his theory. For example, Dalton wrote the formula of water HO, which is not consistent with the observation that two volumes of hydrogen gas combine with one volume of oxygen and yet equal volumes contain the same number of particles. The difficulty was resolved by Avogadro[3] (1811)

---

[1] John Dalton, *A New System of Chemical Philosophy*, 1808.
[2] W. H. Wollaston, *Phil. Trans. Roy. Soc.* **98**, 96 (1808).
[3] Amadeo Avogadro, *J. Phys.* **73**, 58 (1811).

with his suggestion that the "particles" present in the elemental gases are diatoms, O_2, H_2, etc. This was no solution to Dalton, who had a strong intuitive feeling that only unlike atoms have a chemical affinity for each other. The existence of homopolar or covalent bonds, A—A, between like atoms presented great difficulties to theoretical understanding until the advent of quantum mechanics.

Berzelius was an early convert to Dalton's atomistic theory, but he readily accepted Gay-Lussac's laws of gaseous combination and urged Dalton to do the same. Unfortunately, Berzelius, who was to dominate nineteenth-century chemistry, rejected the Avogadro hypothesis. The result was that until 1860 chemistry was disfigured by the use of various incorrect sets of atomic weights, which led to incorrect molecular formulas and prevented a rational discussion of chemical bonding.

Influenced by the electrolytic experiments of Davy, who believed that chemical combination and decomposition were electrical phenomena, Berzelius put forth the first bonding theory of note. Berzelius[4] (1811) proposed that all chemical compounds are divisible into parts of opposite net electrical charges; this was a fruitful idea in much of nineteenth-century chemistry.

The explicit concept of valence seems to have grown with Frankland[5] during the years 1852–1866. He was at first loath to go against the Berzelius theory, but after Cannizzaro[6] (1860) had won acceptance of Avogadro's hypothesis and the correct atomic weights, Frankland began writing down correct chemical formulas using the word "bond." He says,[7] "By this term I do not intend to convey the idea of any material connection, the bonds actually holding together the atoms of a compound being probably much more like those which connect the members of the solar system." The chemists Couper and Brown suggested the notation for these bonds which survives to the present day. They represented the atoms as labeled circles with a number of radiating lines equal to the valence of the atom. The atoms are assembled into molecules by joining the lines. Skeptics objected that the formulas implied (without proof) a real physical arrangement of the atoms in the molecule.

Kekulé became the principal exponent of the concept of valence. A man of wide acquaintance, he organized the Congress of 1860 at which

---

[4] The "Dualistic Theory" of J. J. Berzelius, *La théorie des proportions chimiques et de l'influence de l'électricité dans la nature inorganique*, Paris, 1835. Originally published in Swedish in 1819.

[5] E. Frankland, *Experimental Researches in Pure, Applied and Physical Chemistry*, London, 1877.

[6] S. Cannizzaro, "An Outline of a Course of Chemical Philosophy," *Il Nuovo Cimento*, 1858.

[7] E. Frankland, *J. Chem. Soc.* **19**, 372 (1866).

Cannizzaro acted out his historic role by winning approval of the Avogadro hypothesis. In a series of lucid papers[8] (1858–1876), Kekulé built the foundations of carbon chemistry. He is best remembered for suggesting the ring structure of benzene; he noted that each pair of atoms in benzene is chemically equivalent and by way of explanation suggested that over a short period of time the atoms take on both valence arrangements

⌬ and ⌬

Kekulé[9] postulated a very rapid oscillation of the atoms about their equilibrium positions as the mechanism responsible for valence isomerization in benzene. The existence of the electron and its role in chemical bonding were totally unknown. Thiele's suggestion of a theory of partial valence was almost clairvoyant. He postulated[10] (1899) that neither atom saturates its valence in a double bond, but rather that each preserves a partial valence (dashed line).

$$C=C-C=C \longrightarrow C=C-C=C$$

With all valencies and partial valences saturated, benzene should be stable, but butadiene and such conjugated chains should add at the chain ends. These same observable results are obtained from the molecular orbital theory of the electronic structure of unsaturated molecules, but in a quantitative form (Chapter 8).

From Pasteur's work on optical activity, Le Bel[11] stated the basic principles of optical activity on the molecular level in terms of a postulated *tetrahedral carbon valency*. In that very year, 1874, van't Hoff suggested the same idea: Thus *directional valence* was ushered in by its codiscovers.

In 1887 Arrhenius[12] advanced the notion of electrolytic dissociation of certain molecules into *ions*, specifically, acids, bases, and salts (the electrolytes). The implication is that the molecules exist as stable entities A^+B^- before dissociation (ionic bond). The electrochemical bonding theory of

---

[8] R. Anschutz, *August Kekulé*, vols. I and II, Berlin, 1929.
[9] A. Kekulé, *Liebigs Ann.* **162**, 77 (1872).
[10] J. Thiele, *Liebigs Ann.* **306**, 125 (1899).
[11] Joseph-Achille Le Bel, *Bull. Soc. Chim.* **22**, 337 (1874).
[12] Svante Arrhenius, *Z. Phys. Chem.* **1**, 631 (1887).

Berzelius is thus reborn as the ionic bond and demonstrated by ionic equilibria.

The ideas of Alfred Werner made little headway against this reestablished ionic theory of inorganic bonding. Werner[13] originated the *coordination* bonding theory in 1893. He was concerned with the compounds of metals with various ligands, which others had tried to show had structural formulas like Cl—NH_3—NH_3—M—NH_3—NH_3—Cl, where M is the metal atom. Werner visualized the ligands as directly bonded to the metal; he defined the *coordination number* as "the maximum number of atoms or groups with which a given atom can combine directly." He suggested an octahedral arrangement for coordination number 6, that is, $M(NH_3)_4Cl_2$ for the cited example, which is now established.

In the early twentieth century the increasing atom-mindedness of physics, brought about by the discovery of the electron and the nucleus, led to a coherent if inexact view of atomic structure. At the time of World War I (1914) the following had been established:

1. The atomic number Z of the element is the charge in units of the proton on the nucleus; for neutrality the atom contains Z electrons.
2. The electrons are arranged in shells with a periodic structure related to the Mendeleev table, the outermost electrons being the least bound.
3. There exist at certain places in the periodic table atoms of great chemical stability.

Kossel[14] conceived the valence of the atom to be an aspect of the behavior of the outermost electrons in the shell structure. Kossel noted that atoms tend to lose or gain electrons until they achieve the same number as an inert gas. The resulting chemical affinity he ascribed to the Coulomb attraction between the ions. Kossel noted that eight outer electrons constituted the stable "full ring of electrons." However, this explains neither the covalent bond (homopolar bond of H_2, N_2, etc.) nor directional valency. Although Kossel did employ models of molecules (like N_2) with rings of electrons common to both atoms, G. N. Lewis[15] is generally credited with the theory of the *shared pair* of electrons (later elaborated by Langmuir) which is so familiar to most students of chemistry. The idea of sharing, rather than transferring, electrons to achieve stability makes for a reasonable explanation of the covalent bond. Lewis represented the stable octet of electrons and the covalent and coordinate covalent bonds as follows:

---

[13] Alfred Werner, *Z. Anorg. Chem.* **3**, 267 (1893).
[14] W. Kossel, *Ann. Phys. Leipzig* **49**, 229 (recd. Dec. 1915).
[15] G. N. Lewis, *J. Am. Chem. Soc.* **38**, 762 (recd. Jan. 1916); G. N. Lewis, *Valence and the Structure of Atoms and Molecules*, Dover, New York, 1966. First published in 1923.

$$:\overset{..}{\underset{..}{F}}\cdot + \cdot \overset{..}{\underset{..}{F}}: \;=\; :\overset{..}{\underset{..}{F}}:\overset{..}{\underset{..}{F}}:$$

Homopolar Covalent Bond
(each atom contributes one electron to the shared pair)

$$:\overset{..}{\underset{..}{Na}}\cdot + \cdot \overset{..}{\underset{..}{F}}: \;=\; :\overset{..}{\underset{..}{Na}}:^{+} \;\; :\overset{..}{\underset{..}{F}}:^{-}$$

Ionic Bond
(transfer of an electron)

$$:\overset{..}{\underset{..}{F}}:Be:\overset{..}{\underset{..}{F}}: + 2(:\overset{..}{\underset{..}{F}}:^{-}) \;=\; \left[\begin{array}{c} :\overset{..}{\underset{..}{F}}: \\ :\overset{..}{\underset{..}{F}}:Be:\overset{..}{\underset{..}{F}}: \\ :\overset{..}{\underset{..}{F}}: \end{array}\right]^{2-}$$

Coordinate Bond
(one atom contributes both shared electrons)

Beryllium displays a coordination number of 4 even though it is divalent (e.g., BeF_2, BeH_2). The otherwise puzzling observation of Werner that neutral molecules such as NH_3 (:N:::H_3) and H_2O (H_2::$\overset{..}{O}$:) can replace ions in metallic compounds is easily explained by the unshared pairs of electrons which they have available. The greatest value of the Lewis and Langmuir model of chemical bonding was its unification of the electrochemical, valence, and coordination viewpoints of bonding in one simple pictorial representation.

The following scheme reviews the bonding types, gives some alternative names, and shows the effect of increasing *polarity* on the chemical bonds.

$$A:A \quad \xrightarrow{\text{increasing polarity}} \quad A:B \quad \xrightarrow{\text{increasing polarity}} \quad A^{+}B^{-}$$

Covalent Bond   Polar Covalent Bond   Ionic Bond
Homopolar Bond   Heteropolar Bond   Electrovalent
Homonuclear   Heteronuclear   Bond
  Diatomic Molecule   Diatomic Molecule   Electrostatic
                                                   Bond

$$A: + B \quad \longrightarrow \quad A:B$$

Coordinate Bond
Dative Bond
Acceptor–Donor Complex

Lewis structures are just formal pictures of stationary electrons and do not tell us "why" the atoms are bound. Furthermore, *no stationary distribution of point charges is in stable equilibrium: only a dynamical system can achieve stability.*[16] In general, the stability of the molecule (the "why" of

---

[16] Earnshaw's theorem: *A point charge, acted upon by purely electrical forces, cannot be at rest in stable equilibrium in an electric field due to other charges.* Before the advent of the Bohr model of the atom Earnshaw's theorem provided the basis for one of two conclusions: (1) the inverse square law of electric force does not hold at atomic distances; (2) atoms do not consist of charged particles at *rest*. Rutherford's analysis of experiments on the scattering of α particles by atoms proved that the inverse square law holds down to distances of the order of a nuclear diameter! Thus the structure of the molecule is removed from the realm of electrostatics. After the quantum mechanical problem of dynamic electronic stability has been solved we return to an *electrostatic interpretation* of chemical bonding (see Chapter 5, Section V.3) but not in terms of point charges.

chemical bonding) must arise from the fact that the molecule has an energy which is lower (more negative) than that of the isolated atoms. To see this, it is only necessary to plot the energy of a diatomic molecule against the distance between the atoms (Fig. 1.1.1). The minimum in the curve indicates a stable A—B molecule with an equilibrium distance R_e and a dissociation energy D_e. D_e is the amount of energy necessary to dissociate the molecule into its component atoms $A + B$.

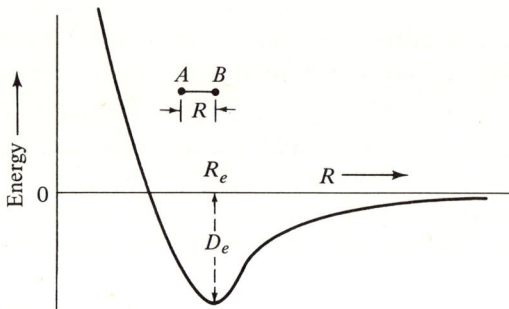

FIG. 1.1.1 Energy of the diatomic molecule AB plotted against the internuclear distance. R_e is the equilibrium distance between the nuclei and D_e is the dissociation energy.

The next step toward fulfilling Gay-Lussac's hope of submitting chemical phenomena to calculation would require us to be able to calculate D_e and R_e. The principles involved in such a calculation are those of quantum mechanics, which was not established until 1926 and which we will explore in the next chapter. Heitler and London[17] applied the new mechanics to the calculation of D_e and R_e for H_2. They found the molecule to be stable, obtaining reasonable estimates of D_e and R_e; historically, this was the first successful treatment of the covalent bond.

The success of Heitler and London on H_2 suggested that similar quantum mechanical studies could be made of the bonding in other diatomic and even polyatomic molecules. In general, great strides have been made in this direction. The situation from the time of Dalton to today is illustrated in Fig. 1.1.2. The *valence bond method* (related to the original Heitler-London work and chiefly used for qualitative descriptions of chemical bonding),[18,19]

---

[17] W. Heitler and F. London, *Z. Phys.* **44**, 455 (1927).
[18] Linus Pauling, *Proc. Nat. Acad. Sci. (U.S.)* **14**, 359 (1928); *J. Amer. Chem. Soc.* **53**, 1367, 3225 (1931).
[19] J. C. Slater, *Phys. Rev.* **37**, 481; **38**, 325, 1109 (1931).

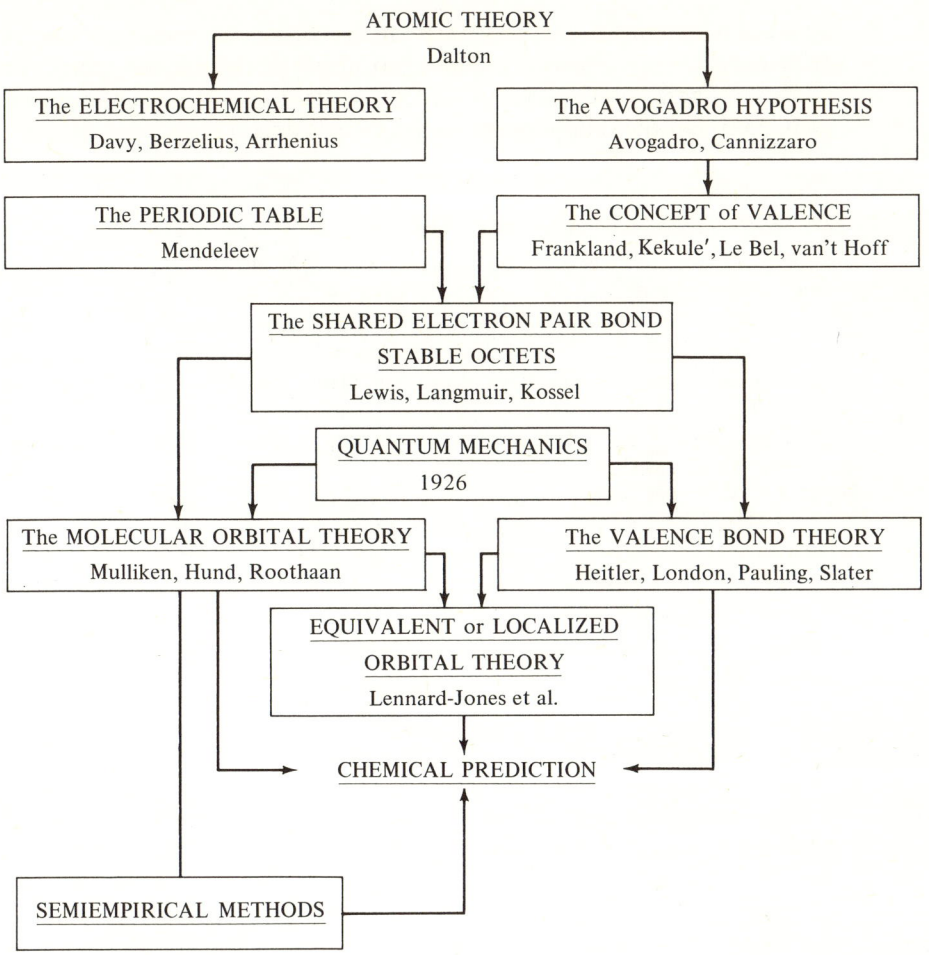

FIG 1.1.2 The history of chemical bonding theory. Semiempirical methods (shown here "coming in the back door") are often the most fruitful for the practicing chemist (Chapter 8).

and the *molecular orbital method*[20] (which has seen very extensive quantitative application) grew up in a competitive atmosphere but have achieved a sort of synthesis through *localized orbital methods*[21] and more exact

[20] R. S. Mulliken, *J. Chem. Phys.* **43**, S2 (1965); historical recollections of the development of molecular orbital theory. C. C. J. Roothaan, *Rev. Mod. Phys.* **23**, 69 (1951), built a rigorous mathematical framework for the molecular orbital method.

[21] J. Lennard-Jones, *Proc. Roy. Soc.* (*London*) **198A**, 1 (1949), first in a series of papers to which G. G. Hall, J. A. Pople, and A. C. Hurley also contributed.

theoretical methods. For very large molecules, and therefore those of principal chemical interest, semiempirical methods (sound theoretical procedures based on empirical parameters) offer the chief hope of chemical prediction and the fulfillment of Gay-Lussac's dream.

REFERENCES

Knight, D. M., ed., *Classical Scientific Papers—Chemistry*, American Elsevier, New York, 1968.

Lewis, G. N., *Valence and the Structure of Atoms and Molecules*, Dover, New York, 1966. This little book, which was first published in 1923, is a fascinating historical insight into the state of our knowledge before the establishment of quantum mechanics.

Mulliken, Robert S., "Molecular Scientists and Molecular Science," *J. Chem. Phys.* **43**, S2 (1965).

Nash, L. K., "The Atomic-Molecular Theory," Case 4 of *Harvard Case Histories in Experimental Science*, J. B. Conant, ed., Harvard Univ. Press, Cambridge, 1950.

Palmer, W. G., *A History of The Concept of Valency to* 1930, Cambridge Univ. Press, Cambridge, 1965.

Slater, J. C., "Molecular Orbital and Heitler-London Methods," *J. Chem. Phys.* **43**, S11 (1965).

2 DEVELOPMENT OF THE WAVE THEORY OF MATTER

The future truths of physical science
are to be looked for in the sixth
decimal place.

>An eminent physicist of the 19th
century quoted by A. A. Michelson
in 1894

Today I have made a discovery as important
as that of Newton.

>**MAX PLANCK** to his son in 1900

QUANTUM CHEMISTRY, like its parent quantum mechanics, is quite firmly based on a rich mechanistic heritage which existed prior to 1900 and came to be called "classical." The first section of this chapter provides an introduction to those aspects of classical mechanics used in quantum chemistry. Later sections outline the development of the quantum theory through the Bohr theory of the hydrogen atom, de Broglie's matter waves, and the Schrödinger equation. The final three sections treat the important mathematical methods of quantum chemistry which will be extensively applied in the following chapters.

2.1 CONSERVATION LAWS AND CLASSICAL MECHANICS

NEWTONIAN MECHANICS

For our purposes Newtonian mechanics consists of the verbal and mathematical statement of Newton's three laws of motion, which, together with a few definitions, allow us to derive the conservation laws.

First Law (*Law of Inertia*). Every body tends to remain at rest or in uniform motion (constant velocity) in a straight line unless acted upon by an external influence (force).

Second Law (*Law of Force*). The rate of change of momentum is proportional to the force producing it.

Third Law (*Law of Action and Reaction*). If body 1 exerts a force on body 2 this is equivalent to body 2 exerting a force of equal magnitude but opposite sign on body 1.

The mathematical statement of the second law

$$\mathbf{F} = \frac{d\mathbf{p}}{dt}$$

leads directly to the important results of Newtonian mechanics. F is the *total force* acting on the body, and p is the *momentum* of the body, defined as

$$\mathbf{p} = m\mathbf{v} = m\frac{d\mathbf{r}}{dt} \qquad (2.1.1)$$

where *m* is the mass of the particle, **v** is the velocity, and **r** the radial vector from the origin to the position of the body.

Conservation of Momentum. If the total force **F** is zero then $d\mathbf{p}/dt$ is zero and the momentum **p** is said to be *conserved*.

The important dynamical quantity, angular momentum, denoted by **L**, is defined as

$$\mathbf{L} = \mathbf{r} \times \mathbf{p} \tag{2.1.2}$$

L is the vector cross product (see Appendix A) of the radial vector and the momentum. The *torque* or twisting force on the body is

$$\mathbf{N} = d(\mathbf{r} \times \mathbf{p})/dt = d\mathbf{L}/dt$$

The last equation yields a second conservation law of the utmost importance.

Conservation of Angular Momentum. If the total torque **N** is zero, then $d\mathbf{L}/dt$ is zero, and the angular momentum **L** is said to be *conserved*.

Like (2.1.1.) this is a vector theorem.

$$\mathbf{N} = \mathbf{i}N_x + \mathbf{j}N_y + \mathbf{k}N_z$$

$$\mathbf{L} = \mathbf{i}L_x + \mathbf{j}L_y + \mathbf{k}L_z$$

So, $dL_x/dt = 0$ if $N_x = 0$, and the angular momentum in the *x* direction is said to be conserved. If N_y and N_z are also zero, then the total angular momentum is conserved, as well as the component in a given direction.

Work and Energy. The definition of physical work is Work = (Force) (Distance), or in vector form, $W = \mathbf{F} \cdot d\mathbf{s}$. The total work done in bringing the body from position 1 to position 2 is the line integral

$$W_{12} = \int_1^2 \mathbf{F} \cdot d\mathbf{s} \tag{2.1.3}$$

where *ds* is the differential element of length. Using previous definitions and a constant mass *m*, (2.1.3) is rewritten

$$W_{12} = m\int_1^2 \frac{d\mathbf{v}}{dt} \cdot \frac{d\mathbf{s}}{dt} dt = m\int_1^2 \frac{d\mathbf{v}}{dt} \cdot \mathbf{v}\, dt$$

or

$$W_{12} = \frac{m}{2}\int_1^2 \frac{d}{dt}(\mathbf{v} \cdot \mathbf{v})\, dt = \frac{m}{2}(v_2^2 - v_1^2)$$

Define a new quantity, the *kinetic energy*, $T = \tfrac{1}{2}mv^2$. We now have

$$W_{12} = T_2 - T_1 \tag{2.1.4}$$

2.1 Conservation Laws and Classical Mechanics

Equation (2.1.4) states that the work done is equal to the change in the kinetic energy.

Suppose the total force is such that the work done in moving around a closed circuit is zero, that is,

$$\int_1^2 \mathbf{F} \cdot d\mathbf{s} + \int_2^1 \mathbf{F} \cdot d\mathbf{s} = \oint \mathbf{F} \cdot d\mathbf{s} = 0 \tag{2.1.5}$$

then the force is said to be a *conservative force* and the system itself is termed *conservative*. In chemistry one deals with systems which are conservative.[1] Now, when $\oint \mathbf{F} \cdot d\mathbf{s} = 0$, the work is independent of the path and can be viewed as the difference between two numbers, one characteristic of the starting point, the other characteristic of the end point, that is,

$$W_{12} = \int_1^2 \mathbf{F} \cdot d\mathbf{s} = V_1 - V_2 \tag{2.1.6}$$

or

$$\mathbf{F} \cdot d\mathbf{s} = -dV = -\left(\frac{\partial V}{\partial x} dx + \frac{\partial V}{\partial y} dy + \frac{\partial V}{\partial z} dz\right)$$

or

$$\mathbf{F} = -\nabla V$$

where

$$\nabla = \mathbf{i}\frac{\partial}{\partial x} + \mathbf{j}\frac{\partial}{\partial y} + \mathbf{k}\frac{\partial}{\partial z}$$

The numbers V_1 and V_2 are the *potential energy* at points 1 and 2, respectively. There is a number V_i characteristic of each point in space: The totality of these numbers is V, the potential, or potential field. The force acting on the body at any point is derivable from the potential $\mathbf{F} = -\nabla V$. From (2.1.4) and (2.1.6), we have for a conservative system

$$T_1 + V_1 = T_2 + V_2$$

Conservation of Energy. If the total force acting is conservative then the *total energy* $T_i + V_i$ is *conserved*. The principle of conservation of energy makes the total energy $E = T_i + V_i$ like the momentum and angular momentum in the absence of external forces, a conserved quantity or *constant of the motion*.

---

[1] An example of a nonconservative system is one subjected to a frictional force, in which case any closed circuit will involve a net loss in energy (given up to the environment as heat).

External Forces. In the treatment of many-particle systems (all chemical systems are many-particle systems) one can distinguish between the internal forces which act between particles of the system and the external forces which originate outside the system but act on it. In response to external forces the many-particle system responds as a whole, that is, as if the entire mass of the system were concentrated at a single point, the *center of mass*. If no external forces (torques) are acting on the system then the total momentum of the center of mass is conserved and the angular momentum about the center of mass is also conserved. Internal forces have no effect on these conservation theorems. Finally, if all the forces (internal and external) are conservative forces, the total energy of the many-particle system is conserved.

HAMILTON'S EQUATIONS

At the next level of sophistication classical mechanics utilizes the properties of the scalar function $T + V$. Scalar equations of motion replace the vector equation which is Newton's second law. In dealing with a many-particle system of N particles there are $3N$ coordinates $q_1, q_2, q_3, \ldots, q_{3N}$ since three coordinates specify the position of each particle. Similarly, there are $3N$ components of the velocity $\dot{q}_1, \dot{q}_2, \ldots, \dot{q}_{3N}$, where $\dot{q}_i = dq_i/dt$, and $3N$ components of the momentum p_i. Defining the *Hamiltonian* function $H = T + V = H(q_i, p_i, t)$ the equations of motion are

$$\frac{\partial H}{\partial q_i} = -\dot{p}_i \quad \text{and} \quad \frac{\partial H}{\partial p_i} = \dot{q}_i \qquad (2.1.7)$$

The Hamiltonian function is the total energy function of the system, p_i and q_i are called canonically conjugate momentum and coordinate (e.g., $m\dot{x}$ and x).

EXERCISE Show that Hamilton's equations for the x, y, and z degrees of freedom of a free particle ($V = 0$) are equivalent to Newton's laws; that is, show that the force on the particle is zero and that it moves with constant velocity.

CONSTANTS OF THE MOTION

We have shown that in classical mechanics the momentum, angular momentum, and energy are conserved under certain conditions; that is, they are constant in time. Is there a general method of finding if a quantity is conserved? Consider the total time derivative of a function of the dynamical variables q_i, p_i, and t. Consider first the Hamiltonian function itself.

$$\frac{dH}{dt} = \frac{\partial H}{\partial t} + \sum_{i=1}^{3N} \left(\frac{\partial H}{\partial p_i} \frac{dq_i}{dt} + \frac{\partial H}{\partial p_i} \frac{dp_i}{dt} \right) = \frac{\partial H}{\partial t} + \sum_{i=1}^{3N} \left(\frac{\partial H}{\partial q_i} \frac{\partial H}{\partial p_i} - \frac{\partial H}{\partial p_i} \frac{\partial H}{\partial q_i} \right)$$

$$\frac{dH}{dt} = \frac{\partial H}{\partial t} \qquad (2.1.8)$$

2.1 Conservation Laws and Classical Mechanics

Equation (2.1.8) says that if the Hamiltonian function is not an explicit function of the time then $dH/dt = 0$ and the energy is conserved. (E is a constant of the motion.)

Next consider any general function of the dynamical variables. $G = G(q_i, p_i, t)$

$$\frac{dG}{dt} = \frac{\partial G}{\partial t} + \sum_i \left(\frac{\partial G}{\partial q_i} \frac{dq_i}{dt} + \frac{\partial G}{\partial p_i} \frac{dp_i}{dt} \right)$$

$$= \frac{\partial G}{\partial t} + \sum_i \left(\frac{\partial G}{\partial q_i} \frac{\partial H}{\partial p_i} - \frac{\partial G}{\partial p_i} \frac{\partial H}{\partial q_i} \right)$$

$$\frac{dG}{dt} = \frac{\partial G}{\partial t} + \{G, H\} \tag{2.1.9}$$

The quantity $\{G, H\}$ is called the Poisson bracket of G and H. If the function G is not an explicit function of the time (i.e., $\partial G/\partial t = 0$), then the time derivative of the function G is given entirely by the Poisson bracket. If the Poisson bracket also vanishes then the function G is conserved (G is a constant of the motion).

To find if a particular momentum is conserved, write

$$\frac{dp_i}{dt} = \frac{\partial p_i}{\partial t} + \sum_j \left(\frac{\partial p_i}{\partial q_j} \frac{\partial H}{\partial p_j} - \frac{\partial p_i}{\partial p_j} \frac{\partial H}{\partial q_j} \right)$$

The first two terms vanish because p_i, q_i, t form an independent set of variables; only $i = j$ survives in the last term, so $\dot{p}_i = -\partial H/\partial q_i$. This is just Hamilton's equation of motion, but it emphasizes that the momentum is conserved if the Hamiltonian has no explicit dependence on the canonically conjugate coordinate q_i.

The significance of the constants of motion lies in the following. There are $6N$ Hamiltonian equations of motion, and these equations must be solved to obtain the complete solution to the mechanics of the system in time and space (e.g., the trajectories of the particles). Each equation, if integrable, will have one constant of integration which is determined by initial conditions, if known. However, in any many-particle problem of importance, the equations of motion are so inextricably coupled by potentials of the form $\sum_{i>j}^{3N} V_{ij}$ that exact solution in terms of a finite number of known functions is impossible. In this realistic situation each equation that can be solved by $\partial H/\partial q_i = 0$ gives some information about the system which is very valuable. Also, the energy is a constant of the motion if H has no explicit time dependence.

This is enough mathematical physics until we come to the Bohr theory of the hydrogen atom. In the next two sections we summarize the developments in atomic physics which led to the discovery of the quantum theory and the development of Bohr's theory.

2.2 ATOMIC PHYSICS

In the period of over two hundred years from the time of Galileo to the twentieth century the science of physics developed a great body of theory and experiment. Most of the development was in mechanics, which reached a high level of sophistication. Newton set down the fundamental principles and invented the necessary mathematical tool, the calculus. In the next century celestial mechanics attracted the greatest mathematicians of the period: Laplace, Euler, Lagrange, and Gauss. Finally, Hamilton put the science of mechanics into elegant form. Progress was slower in optics, electricity, and magnetism, yet Maxwell achieved a unification of these fields under his equations of electromagnetic radiation well before the twentieth century. This led to the "all is in its place" feeling of physicists in the last twenty years of the nineteenth century which is expressed in the opening quotation of this chapter. It was generally believed that the fundamental laws had been discovered and only a mopping-up action in the form of more precise measurements was necessary.

Trouble arose from the fact that physics had not yet come to grips with the atomic and subatomic structure of matter. Indeed there was strong resistance to the atomic concept among certain scientists who smelled dogma in these unseen entities.[2] Although the idea that matter is composed of elementary particles is almost as old as western civilization, until this century the atomic hypothesis was more useful to chemists than to physicists. Chemists offered strong experimental evidence for the atom during an entire century which began with Dalton (1808). Physicists of this period searched for a theory of matter and its transformation which was independent of a structural model. (The product of this quest was thermodynamics.) In this section we describe how physics finally came to grips with the world of the atom.

THE ELECTRON

Roentgen's discovery of the x-ray in 1895 removed the "Crookes tube" from the class of scientific curiosities. Experimenting with a Crookes tube, he noted that coming from the tube were invisible rays capable of blackening photographic plates. A Crookes tube consists of a cathode and an anode in a glass envelope containing a gas at very low pressure. When voltage of several kilovolts is applied between the electrodes an electrical discharge occurs between them, that is, the gas becomes conducting. At sufficiently low pressure the glow is lessened but a beam emanates from the cathode (negative) and upon striking the glass wall of the tube produces a phosphoresence. This beam is called the *cathode ray*.

---

[2] As recently as 1910 Wilhelm Ostwald, the eminent physical chemist, would gladly have done away with much atomic theory.

Crookes (1878) performed many experiments with cathode rays, including the deflection of the ray by magnets, the casting of shadows by objects placed in the ray, and the exertion of pressure by the ray on objects in its path. Crookes concluded that the cathode ray consists of a high-velocity stream of electrified particles.[3] The x-rays, on the other hand, were shown by Roentgen to be undeflected by magnetic fields and therefore do not consist of electrified particles. Roentgen also showed that the x-rays have amazing penetrating power and constitute an important medical tool. The resulting popular and scientific furor over x-rays led to many experiments, but our interest centers on those of J. J. Thomson.

Thomson performed careful experiments on cathode rays using electric and magnetic fields. He proved that the cathode ray is a stream of negative charges (electrons) and that these electrons are one and the same regardless of the gas used in the Crookes tube. These facts constitute the discovery of the electron and resulted in the award of the Nobel Prize to Thomson in 1906. One of his experiments,[4] the deflection of a cathode ray by an electric field, is represented in Fig. 2.2.1. By performing such deflection experiments in electric and magnetic fields Thomson arrived at the ratio of electronic charge to mass, $e/m = 1.7589 \times 10^8$ coulombs/gm.

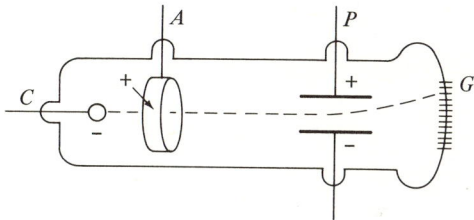

FIG. 2.2.1 Deflection of a cathode ray by an electric field. Emanating from the cathode (C) the ray passes through the hollow anode (A) is deflected by the electric field between the plates (P) and is displayed on the graduated phosphorescent screen (G).

The charge of the electron was determined by the American scientist R. A. Millikan[5] (1909) in his famous oil drop experiments. Figure 2.2.2 illustrates how oil droplets are sprayed into the air between two parallel plates. The droplets fall as free bodies, achieving a terminal velocity v_1 under the influence of gravity and the resistance of the air. Now the plates are charged so that the upper has positive polarity. Most droplets have acquired a charge either through friction

---

[3] Certainly true was Crookes's comment, "the phenomena in these tubes reveal to physical science a new world," W. Crookes, *Phil. Trans.* **170**, 135 (1879).

[4] J. J. Thompson, *Phil. Mag.* **44**, 293 (1897).

[5] R. A. Millikan, *The Electron*, 2nd ed., Univ. of Chicago Press, Chicago, 1924.

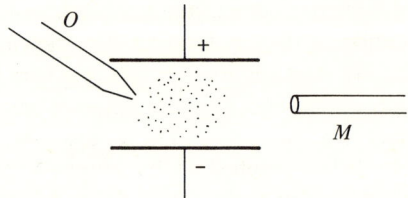

FIG. 2.2.2 Oil drop experiment. The droplets are sprayed between the charged plates by the sprayer (O) and their motion is observed by means of the microscope (M).

in the spraying process or by the background ionizing radiation which is always present on the surface of the earth. If this charge on the droplet is negative the electric field acts to push the droplet upward; the net upward force is

$$Ee_n - Mg = \text{net upward force}$$

where g is the acceleration due to gravity, E is the field strength, M is the mass of the droplet, and e_n is the net charge on the droplet. This force then determines an upward terminal velocity against air resistance, v_2.

$$v_1/v_2 = Mg/(Ee_n - Mg)$$

or

$$e_n = Mg(v_1 + v_2)/Ev_1$$

If the droplet now changes its net charge by further acquisition of charge there is a new velocity v_2' and

$$e_m = Mg(v_1 + v_2')/Ev_1$$

where e_m is the new net charge.

$$e_m - e_n = e_i = \frac{Mg}{Ev_1}(v_2' - v_2) \qquad (2.2.1)$$

In a given experiment Mg/Ev_1 is a constant and extensive observations yield various values for $(v_2' - v_2)$. Millikan found that there was a minimum value of $(v_2' - v_2)$ of which all others are multiples. This established that e_i has a certain minimum value and all other charges are multiples of the minimum charge. The value of the electronic charge was found[6] to be 1.6×10^{-19} coulomb. From the charge the electronic mass is determined by the known e/m ($m = 9.1 \times 10^{-28}$ gm) and Avogadro's number follows from $F = Ne$, where F is Faraday's constant.

ELECTROMAGNETIC RADIATION

Before the discovery of *interference* most scientists favoured a theory in which light consisted of particles (corpuscular theory of light). Huygens had suggested a wave theory of light in 1690, but the corpuscular theory

---

[6] E, g, and v_1 are known. From the observed v_1, Stokes's law yields the radius of the drop, and, with the density, its mass; e then follows from (2.2.1).

originated with Newton, so the wave theory made little headway against the overwhelming scientific prestige of Sir Isaac.

Interference is uniquely a phenomenon associated with waves. A wave is a periodic disturbance which travels through space. Waves have the ability to superimpose themselves one upon the other to yield a net disturbance which may be zero (complete cancellation). The identification of light as an electromagnetic wave resulted from the investigations of James Clerk Maxwell in the nineteenth century. Before Maxwell the concept of "action at a distance" dominated physics. Two charges, q_1 and q_2, need not be in contact to affect each other, but will be attracted or repelled at a distance r_{12} according to the Coulomb force law, $F_{12} = q_1 q_2 / r_{12}^2$. One observes a charge exerting a force on another with nothing between them, that is, action at a distance. Maxwell[7] proposed a theory of the *electromagnetic field*, which effectively deals with this problem. In Maxwell's theory oscillating charges set up an electrical disturbance (electromagnetic wave) which propagates through space with a finite velocity. Upon reaching another charge, that charge will be set in motion. Thus the electromagnetic wave (a periodic disturbance in the electric and magnetic fields) exerts force on the charge and transmits energy to it (see Fig. 2.2.3).

Maxwell showed that the velocity of propagation of the electromagnetic waves is equal to the velocity of light. In Maxwell's words, "This velocity is so nearly that of light, that it seems we have strong reason to conclude

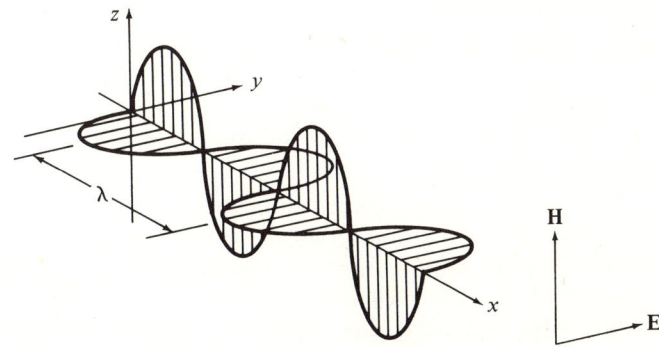

FIG. 2.2.3 Plane-polarized electromagnetic wave. The electric field strength vector **E** and the magnetic field strength vector **H** are perpendicular to each other and the direction of propagation x. The wavelength, λ (cm) is the distance crest to crest. The number of crests per second passing a given point is the frequency ν (sec^{-1})

[7] J. C. Maxwell, *Phil. Trans.* **155**, 459 (1865).

that light itself ... is an electromagnetic disturbance in the form of waves propagated through the electromagnetic field according to electromagnetic laws." He thus effected the unification of optics, electricity, and magnetism in one theory.

Today the propagation of electromagnetic waves in invisible portions of the *spectrum* is an everyday experience (see Table 2.2.1); wave propagation was first demonstrated by Hertz (1888) by discharging a condenser across a spark gap and simultaneously observing a spark across a loop of wire (antenna) in an evacuated bell jar some distance away.

An electromagnetic wave is characterized by its *wavelength* or *frequency*, and its *polarization* (direction of **E** and **H** vectors, e.g., the wave in Fig. 2.2.3 has the electric vector in the y direction, therefore, it is y-polarized). The wavelength λ is expressed in meters or fractions thereof: centimeters (cm), millimeters (mm), microns ($\mu = 10^{-4}$ cm), millimicrons (m$\mu = 10^{-7}$ cm), and angstrom units (A = 10^{-8} cm). The *frequency* is the number of cycles per second, v (sec^{-1}); v and λ are related to the velocity of the wave, c ($c = 3 \times 10^{10}$ cm/sec), by

$$v\lambda = c \qquad (2.2.2)$$

Another useful quantity is the "frequency" expressed in *wavenumbers*, $\bar{v}$, that is, units of cm^{-1}.

$$\bar{v} = \frac{1}{\lambda} \qquad (2.2.3)$$

QUANTUM THEORY

From the thermodynamic point of view a material body emits and absorbs electromagnetic radiation to establish equilibrium with its surroundings. At equilibrium the intensity of this radiation at a given wavelength depends upon the temperature of the body. Radiation in equilibrium with matter is called *blackbody radiation*.[8]

For purposes of experiment a heated furnace with a small hole drilled in its side for observation serves as a convenient source of blackbody radiation. At "low" temperature the furnace glows cherry red; as the temperature is raised it glows yellow and then white (see Fig. 2.2.4).

Since the radiation distribution is independent of the nature of the material[8] there must exist a universal function relating I_λ to the wavelength and temperature. Previous attempts (Rayleigh, Wien) had not been

---

[8] The name "blackbody radiation" arises from the fact that a perfectly black body (i.e., a perfect absorber of radiation) heated to temperature T has a distribution of emitted radiation identical to the above. However, at *thermodynamic equilibrium* between radiation and matter the radiation distribution is independent of the nature of the body (Kirchhoff's law).

TABLE 2.2.1 The Electromagnetic Spectrum

WAVELENGTH	10^{-3} Å	50 Å	2000 Å	4000 Å	8000 Å	30 μ	600 μ	50 cm
					0.8 μ		0.06 cm	
NAME OF SPECTRAL REGION	γ-rays	Vacuum ultraviolet		Visible	Near infrared	Far infrared		Radio waves
	X-rays		Ultraviolet				Microwave	
ORIGIN	Nuclear transitions		Electronic transitions		Molecular vibrations and rotations			
ENERGY[a]								
Electron volts (eV)[b]	250	5.7 × 10³		3	1.5	0.04	2 × 10⁻³	
Kcal/mole				72	36	0.94	0.46	

[a] Energy of a photon of frequency ν from Planck's law, $E = h\nu$.
[b] 1 eV is the amount of energy gained by an electron upon falling through a potential difference of 1 volt.

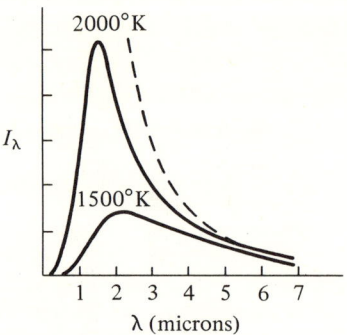

FIG. 2.2.4 Distribution of blackbody radiation as a function of temperature. I_λ is the intensity at wavelength λ. The dotted line is the classical prediction at 2000°K (Rayleigh-Jeans).

successful in accounting for the distribution in its entirety. Max Planck was able to fit the entire curve with an "interpolation formula found by happy guesswork."[9] Planck was now faced with the problem of giving real physical meaning to his interpolation formula. He accomplished this[10] by considering the equilibrium distribution of a system of oscillators (emitters and absorbers of electromagnetic radiation) at temperature T. This gave a formula of the same form as that obtained previously by "happy guesswork."

$$I_\lambda = \frac{2\pi c^2 h}{\lambda^5 [\exp(hc/kT\lambda) - 1]} \tag{2.2.4}$$

Of the constants in the Planck equation two were well known: the velocity of light, c, and the Boltzmann constant, k. The third constant resisted all attempts by Planck to fit it into a classical interpretation. This constant, h, having units of erg-seconds (energy × time = action), was called by Planck the "elementary quantum of action." Today h is just called *Planck's constant*, $h = 6.626 \times 10^{-27}$ erg-sec. Quite reluctantly, Planck concluded that the existence of h represented a revolutionary departure from classical physics. Specifically, h implied the existence of an atomism in radiation. Radiation of frequency ν is not absorbed or emitted continuously in arbitrary amounts but only in *quanta* of energy, $E = h\nu$.

$$E = h\nu \qquad \text{PLANCK'S LAW} \tag{2.2.5}$$

This is a serious departure from classical mechanics because it introduces a

---

[9] M. Planck, Nobel Prize in Physics Award Address, 1919.
[10] M. Planck, *Ann. Phys.* **4**, 553 (1901).

granularity, a discreteness, or a quantization into the nature of a fundamental and classically continuous quantity.

Planck's law says that at high frequency (short wavelength) a large amount of energy is packed into a quantum. As a result, at a given temperature the high-frequency material oscillators of the blackbody are not involved in the matter–radiation equilibrium since a very high temperature is necessary to excite them (very high thermal energy would be necessary to excite the high-frequency oscillators since according to Planck's law they can only emit or absorb amounts of energy $h\nu$). Thus Planck's quantum theory accounts for the I_λ curves of Fig. 2.2.4, including the falloff at high frequency (short wavelength).

The small numerical value of h indicates that the discontinuities in nature are very minute, and therefore mechanical and radiative phenomena appear continuous to the unaided eye. However, the quantization of the absorption and emission radiation should be evident in all such interactions and not just blackbody radiation. After a 5-year pause, Einstein closed the gap.[11]

> It appears to me that the observations on "black body" radiation, photoluminescence, the generation of cathode rays by ultraviolet radiation, and other phenomena related to absorption and emission of light are better understood on the assumption that energy in light is . . . not distributed continuously over larger and larger volumes of space but consists of a finite number of energy quanta, localized at points of space, which move without subdividing and which are absorbed and emitted only as units.

These *quanta of light* proposed by Einstein came to be called *photons*. Einstein[11] was able to explain the photoelectric effect in terms of photons of light.

The photoelectric experiment is illustrated in Fig. 2.2.5. The cathode is illuminated by light in a highly evacuated tube. With certain choices of voltage

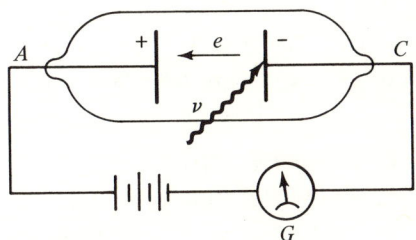

FIG. 2.2.5 Photoelectric effect. Light of frequency ν impinges on the cathode. The photocurrent e is measured with the galvanometer G.

[11] A. Einstein, *Ann. Phys.* **17**, 132 (1905)

drop between electrodes and frequency v a current is observed to flow, that is, electrons are emitted from the cathode. It was found that for a given type of cathode surface the frequency of light must be greater than a certain value (the threshold frequency) in order to observe the photocurrent. At frequencies above threshold the electrons come off the cathode with excess kinetic energy. The excess kinetic energy can be determined by reversing the polarity of the electrodes at constant illumination until the current flow is stopped. It is found that the excess energy of the electron is independent of the intensity of light and depends only on the frequency, although the photocurrent is proportional to the light intensity. Einstein concluded that when light falls on the cathode surface the total energy of the photon, hv, is given to a single electron within the metal. If this energy is greater than some minimum value hv_t (v_t is the threshold frequency), the electron surmounts the potential barrier at the surface and leaves the metal with excess kinetic energy $\tfrac{1}{2}mv^2$.

$$\tfrac{1}{2}mv^2 = hv - hv_t = eV$$

V is the voltage necessary to reduce the flow of electrons to zero and e is the electronic charge. The intensity of light is proportional to the number of photons present in the beam so the photocurrent is proportional to the intensity. Finally, a plot of V versus v has intercept v_t and slope h/e and offers a means of obtaining h (Fig. 2.2.6). Complete experimental verification of Einstein's explanation of the photo-electric effect[12] was not obtained until 10 years later, by Millikan.

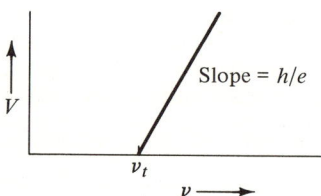

FIG. 2.2.6 Einstein's explanation of the photoelectric effect.

Einstein's theory of light photons seems to be a reversion to the corpuscular theory of light, but is actually a synthesis of wave and corpuscular theories. Light is considered to act as a photon when interacting with matter, but to propagate through space as a wave. Einstein thus introduced a *wave–particle duality* into the description of light.

SPECTROSCOPY AND THE ATOM

Although blackbody radiation is continuous over a certain region of the spectrum, the light given off by the gases in a Crookes tube or in other

---

[12] The Nobel Prize in Physics was awarded to Einstein (1921) not for the theory of relativity, but for the explanation of the photoelectric effect. While Einstein's fame principally rests on his theory of relativity, Nobel's will was being strictly interpreted to honor only work conferring "great benefits on mankind."

electrical discharges is far from continuous. In general, it consists of a number of emissions at well-specified wavelengths, that is, a discrete line spectrum for atoms or a band spectrum for molecules (see Fig. 2.2.7). Furthermore the spectrum of each element, ion, or molecule, is characteristic of that particular species.[13]

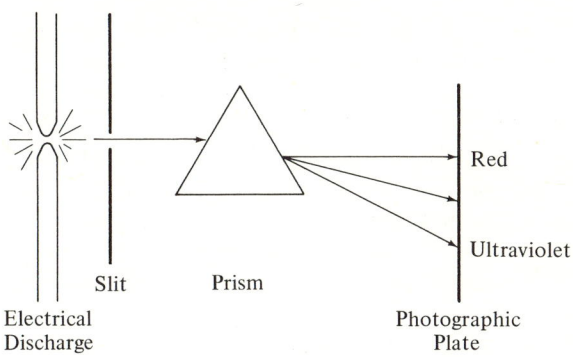

FIG. 2.2.7 Schematic of the emission spectrograph.

Although the observation of these line spectra had been going on for years, they resisted all attempts at explanation and organization. Physicists knew that in these discharges the spectra were due to the emission by isolated atoms and ions, but according to Maxwell's theory (and any classical atomic model) the atoms should emit continuously. The first important advance was a suggestion made by Balmer[14] in 1885. Balmer suggested that certain lines of the H-atom spectrum (now the Balmer series) could be organized according to their wavelength in a series of integers. If λ is the wavelength of the line, the Balmer relation can be written

$$\frac{1}{\lambda} = \bar{v} = R\left(\frac{1}{n_1^2} - \frac{1}{n_2^2}\right) \quad \text{RYDBERG-RITZ EQUATION} \quad (2.2.6)$$

---

[13] The spectrum of a gas or liquid may also be observed by absorption of radiation, i.e., pass the continuous radiation of a solid through the sample and then analyze the transmitted light in the spectrograph.

[14] J. Balmer, *Ann. Phys. Chem.* **25**, 80 (1885). It is interesting to note that Balmer was said to be a devoted Pythagorean from boyhood. That is, he believed that the mystery of the universe could be understood by correlating phenomena with integers. While such a belief can lead to the pseudoscience of numerology, it is difficult to imagine a more appropriate field for a Pythagorean than the organization of atomic spectra!

where $n_1 = 2$ and $n_2 = 3, 4, 5, 6, \ldots$, and R is the Rydberg constant, $R = 109{,}737$ cm^{-1}.

Balmer's discovery gave impetus to the study of atomic spectra, and from then on spectroscopists tried to represent the frequency of the line by numerical relations involving the square of integers. In 1908 Ritz suggested a combination principle by which the frequency of any line in the H-atom spectrum can be expressed as the difference between *terms* like R/n^2. Balmer's relation (2.2.6) has been written in this Rydberg-Ritz form: $n_1 = 1$, $n_2 > 1$ is the Lyman series: $n_1 = 2$, $n_2 > 2$ the Balmer series: $n_1 = 3$, $n_2 > 3$ the Paschen series, etc. In this way a number of terms account for the observed line spectrum of hydrogen.

THE BOHR MODEL OF THE ATOM

In 1885, the very year that Balmer published his paper on the spectrum of the H atom, Niels Bohr was born in Denmark. Bohr was to bring that order to the mechanical description of the H atom which would give a quantitative explanation for the spectral regularities discovered by Balmer. We will examine what Bohr had to work with in building an atomic theory.

The initial step in the clarification of atomic structure was due to Ernest Rutherford[15] (1911). Rutherford observed that when gold foil is bombarded with a beam of α particles (helium nuclei) almost all the particles pass through undeflected. One can conclude that the passing α particles see an "atom" which is almost entirely empty space. Analysis of the scattering of the deflected particles showed that the atomic mass is concentrated in a tiny, positively charged body called the nucleus. Negative charge is required in the atom for neutrality, but these experiments give no information about its distribution.

To treat the problem of the distribution of negative charge in an atom, Niels Bohr[16] (1913) proposed the following model for the simplest atom,

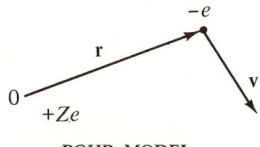

BOHR MODEL

hydrogen. He combined the Rutherford atom with Planck-Einstein quantization of the energy and a novel assumption of his own to obtain the first quantum theory of matter. Bohr's model consists of a very light

---

[15] E. Rutherford, *Phil. Mag.* **21**, 669 (1911); **27**, 488 (1914).
[16] N. Bohr, *Phil. Mag.* **26**, 1, 476, 875 (1913).

electron of mass m and charge $-e$ in circular orbit about a massive nucleus of charge $+Ze$. Bohr's hypothesis consists of quantizing the angular momentum as follows: $L = rmv = n\hbar$, where $\hbar = h/2\pi$. As a condition for a stable orbit the centripetal force pulling the electron into the nucleus (Coulomb attractive force Ze^2/r^2) must be balanced by the centrifugal force (mv^2/r).

$$Ze^2/r^2 = mv^2/r$$

or

$$v = \frac{Ze^2}{mvr} = \frac{Ze^2}{n\hbar} \qquad (2.2.7)$$

from the Bohr Hypothesis. Also

$$r = \frac{n\hbar}{mv} = \frac{n^2\hbar^2}{Ze^2 m} \qquad (2.2.8)$$

From (2.2.7) and (2.2.8) it's simple to determine the total energy, $E =$ kinetic $+$ potential energy, $E = \tfrac{1}{2}mv^2 - Ze^2/r$.

$$E = -\frac{e^4 m Z^2}{2n^2 \hbar^2} \qquad (2.2.9)$$

This expression for the energy of one-electron atoms and ions is essentially correct and is in *complete agreement with the Rydberg-Ritz equation* (2.2.6). The lowest-energy state has $n = 1$. This is the *ground state* (largest negative energy). States with $n > 1$ have higher energies (less negative energy); these are called the *excited states* of the atom or ion. An *electronic transition* is said to occur when the electron gains sufficient energy to jump from n_1 to n_2, $n_2 > n_1$, or loses energy in jumping from n_2 to n_1. The energy in each case is supplied or given up as a photon of energy $h\nu$, that is, *absorption or emission* of light. From Bohr's energy expression, (2.2.9), the spectrum of an H atom is predictable as follows: Consider the resonance transition[17] $n_1 = 1$ to $n_2 = 2$. The energy change is

$$E_{n_2} - E_{n_1} = \frac{e^4 m}{2\hbar^2}\left(\frac{1}{n_1^2} - \frac{1}{n_2^2}\right) = \frac{e^4 m}{2\hbar^2}(1 - \tfrac{1}{4}) \qquad (Z=1)$$

so the transition requires a photon of frequency[18]

$$\nu = \frac{(E_2 - E_1)}{h} = \frac{3}{4h}\left(\frac{e^4 m}{2\hbar^2}\right)$$

---

[17] The resonance transition is the lowest-energy transition from the ground state.
[18] Where frequency is often used as a spectroscopic unit of measure, see Section 6.9, it is preferable to adopt *hertz* as the unit, for example, 9×10^9 sec^{-1} = 9 kilomegahertz.

Development of the Wave Theory of Matter

EXAMPLE Find the frequency of the photon absorbed in the resonance transition. Find the wavelength of light absorbed, in centimeters (cm) and in angstroms (A). Find the energy of the transition, in ergs and in cm^{-1} (wavenumbers)

$$\nu = \frac{3(4.80298 \times 10^{-10} \text{ esu})^4 (9.1091 \times 10^{-28} \text{ gm})}{4(6.6256 \times 10^{-27} \text{ erg-sec}) \times 2 \times (1.0544 \times 10^{-27} \text{ erg-sec})^2}$$

$$= 2.467 \times 10^{15} \text{ sec}^{-1}$$

To convert to wavelength units the fundamental relation is $\lambda \nu = c$, where λ is the wavelength in centimeters and c is the velocity of light, $c = 2.9979 \times 10^{10}$ cm sec^{-1}.

$$\lambda = \frac{2.9979 \times 10^{10} \text{ cm sec}^{-1}}{2.467 \times 10^{15} \text{ sec}^{-1}} = 1.215 \times 10^{-5} \text{ cm}$$

Or in angstrom units

$$\lambda = 1.215 \times 10^{-5} \text{ cm} \times 10^8 \text{ A/cm} = 1215 \text{ A}$$

The transition energy in ergs is

$$(E_2 - E_1) = 6.6256 \times 10^{-27} \text{ erg-sec} \times 2.467 \times 10^{15} \text{ sec}^{-1}$$

$$= 1.635 \times 10^{-11} \text{ erg}$$

Frequencies are often expressed in wavenumbers, cm^{-1}.

$$\bar{\nu} = (E_2 - E_1)/hc = 1/\lambda = 8.229 \times 10^4 \text{ cm}^{-1}$$

EXERCISE Calculate the Rydberg constant R in the experimental expression for the observed transition frequencies of the H atom.

$$\bar{\nu}(\text{cm}^{-1}) = R \left(\frac{1}{n_1^2} - \frac{1}{n_2^2} \right)$$

Compare to the experimental value, 109.7×10^3 cm^{-1}.

The energy required to remove the electron to infinite distance from the nucleus can be calculated by letting the quantum number of the upper state go to infinity (see the radius expression in Table 2.2.2). The *ionization potential* of the atom is then $(E_{n_1} - E_{n_2})(n_1 = 1, n_2 = \infty) = -E_1$.

EXAMPLE The ionization potential of the H atom.

$$\text{I.P.} = -E_1 = \frac{e^4 m}{2\hbar^2} = \frac{(4.80298 \times 10^{-10} \text{ esu})^4 (9.1091 \times 10^{-28} \text{ gm})}{2(1.0544 \times 10^{-27} \text{ erg-sec})^2}$$

$$= 2.180 \times 10^{-11} \text{ erg}$$

Since ionization potentials are usually expressed in electron volts we convert by 1 erg = 6.24×10^{11} eV,

$$\text{I.P.} = 13.6 \text{ eV}$$

2.2 Atomic Physics

EXERCISE Calculate the second ionization potential of helium, i.e., the ionization potential of He$^+$. Express the answer in electron volts.

Table 2.2.2 gives the characteristic values and formulas for the interesting properties of the Bohr hydrogenic atom.

TABLE 2.2.2[a]

Dynamical property	Formula	Numerical value
Energy, E_n	$-\dfrac{e^4 m}{2\hbar^2}\left(\dfrac{Z^2}{n^2}\right)$	$\left.\begin{array}{l}-109{,}737 \text{ cm}^{-1}\\-13.60 \text{ eV}\\-2.180 \times 10^{-11} \text{ erg}\end{array}\right\} \times \left(\dfrac{Z^2}{n^2}\right)$
Ionization potential	$-E_1$	$13.6 \text{ eV} \left(\dfrac{Z^2}{n^2}\right)$
Velocity, v	$\dfrac{e^2}{\hbar}\left(\dfrac{Z}{n}\right)$	$2.188 \times 10^8 \text{ cm sec}^{-1} \times \left(\dfrac{Z}{n}\right)$
Radius, r	$\dfrac{\hbar^2}{e^2 m}\left(\dfrac{n^2}{Z}\right)$	$\left.\begin{array}{l}0.529 \times 10^{-8} \text{ cm}\\0.529 \text{ A}\end{array}\right\} \times \left(\dfrac{n^2}{Z}\right)$
Period, t	$\dfrac{2\pi\hbar^3}{e^4 m}\left(\dfrac{n^3}{Z^2}\right)$	$1.52 \times 10^{-16} \text{ sec} \times \left(\dfrac{n^3}{Z^2}\right)$

[a] R. S. Berry, *J. Chem. Educ.* **43**, 283 (1966). Used by permission.

It is illuminating to note the dependence of these properties on n and Z. For example, the radius of the orbit increases as the square of the principal quantum number, but its energy increases (becomes less negative) as $1/n^2$. Therefore, even as the orbits get very, very large for large n, the energies all squeeze into a small energy range below the ionization limit, as in Fig. 2.2.8, where part of the Lyman series is also indicated.

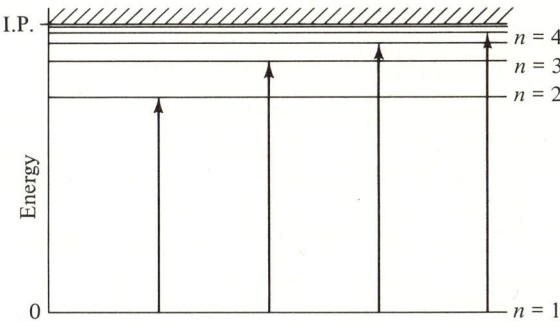

FIG. 2.2.8 Energy levels of H atom. The Lyman series in absorption.

To get a feeling for the velocity of the electron (and therefore perhaps for its localization) compare its velocity (Table 2.2.2) to the velocity of light. The period is the time required for one traversal of the orbit. In 1 sec the electron goes around the nucleus about 10^{16} times for $Z = n = 1$.

EXERCISE   In the Lyman series of the H-atom spectrum seen in absorption the initial state is $n_1 = 1$, and the final states are $n_2 = 2$, 3, 4, ... Calculate the frequencies at which a few of these transitions are to be observed. Give the results in $\sec^{-1}$, cm^{-1}, and in electron volts.

2.3 MATTER WAVES

The Bohr theory of one-electron atoms gives the Rydberg-Ritz equation for the transition energies and the correct value of the Rydberg constant, but it is clearly wrong. For example, this theory incorrectly predicts a nonzero angular momentum for the ground state. But the greatest failure of the theory was that it proved incapable of extension to atoms containing more than one electron. A new start is necessary, as was evident by the early 1920s.

The key to a new start is the wave–particle duality implied in Planck-Einstein quantization. Louis de Broglie (1924), noting the wave–particle duality for light, asked if there could be such a duality in matter. Could an electron be a classical particle of momentum p which for quantum interactions behaves like a wave of wavelength λ? Two questions arise:

1. First and most important, where is the experimental evidence that the electron can behave like a wave?
2. What is the relation, if any, between the properties of the electron (i.e., energy, mass, momentum, charge, etc.) and its wavelength?

De Broglie, being a theoretician, could not answer question 1, but he could offer a suggestion for question 2. In fact there are two ways to proceed to answer 2.

1. Find the wave equation of the electron (i.e., construct the quantum theory of matter).
2. Find a direct relation between some classical property of a particle and its quantum wavelength.

Just as Planck found a law which directly connects the classical frequency of light to the quantum mechanical energy of the photon without building the complete theory of quantum electrodynamics, de Broglie suggested[19] a relation of type 2 which is dimensionally correct.

---

[19] De Broglie used a relativistic argument as justification for this relation. According to the well-known relativistic relation of Einstein, $E = mc^2$, which combined with Planck's law, $E = h\nu$, gives the momentum of a photon, $mc = h\nu/c = h/\lambda$.

2.3 Matter Waves

$$\lambda = h/p \qquad \text{DE BROGLIE RELATION} \qquad (2.3.1)$$

The wavelength λ and the momentum p are related through Planck's constant.

EXERCISE   Show that h/p has dimensions of length.

The de Broglie relation introduces a wave–particle duality into the concept of matter by associating a quantum wavelength λ with each classical particle of momentum p.

THE DAVISSON–GERMER EXPERIMENT

Equation (2.3.1) is an interesting relation, but it requires strong experimental corroboration. Is there experimental evidence that the electron can behave as a wave? This question was answered by the Davisson-Germer experiment[20] (1927). While bombarding a specimen of crystalline nickel with an electron beam, these scientists unexpectedly obtained an interference pattern from the scattered beam. The electrons behaved like x-rays in a diffraction experiment. Interference is inherently a *wave* phenomenon. Using Bragg's law one calculates the wavelength of electrons necessary to produce this interference pattern. From the known momentum of the incident electrons a verification of the de Broglie relation is obtained.

Comparison of an x-ray experiment with the observed results of Davisson and Germer is made in Fig. 2.3.1. Following the Bragg law, the intensity of a beam of x-rays reflected from a crystal lattice is a maximum when

$$\frac{1}{\lambda} = \frac{n}{2d \sin \theta}$$

where λ is the wavelength of the x-ray, θ the angle of reflection, d, the lattice spacing, and $n = 1, 2, 3, 4, \ldots$. Thus, if one holds the angle θ constant and varies the wavelength of the x-rays, one will observe a series of maxima in the intensity of reflected waves. A plot of $1/\lambda$ versus intensity will show the x-ray maxima as a series of lines corresponding to $n = 1, 2, 3, 4, \ldots$. Davisson and Germer found corresponding maxima in a plot of the intensity of *reflected electrons* versus $(E)^{1/2}$. The electronic energy is $E = \frac{1}{2}p^2/m$, $(E)^{1/2} \propto p$, or $(E)^{1/2} \propto 1/\lambda$ according to the de Broglie relation, which is thereby verified.

EXERCISE   Electrons of energy 20 eV are incident on a crystal of lattice spacing $d = 2$ A. Find the angles of maximum reflectance.

[20] C. J. Davisson and L. H. Germer, *Nature* 119, 558 (1927).

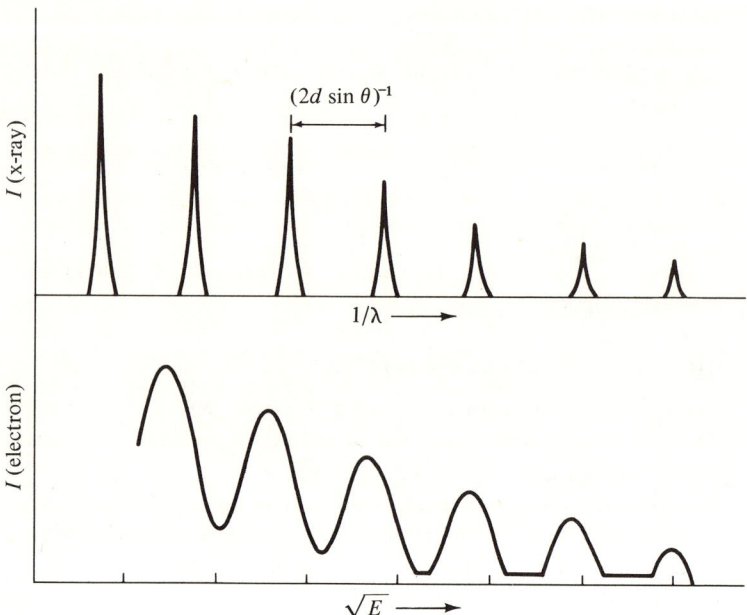

FIG. 2.3.1 Selective reflection of x-rays and electrons from nickel crystal. I(x-ray) is the theoretical intensity of a reflected beam of x-rays. I(electron) is the observed intensity of the reflected beam of electrons. After C. J. Davisson, *Franklin Inst. J.* **205**, 597 (1928). Used by permission.

EXERCISE Show that the Bohr assumption of quantized angular momentum, $L = mvr = m\hbar$, leads to a *standing electron wave* around the nucleus. That is, show that the circumference of the circular orbit contains an integral number of wavelengths. What would be the result if n were not an integer?

PRINCIPLE OF COMPLEMENTARITY

If an elementary particle is a wave then where is the idea of location and trajectory? Bohr[21] stated a *principle of complementarity* which clarifies the wave–particle duality. Bohr says, essentially, that wave and particle properties are complementary, exclusive, and cannot be exhibited simultaneously. This implies that there are conditions under which we can expect matter to display wave properties and other conditions under which a classical particle with classical trajectory is the correct description. In general, the particle of momentum p behaves like a wave of wavelength

---

[21] N. Bohr, *Nature* **121**, 580 (1928).

2.3 Matter Waves

h/p when the *potential* $V(x, y, z)$ changes over a distance of the same order as h/p. This is certainly the case for electrons in atoms, molecules, and crystals.

EXERCISE A cathode ray tube consists of a beam of electrons deflected over the face of a phosphorescent screen by applied potentials. If the electrons have energies of order 1 eV and the potentials vary appreciably over distances of the order of millimeters, show that the electrons will follow classical trajectories.

THE UNCERTAINTY PRINCIPLE

If the position and momentum of a particle could be determined very precisely under quantum conditions then the particle–wave would have both a well-defined wavelength *and* a well-defined position. This seems contradictory. Are there not limits on the precision with which x and p_x can be simultaneously determined? The easiest way to show there must be such a relation between x and p_x in practice is to try to imagine an experiment which simultaneously determines the position and momentum of an electron. Say two precise measurements of position at different times are made. From the elasped time between measurements, the velocity (i.e., momentum) of the electron follows at once. However, in order to locate a particle with such precision it is necessary to "see" the particle by using radiation of very short wavelength, that is, photons of large momentum. But the interaction of photon and electron changes the momentum of the electron.[22] Thus the very interaction of the particle with the measuring device ensures that both position and velocity cannot be measured simultaneously.

Werner Heisenberg[23] (1927) stated the *uncertainty principle*, which gives quantitative limits to the wave–particle duality. The relation can be derived from quantum theory

$$\Delta p_x \, \Delta x \geq \hbar/2 \qquad \text{HEISENBERG UNCERTAINTY PRINCIPLE} \qquad (2.3.2)$$

where Δp_x is the "uncertainty" (root-mean-square deviation of statistics) in the linear momentum corresponding to coordinate x, and Δx is the uncertainty in x. The other conjugate dynamical variables have similar uncertainty relations, that is, Δp_y and Δy, Δenergy and Δtime, ΔL, and $\Delta \theta$, where L and θ are angular momentum and angular coordinate. The

---

[22] The Compton effect, A. H. Compton, *Phys. Rev.* **21**, 483 (1923). Simply stated, Compton found that x-rays are inelastically scattered by electrons (i.e., the incident and scattered x-rays have different wavelengths). Conservation of energy and conservation of momentum lead to an expression which correctly describes the observations.

[23] W. Heisenberg, *Z. Phys.* **43**, 172 (1927).

Heisenberg uncertainty principle is a fundametal limitation on observation. For macroscopic bodies the uncertainty principle is meaningless because of their large size, but for particles it is a severe limitation. One can no longer talk of electron orbits as in the Bohr theory of the H atom. The definite orbits of the Bohr theory are replaced by *wavefunctions* $\Psi(x, t)$ which by their very nature imply a diffuseness of specification in the coordinates.

2.4 THE SCHRÖDINGER EQUATION

The classical Hamiltonian equations of motion (2.1.7) certainly have to be discarded when the position and momentum of a particle cannot be simultaneously specified at time t. Given that de Broglie's matter waves replace the classical trajectory of the particle, what we now require is a *wave equation*: a differential equation which generates the *wavefunction* $\Psi(x, t)$ of the quantum particle as a function of time and position. A possible candidate for a wavefunction is a cosine wave, $\psi(x) = \cos 2\pi x/\lambda$ (λ is the wavelength so x/λ is dimensionless). As in Fig. 2.4.1, the cosine

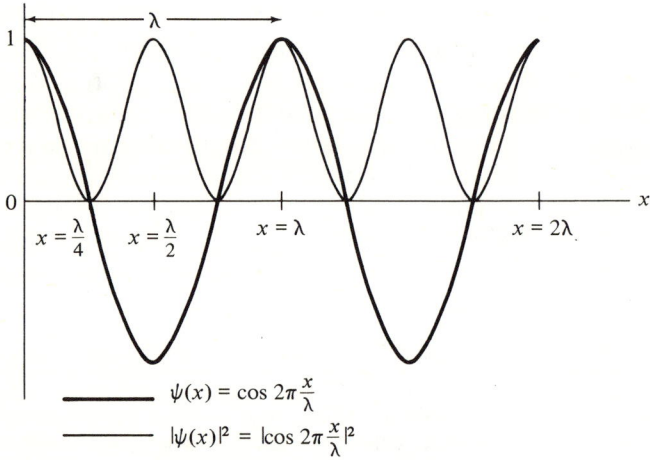

——— $\psi(x) = \cos 2\pi \frac{x}{\lambda}$

——— $|\psi(x)|^2 = |\cos 2\pi \frac{x}{\lambda}|^2$

FIG. 2.4.1 Cosine wave.

wave stretches from minus infinity to plus infinity and oscillates in value between $+1$ and -1 with wavelength λ. However, the infinite extent of this wave makes it a bad choice for an electron bound to an atom! In this case we need some degree of localization for physical reasons. In order to proceed, however, we will find the wave equation for the free (unbound) particle, that is, a particle possessing only kinetic energy. Like other wave

2.4 The Schrödinger Equation

equations of mathematical physics, the equation we seek is of the general form

$$\frac{\partial^n \Psi(x, t)}{\partial t^n} \propto \frac{\partial^m \Psi(x, t)}{\partial x^m}$$

where n and m are integers.

To help find the wave equation, combine Planck's law with the de Broglie relation and the known energy of the free particle.

$$E = h\nu = p^2/2m, \quad \text{but} \quad p = h/\lambda$$

So

$$\nu = h/2m\lambda^2 \qquad (2.4.1)$$

Equation (2.4.1) is the $\lambda\nu$ relation for quantum waves. (Note that it differs from the $\lambda\nu$ relation for light waves, $\lambda\nu = c$.)

In the wavefunction $\Psi(x, t)$ the frequency ν must be associated with the time t to give a dimensionless variable νt. Similarly, the wavelength λ must be associated with x to give a dimensionless variable x/λ. Since ν appears to the first power in (2.4.1) the wave equation involves only the first derivative with respect to t, but because $1/\lambda^2$ appears in (2.4.1) the wave equation contains the second derivative of Ψ with respect to x. Although this is just a dimensionality argument, it leads immediately to the wave equation[24]

$$\frac{\partial \Psi(x, t)}{\partial t} \propto \frac{h}{2m} \frac{\partial^2 \Psi(x, t)}{\partial x^2} \qquad (2.4.2)$$

consistent with the relation (2.4.1). The general solution to (2.4.2) is the wavefunction $\Psi(x, t) = K \exp(\nu t - x/\lambda)$. Substituting into (2.4.2) (K is an arbitrary constant)

$$\frac{\partial}{\partial t} \exp\left(\nu t - \frac{x}{\lambda}\right) = \frac{h}{2m} \frac{\partial^2}{\partial x^2} \exp\left(\nu t - \frac{x}{\lambda}\right)$$

$$\nu \exp\left(\nu t - \frac{x}{\lambda}\right) = \frac{h}{2m} \frac{1}{\lambda^2} \exp\left(\nu t - \frac{x}{\lambda}\right)$$

or

$$\nu = \frac{h}{2m\lambda^2}$$

---

[24] R. S. Berry, *J. Chem. Educ.* **43**, 283 (1966). This paper contains an excellent informal discussion of quantum waves, atomic orbitals, and many-electron atoms. The annotated bibliography of nearly 50 papers and books is unique.

However, since the magnitude of the wavefunction, $|\exp(vt - x/\lambda)|$, has the unfortunate property of going to infinity at large t, this solution must be rejected. The infinity is removed if the constant of proportionality in (2.4.2) is taken to be $ih/2m$ (or $i\hbar/2m$ in radians) where $i = \sqrt{-1}$. The resulting equation is the *time-dependent Schrödinger equation*[25] for the free particle.

$$\frac{\partial \Psi(x, t)}{\partial t} = \frac{i\hbar}{2m} \frac{\partial^2 \Psi(x, t)}{\partial x^2} \quad (2.4.3)$$

The general solution of (2.4.3) is the free-particle wavefunction

$$\Psi(x, t) = K \exp[i(2\pi x/\lambda - 2\pi vt)] \quad (2.4.4)$$

EXERCISE Show that (2.4.4) is a solution of the Schrödinger equation (2.4.3) and yields the relation (2.4.1).

The magnitude of the wavefunction is finite everywhere in space and time.

$$|\Psi(x, t)|^2 = \Psi^*(x, t)\Psi(x, t) = K^2$$

(Note that the magnitude squared of a complex function is the function times its complex conjugate. The complex conjugate, indicated by the star, is obtained from the function by changing the sign in front of each i in the function.)

In order to generalize this result we note that the Schrödinger equation and its free-particle solution imply the following: The *operator*

$$i\hbar \frac{\partial}{\partial t}$$

operates on $\Psi(x, t)$ to give the result

$$i\hbar \frac{\partial \Psi(x, t)}{\partial t} = hv\Psi(x, t) = E\Psi(x, t) \quad (2.4.5)$$

and the operator

$$\frac{\hbar}{i} \frac{\partial}{\partial x}$$

operates on $\Psi(x, t)$ to give the result

$$\frac{\hbar}{i} \frac{\partial \Psi(x, t)}{\partial x} = \frac{h}{\lambda} \Psi(x, t) = p\Psi(x, t) \quad (2.4.6)$$

---

[25] E. Schrödinger, *Ann. Phys.* **81**, 109 (1926).

2.4 The Schrödinger Equation

From (2.4.3) and (2.4.5) the operator relation follows.

$$i\hbar \frac{\partial \Psi(x, t)}{\partial t} = -\frac{\hbar^2}{2m}\frac{\partial^2 \Psi(x, t)}{\partial x^2} = E\Psi(x, t) \qquad (2.4.7)$$

Furthermore,

$$-\frac{\hbar^2}{2m}\frac{\partial^2 \Psi}{\partial x^2} = \frac{p^2}{2m}\Psi$$

and $E = p^2/2m$ is the energy of the free particle. It is natural to say that

$$-\frac{\hbar^2}{2m}\frac{\partial^2}{\partial x^2}$$

is the *operator for the energy* of a free particle and

$$\frac{\hbar}{i}\frac{\partial}{\partial x}$$

is the *operator for the momentum*. By analogy with classical mechanics, where the Hamiltonian H is the energy function,

$$\hat{H} = -\frac{\hbar^2}{2m}\frac{\partial^2}{\partial x^2}$$

is the *Hamiltonian operator* for the free particle in quantum mechanics.

The time-dependent Schrödinger equation can now be written

$$i\hbar \frac{\partial \Psi}{\partial t} = \hat{H}\Psi \qquad (2.4.8)$$

From $2\pi v = E/\hbar$ we obtain for the free-particle wavefunction (2.4.4)

$$\Psi(x, t) = K \exp(-iEt/\hbar)\psi(x)$$

where $\psi(x) = \exp(i2\pi x/\lambda)$. This is called a *stationary-state wavefunction* because, as shown previously, $\Psi^*(x, t)\Psi(x, t)$ is independent of the time (i.e., stationary). The time dependence can be removed from the Schrödinger equation by combining (2.4.7) and (2.4.8).

$$\hat{H}\psi(x) = E\psi(x) \qquad (2.4.9)$$

THE TIME-INDEPENDENT SCHRÖDINGER EQUATION

EXERCISE   Derive (2.4.9).

The time-independent Schrödinger equation is the working equation of quantum chemistry. At this point the basic equations and relations of quantum mechanics have been exhibited and made plausible. Guided by the theoretical suggestions of Planck, Einstein, and de Broglie, as well as the experimental observation of electron wave phenomena, we have been led to a wave equation, the Schrödinger equation. Time dependence was removed from the Schrödinger equation because we will work with conservative systems and systems in which energy is a precisely known constant of the motion. In the resulting equation $\hat{H}\psi = E\psi$, E is the energy (a number) and $\hat{H}$ is the Hamiltonian (an operator). Although the classical Hamiltonian is just $p^2/2m$, the particle in a potential $V(x)$ has the classical Hamiltonian $p^2/2m + V(x)$: Analogously, the quantum mechanical Hamiltonian operator is

$$\hat{H} = -\frac{\hbar^2}{2m}\frac{\partial^2}{\partial x^2} + V(x)$$

We also found an operator for the momentum

$$\hat{p} = \frac{\hbar}{i}\frac{\partial}{\partial x}$$

This important relation between the properties of the system and operators will be pursued further in the next section.

It is not clear at this point how one is to use the Schrödinger equation, an equation which contains two unknowns, E and $\psi(x)$. The clarification of this point requires us to look more closely at differential equations of this type in the next section. Then in the following two sections we examine two practical methods of obtaining approximate solutions of the Schrödinger equation: the variation method and the perturbation method. These methods will be applied in later chapters. The student who feels his mathematical skill is not sufficient will be assured to know that only the usual minimum knowledge of calculus is assumed. Although many new concepts and notations must be introduced, the basic mathematical manipulations are completely familiar.

2.5 EIGENVALUES, EIGENFUNCTIONS, AND OPERATORS

Our interest in eigenvalue equations stems from the fact that they yield the constants of the motion in quantum mechanics.

$$\hat{O}\phi = k\phi \quad \text{EIGENVALUE EQUATION} \quad (2.5.1)$$

$\hat{O}$ is an *operator*,[26] ϕ is an *eigenfunction* to said operator, and k is the *eigenvalue*. The operator is a formal representation of a *procedure to be carried out on the function written next to it*. In this vein we saw

$$\hat{p}\phi(x) = \frac{\hbar}{i}\frac{\partial}{\partial x}\phi(x) = p\phi(x) \quad (2.5.2)$$

---

[26] Specifically, our interest lies solely in *linear* operators, that is, $\hat{O}(\phi_1 + \phi_2) = \hat{O}\phi_1 + \hat{O}\phi_2$ (see Appendix A).

2.5 Eigenvalues, Eigenfunctions, and Operators

where $\hat{p}$ is the momentum operator, p is the momentum eigenvalue (i.e., the numerical value of the momentum), and $\phi(x)$ is the eigenfunction of $\hat{p}$. In this case the operation to be carried out is differentiation.

EXAMPLE Find the eigenfunctions of the operator d/dx.

$$\frac{d\phi(x)}{dx} = k\phi(x) \quad \text{or} \quad \frac{d\phi(x)}{\phi(x)} = d\ln\phi(x) = k\,dx$$

$$\ln\phi(x) = kx + c \quad \text{or} \quad \phi(x) = c'\exp(kx)$$

where c is the constant of integration.

As in the example, finding the eigenfunctions of an operator often involves solving a differential equation. The differential equation has a set of solutions, each solution characterized by the value of k, $\phi_k(x)$, the subscript indicating this is the eigenfunction belonging to the eigenvalue k.

Among differential equations, there is one general equation of great importance in quantum mechanics, the *Sturm-Liouville* equation. The differential equations of Chapters 3 and 4 will all be of this form, so it's necessary to obtain an appreciation of the properties of Sturm-Liouville eigenfunctions and eigenvalues.

$$\left\{\frac{d}{dx}\left[r(x)\frac{d}{dx}\right] - s(x)\right\}\phi(x) = -k\phi(x) \tag{2.5.3}$$

STURM-LIOUVILLE EQUATION

The Sturm-Liouville equation is an eigenvalue equation with the operator

$$\hat{L} = \frac{d}{dx}\left[r(x)\frac{d}{dx}\right] - s(x)$$

and r and s are ordinary functions of x, or constants. Thus, (2.5.3) can be rewritten as

$$\hat{L}\phi(x) = -k\phi(x)$$

EXERCISE What choice of r, s, and k yields (1) the Schrödinger equation of the free particle and (2) the Schrödinger equation of a particle in a potential $V(x)$?

The eigenfunctions $\phi(x)$ are the solutions of the Sturm-Liouville differential equation. The unique properties of these eigenfunctions derive from the Sturm-Liouville *boundary conditions*. For simplicity, consider the Sturm-Liouville (S-L) boundary conditions on the region of interest $a \leq x \leq b$, in which r and s are integrable functions or constants. In certain cases a and b may be allowed to go to $+\infty$ and $-\infty$.

$$\phi(x = a) = \phi(x = b) = 0 \quad \text{BOUNDARY CONDITIONS} \tag{2.5.4}$$

Although these are not the most general S-L boundary conditions, they enable us to proceed to prove some of the mathemtical properties of these eigenfunctions. The eigenfunctions are presumed to be *nondegenerate*, that is, the eigenvalues k are presumed to be *different* for each eigenfunction.

Statements A through J give some of the properties of nondegenerate S-L eigenfunctions and eigenvalues, thereby introducing several new words, definitions, and a notation used throughout the text. Appendix A summarizes the definitions introduced here.

A. The *Dirac braket* notation for the definite integral (including multiple integrals). Functions to the left of the comma in a braket are understood to be complex conjugate functions.

$$\langle \phi_i, \hat{O}\phi_j \rangle = \int_a^b \phi_i^* \hat{O}\phi_j \, dx$$

B. The *norm* N of a function. N is just a number.

$$\langle \phi_i, \phi_i \rangle = \int_a^b \phi_i^* \phi_i \, dx = N$$

EXAMPLE   Find the norm of $\cos \theta$ on the interval 0–2π.

$$N = \int_0^{2\pi} \cos^2 \theta \, d\theta = \left(\frac{\sin \theta \cos \theta}{2} + \tfrac{1}{2}\theta \right) \Big|_0^{2\pi} = \pi$$

C. *Normalization* of the function. If the norm is unity the function is said to be normalized. However, it is very easy to normalize a function which has $N \neq 1$. Let $\langle X_i, X_i \rangle = N$, then $\phi_i = X_i N^{-1/2}$ is the normalized form of X_i.

$$\langle \phi_i, \phi_i \rangle = \frac{1}{N} \langle X_i, X_i \rangle = \frac{N}{N} = 1$$

EXERCISE   Normalize $\cos \theta$ on the interval 0–2π.

D. The Sturm-Liouville operator $\hat{L}$ is *Hermitian*. That is,

$$\int_a^b \phi_i^* \hat{L}\phi_j \, dx = \int_a^b \phi_j \hat{L}^* \phi_i^* \, dx$$

or, in the braket notation,

$$\langle \phi_i, \hat{L}\phi_j \rangle = \langle \hat{L}\phi_i, \phi_j \rangle \tag{2.5.5}$$

2.5 Eigenvalues, Eigenfunctions, and Operators

PROOF

$$\int_a^b \phi_i^* \hat{L} \phi_j \, dx = \int_a^b \phi_i^* \frac{d}{dx}\left[r(x)\frac{d\phi_j}{dx}\right] dx - \int_a^b \phi_i^* s \phi_j \, dx$$

Integrating by parts

$$\int_a^b u\, dv = uv\Big|_a^b - \int_a^b v\, du \quad \text{where} \quad dv = \frac{d}{dx}\left(r\frac{d\phi_j}{dx}\right)$$

$$= \phi_i^* r \frac{d\phi_j}{dx}\Big|_a^b - \int_a^b \left(\frac{d\phi_i^*}{dx}\right) r \frac{d\phi_j}{dx} dx - \int_a^b \phi_i^* s \phi_j \, dx \quad (2.5.6)$$

The first term in equation (2.5.6) vanishes as a result of the S-L boundary conditions (2.5.4). Integrate the second term by parts, with $dv = d\phi_j/dx$,

$$= -\left(\frac{d\phi_i^*}{dx}\right) r \phi_j\Big|_a^b + \int_a^b \frac{d}{dx}\left(r\frac{d\phi_i^*}{dx}\right)\phi_j \, dx - \int_a^b \phi_i^* s \phi_j \, dx \quad (2.5.7)$$

The first term in (2.5.7) vanishes by (2.5.4), so the remainder is

$$= \int_a^b (\hat{L}^* \phi_i^*)\phi_j \, dx = \langle \hat{L}\phi_i, \phi_j \rangle$$

which proves the assertion that $\hat{L}$ is Hermitian. The proof is given with a real $\hat{L}$, so it is identical to its complex conjugate, but in general $\hat{L}^*$ is needed (see the following exercise). Because of property E, below, we require all our operators for observables to be Hermitian.

EXERCISE Given that the functions ϕ_i and ϕ_j satisfy the SL boundary conditions at $x = -\infty$ and $x = \infty$, prove that the momentum operator $\hat{p} = (\hbar/i)\, d/dx$ is Hermitian.

E. *Hermitian operators have only real eigenvalues.* (The eigenvalues of Hermitian operators do not contain $\sqrt{-1}$ and therefore are always real: a necessary property for a physical observable.)

PROOF $\hat{O}\phi_k = k\phi_k$ so $\hat{O}^*\phi_k^* = k^*\phi_k^*$; multiply the first by ϕ_k^* and integrate, the second by ϕ_k and integrate over normalized functions.

$$\langle \phi_k, \hat{O}\phi_k \rangle = k \quad \text{and} \quad \langle \hat{O}\phi_k, \phi_k \rangle = k^* \quad (2.5.8)$$

So by the Hermitian character of $\hat{O}$, $k = k^*$

F. The independent solutions[27] ϕ_k and ϕ_j of the S-L equation are *orthogonal*.

---

[27] By "independent solutions ϕ_k and ϕ_j" we mean that ϕ_k cannot be expressed as a linear combination of functions including ϕ_j (see Appendix A).

$$\langle \phi_j, \phi_k \rangle = \int_a^b \phi_j^* \phi_k \, dx = \begin{cases} 0 & k \neq j \text{ orthogonal} \\ 1 & k = j \text{ normalized} \end{cases} \quad \delta_{kj} \quad \text{(Kronecker delta)} \tag{2.5.9}$$

PROOF From the eigenvalue equations $\hat{L}\phi_k = -k\phi_k$ and $\hat{L}^*\phi_j^* = -j^*\phi_j^*$ it follows from multiplication by ϕ_j^* and ϕ_k, respectively, and integration that $\langle \phi_j, \hat{L}\phi_k \rangle = -k\langle \phi_j, \phi_k \rangle$ and $\langle \hat{L}\phi_j, \phi_k \rangle = -j^*\langle \phi_j, \phi_k \rangle$. From the Hermitian character of $\hat{L}$, $\langle \phi_j, \hat{L}\phi_k \rangle = \langle \hat{L}\phi_j, \phi_k \rangle$, so

$$\langle \phi_j, \hat{L}\phi_k \rangle - \langle \hat{L}\phi_j, \phi_k \rangle = -(k - j^*)\langle \phi_j, \phi_k \rangle = 0$$

Now $j^* = j$, and because $k \neq j$ then $(k - j) \neq 0$, so $\langle \phi_j, \phi_k \rangle$ must vanish.

G. The set of eigenfunctions to $\hat{L}$, ϕ_k, has *discrete*[28] eigenvalues n_k. The nondegenerate eigenvalues form an ordered set of numbers.

H. The set of eigenfunctions to $\hat{L}$, ϕ_k, is a *complete set*. The property of completeness means that if $\psi(x)$ is an arbitrary integrable function on the interval $a \leq x \leq b$ then $\psi(x)$ can be expanded in a complete set of functions on that interval.

$$\psi(x) = \sum_{i=1}^{\infty} c_i \phi_i(x) \quad \text{EXPANSION THEOREM} \tag{2.5.10}$$

The c_i are called *expansion coefficients*. They are just numbers to be determined so that the sum converges to $\psi(x)$ everywhere on the interval. Properties G and H are quite important in achieving practical solutions to the Schrödinger equation.

EXERCISE Find the definition of the coefficient c_k in the expansion (2.5.10) if $\psi(x)$ is known. Hint: multiply by ϕ_k, integrate.

I. If the eigenvalues are arranged in ascending order the corresponding eigenfunctions have *increasing numbers of nodes* between a and b (at a node $\phi(x) = 0$); ϕ_1 has no nodes between a and b; ϕ_2 has one node between a and b; etc.

J. A general property of eigenfunctions which is not limited to Sturm-Liouville eigenfunctions is that if $\phi(x)$ is *simultaneously* an eigenfunction to *two* operators then the operators will *commute*.

PROOF Let the two operators be $\hat{O}$ and $\hat{P}$ with eigenvalues k and m respectively; then $\hat{O}\phi = k\phi$ and $\hat{P}\phi = m\phi$.

$$\hat{P}\hat{O}\phi = \hat{P}k\phi = km\phi \quad \text{and} \quad \hat{O}\hat{P}\phi = \hat{O}m\phi = mk\phi$$

or

$$\hat{O}\hat{P}\phi = \hat{P}\hat{O}\phi \quad \text{that is,} \quad \hat{O}\hat{P} = \hat{P}\hat{O}$$

When $\hat{O}\hat{P} = \hat{P}\hat{O}$ the operators are said to commute.

---

[28] The eigenfunctions and eigenvalues have been presumed to be nondegenerate, and $r(x)$ and $s(x)$ are presumed positive on the *finite* interval (a, b).

2.5 Eigenvalues, Eigenfunctions, and Operators

EXAMPLE The properties of the Sturm-Liouville equation, its eigenfunctions, and eigenvalues will be demonstrated on a simple example. We choose the differential equation $d^2\phi(x)/dx^2 = -m^2\phi(x)$ and ask that $\phi(x)$ satisfy the S-L boundary conditions at $x = 0$ and $x = \pi$. The $\phi(x)$ are then the sine functions on the interval $0 \leq x \leq \pi$ (Fourier series). These functions, on such an interval, are later used in the quantum mechanical problem of a particle in a box.

1. *The S-L equation* is $d^2\phi(x)/dx^2 = -m^2\phi(x)$, which is equivalent to (2.5.3) with $r(x) = 1$, $s(x) = 0$ and $k = m^2$.
2. The *eigenfunctions* to this S-L operator are sin mx, or cos mx, or any linear combination of the two.

PROOF $d^2 \sin mx/dx^2 = m \, d \cos mx/dx = -m^2 \sin mx$; similarly, $d^2 \cos mx/dx^2 = -m^2 \cos mx$.

3. However, only sin mx satisfies the S-L *boundary conditions* at $x = 0$ and $x = \pi$. Thus sin mx is the eigenfunction. The eigenvalues m^2 form an *ordered set* of *discrete numbers*.

PROOF sin $m(x = 0) = 0$ for all m, but sin $m(x = \pi) = 0$ for $m = 1, 2, 3, 4, 5, \ldots$. (The set of functions sin $-mx$ are not only degenerate with sin mx but because sin $-mx = -\sin mx$ they are not independent functions.)

4. The *norm* of sin mx on the interval is $\int_0^\pi \sin^2 mx \, dx = \pi/2$.
5. The *normalization* of sin mx is achieved by $(2/\pi)^{1/2} \sin mx$.

PROOF

$$\int_0^\pi \left(\frac{2}{\pi}\right) \sin^2 mx \, dx = \left(\frac{2}{\pi}\right)\left(\frac{\pi}{2}\right) = 1$$

6. The eigenfunctions are *orthogonal*.

PROOF

$$\int_0^\pi \sin mx \sin kx \, dx = \left[\frac{\sin(m-k)x}{2(m-k)} - \frac{\sin(m+k)x}{2(m+k)}\right]_0^\pi = 0 \quad (m \neq k)$$

7. The S-L operator d^2/dx^2 is *Hermitian*. This follows from 1 and the boundary conditions.
8. The eigenvalues are *real*, that is, $m^2 = 1, 4, 9, 16, \ldots$.
9. The set of eigenfunctions sin mx form a *complete set* of functions on the interval $0 \leq x \leq \pi$. This is a well-known property of these functions and the resultant expansions are called Fourier series.

That is, the arbitrary integrable function $\psi(x)$ may be expanded as follows:

$$\psi(x) = \sum_{m=1}^{\infty} c_m \sin mx$$

The c_m are coefficients to be determined from

$$c_m = \frac{2}{\pi} \int_0^{\pi} \psi(x) \sin mx \, dx$$

Figure 2.5.1 gives a few terms in the expansion for $\psi(x) = x$; note how the series converges towards the representation of the line $\psi(x) = x$.

10. Finally note that $\sin x$ has no *nodes* (zeros) between 0 and π, while $\sin 2x$ has one node on this interval, $\sin 3x$ has two nodes on the interval, etc.

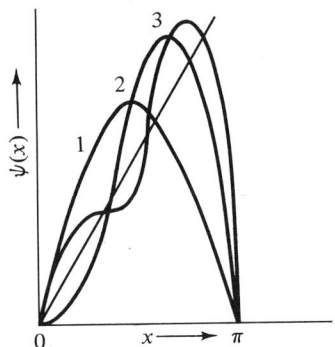

FIG. 2.5.1 Fourier series approximations to $\psi(x) = x$, with 1, 2, and 3 terms in the expansion $\psi(x) = \sum_{m=1}^{\infty} c_m \sin mx$.

2.6 THE VARIATIONAL PRINCIPLE

The properties of solutions to Sturm-Liouville type equations permit us to derive a useful and practical method for finding wavefunctions. We deal directly with the S-L equation which is the Schrödinger equation.

$$\hat{H}\phi_k = E_k \phi_k \quad \text{and} \quad E_k = \langle \phi_k, \hat{H}\phi_k \rangle / \langle \phi_k, \phi_k \rangle \tag{2.6.1}$$

Property G assures us that the energies E_k form an ordered discrete set of numbers, $E_1, E_2, E_3, \ldots$ such that $E_1 < E_2 < E_3 \ldots$. Property H assures us that an arbitrary integrable function can be expanded in the eigenfunctions of the Hermitian operator $\hat{H}$.

$$\tilde{\psi}(x) = \sum_{i=1}^{\infty} c_i \phi_i \tag{2.6.2}$$

2.6 The Variational Principle

The number W is defined as follows:

$$W = \frac{\int_a^b \tilde{\psi}^*(x)\hat{H}\tilde{\psi}(x)\,dx}{\int_a^b \tilde{\psi}^*(x)\tilde{\psi}(x)\,dx} = \frac{\langle \tilde{\psi}, \hat{H}\tilde{\psi}\rangle}{\langle \tilde{\psi}, \tilde{\psi}\rangle} \qquad (2.6.3)$$

Substituting the expansion (2.6.2) into (2.6.3)

$$W = \frac{\sum_{i,j} c_j^* c_i \langle \phi_j, \hat{H}\phi_i\rangle}{\sum_{i,j} c_j^* c_i \langle \phi_j, \phi_i\rangle} \qquad (2.6.4)$$

Using (2.6.1) and the orthogonality property in the numerator and denominator

$$W = \frac{\sum_i |c_i|^2 E_i}{\sum_i |c_i|^2} \qquad (2.6.5)$$

Subtracting E_1 [the lowest eigenvalue of (2.6.1)] from both sides of (2.6.5)

$$W - E_1 = \frac{\sum_i |c_i|^2 (E_i - E_1)}{\sum_i |c_i|^2} \qquad (2.6.6)$$

Since $(E_i - E_1) \geq 0$ for every value of i and $|c_i|^2 \geq 0$, therefore $W - E_1 \geq 0$ or

$$W \geq E_1 \qquad \text{VARIATIONAL PRINCIPLE} \qquad (2.6.7)$$

The variational principle says that in seeking a wavefunction $\tilde{\psi}$ as close as possible to the *unknown* ground-state wavefunction ϕ_1, we have a condition that $\tilde{\psi}$ will satisfy. Notice that this condition does not require us to know the value of E_1 beforehand. *The variational principle assures us that W is always above E_1, and $W = E_1$ when $\tilde{\psi} = \phi_1$.*

NONLINEAR PARAMETER

The manner in which the variational principle determines the best wavefunction of a given form is easily illustrated by variation of a nonlinear parameter. Consider the trial wavefunction $\tilde{\psi}(x, a) = \exp(-ax)$; a is a nonlinear parameter because it appears in the exponential, it is a parameter because W depends on the value of a.

$$W(a) = \frac{\langle \tilde{\psi}(x, a)\hat{H}\tilde{\psi}(x, a)\rangle}{\langle \tilde{\psi}(x, a), \tilde{\psi}(x, a)\rangle} \geq E_1 \qquad (2.6.8)$$

Imagine a Schrödinger equation $\hat{H}\phi_1 = E_1 \phi_1$ with $\phi_1 = \exp(-\tfrac{1}{2}x)$. Clearly, $W(a = \tfrac{1}{2}) = E_1$ and $\tilde{\psi}(x, a = \tfrac{1}{2}) = \phi_1$. How would the variational principle give these correct answers without prior knowledge of E_1? Plot $W(a)$ versus a. The plot gives the values of $W(a)$ versus a over a range of a.

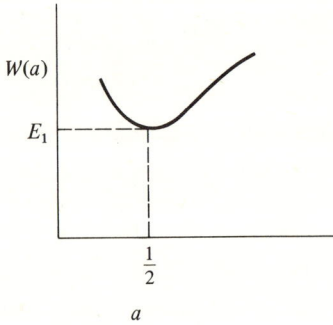

Note that $W(a)$ is always above E_1 and equals E_1 for $a = \frac{1}{2}$. For this simple case $\tilde{\psi}(x, a)$ becomes identical to ϕ_1, but in general the minimum W just brings $\tilde{\psi}$ as close as possible to ϕ_1.

EXERCISE   Suggest a use for $\partial W(a)/\partial a$.

LINEAR PARAMETERS

This is the basic technique of quantum chemistry. In this case the trial wavefunction is written as a linear combination of *known* functions X_i, for example, $\tilde{\psi}(x) = c_1 X_1 + c_2 X_2 + \ldots$. The *unknown* coefficients c_i are the linear parameters to be determined.

Of course, the formal justification for this form of trial wavefunction is the *expansion theorem* with a complete set of functions; in that case it takes the form $\phi_1 = \sum_{i=1}^{\infty} c_i X_i$. In practice, however, the trial wavefunction is a very truncated expansion, say, $\tilde{\psi}(x) = c_1 X_1 + c_2 X_2$. Such a trial wavefunction can only crudely approximate $\phi_1(x)$. The problem then is to choose c_1 and c_2 such that $\tilde{\psi}(x)$ is as close to $\phi_1(x)$ as possible in the *energetic sense*. The answer to this energetic problem is supplied by the variational principle.

$$W(c_1, c_2) = \frac{\langle (c_1 X_1 + c_2 X_2), \hat{H}(c_1 X_1 + c_2 X_2) \rangle}{\langle (c_1 X_1 + c_2 X_2), (c_1 X_1 + c_2 X_2) \rangle} \qquad (2.6.9)$$

Letting $H_{ij} = \langle X_i, \hat{H} X_j \rangle$, using the Hermitian property of $\hat{H}$, $H_{ij} = H_{ji}$, and the orthonormality of the X_i, we get

$$W(c_1, c_2) = \frac{c_1^2 H_{11} + c_2^2 H_{22} + 2 c_1 c_2 H_{12}}{c_1^2 + c_2^2} \qquad (2.6.10)$$

If E_1 is the ground-state energy of the system having Hamiltonian $\hat{H}$, and ϕ_1 is the ground-state wavefunction, then $\hat{H}\phi_1 = E_1 \phi_1$; the variational principle says that $W \geqslant E_1$. Since W will be greater than E_1 we can *minimize* W with respect to c_1 and c_2, that is, find the values of c_1 and c_2 such that $\partial W/\partial c_i = 0$.

$$\frac{\partial W}{\partial c_1} = 0 = \frac{2c_1 H_{11} + 2c_2 H_{12}}{c_1^2 + c_2^2} - \frac{2c_1 W}{c_1^2 + c_2^2}$$

$$\frac{\partial W}{\partial c_2} = 0 = \frac{2c_2 H_{22} + 2c_1 H_{12}}{c_1^2 + c_2^2} - \frac{2c_2 W}{c_1^2 + c_2^2}$$

(2.6.11)

The equations (2.6.11) are just

$$c_1 H_{11} + c_2 H_{12} - c_1 W = 0$$
$$c_2 H_{22} + c_1 H_{12} - c_2 W = 0$$

(2.6.12)

Equations (2.6.12) can be solved simultaneously for the ratio c_1/c_2 by eliminating W.

EXERCISE Solve for the ratio c_1/c_2 from equations (2.6.12).

ANSWER $\dfrac{c_1}{c_2} = \dfrac{(H_{11} - H_{22}) \pm \sqrt{(H_{11} - H_{22})^2 + 4H_{12}^2}}{2H_{12}}$

Having found the ratio of coefficients which minimizes W, the individual magnitudes of c_1 and c_2 can be found by imposing the condition that $\tilde{\psi}(x)$ be normalized to unity. In which case, $\langle \tilde{\psi}(x), \tilde{\psi}(x) \rangle = 1 = c_1^2 + c_2^2$, or $1/c_2 = \sqrt{c_1^2/c_2^2 + 1}$ and $1/c_1 = \sqrt{c_2^2/c_1^2 + 1}$. The problem is now *solved* in the very restricted sense that the limited expansion $c_1 X_1 + c_2 X_2$ has been made as energetically close to $\phi_1(x)$ as is feasible. A better trial wavefunction would include other X_i and would give a better representation of $\phi_1(x)$ and, of course, a better energy. If there are n functions X_i in the expansion, that is, $\tilde{\psi}(x) = \sum_{i=1}^n c_i X_i$, then the n variational conditions like equation (2.6.11) yield n simultaneous equations like (2.6.12) which must be solved by matrix techniques (see Appendix A).

2.7 THE PERTURBATION APPROACH

The perturbation approach to the solution of quantum chemical problems is a basic and simplifying component of quantum chemical technique. When faced with a *insoluble* Schrödinger equation, $\hat{H}\phi = E\phi$, we find a physically related model system which has a *soluble* Schrödinger equation, $\hat{H}_0 \psi^{(0)} = \varepsilon^{(0)} \psi^{(0)}$; all the *zero-order* quantities, $\hat{H}_0$, $\psi^{(0)}$, and $\varepsilon^{(0)}$ are known. Let $\lambda \hat{V} = \hat{H} - \hat{H}_0$; then the exact Hamiltonian is

$$\hat{H} = \hat{H}_0 + \lambda \hat{V} \qquad (2.7.1)$$

$\hat{V}$ is the *perturbation* to the known Hamiltonian $\hat{H}_0$ and λ is the perturbation parameter. Our hope is that $\lambda \hat{V}$ is in fact small, that is, a perturbation; then the following expansions converge:

$$\phi = \psi^{(0)} + \lambda\psi^{(1)} + \lambda^2\psi^{(2)} + \lambda^3\psi^{(3)} + \cdots \qquad (2.7.2)$$

$$E = \varepsilon^{(0)} + \lambda\varepsilon^{(1)} + \lambda^2\varepsilon^{(2)} + \lambda^3\varepsilon^{(3)} + \cdots \qquad (2.7.3)$$

$\psi^{(1)}, \psi^{(2)}, \ldots$, are the *first-order* wavefunction, *second-order* wavefunction, etc., and $\varepsilon^{(1)}, \varepsilon^{(2)}, \ldots$, are the *first-order energy*, *second-order* energy, etc.

Substitute (2.7.1)–(2.7.3) into $\hat{H}\phi = E\phi$, and equate coefficients of equal powers of λ.

$$\hat{H}_0\psi^{(0)} = \varepsilon^{(0)}\psi^{(0)} \qquad (2.7.4)$$

$$(\hat{H}_0 - \varepsilon^{(0)})\psi^{(1)} = (\varepsilon^{(1)} - \hat{V})\psi^{(0)} \qquad (2.7.5)$$

$$(\hat{H}_0 - \varepsilon^{(0)})\psi^{(2)} = (\varepsilon^{(1)} - \hat{V})\psi^{(1)} + \varepsilon^{(2)}\psi^{(0)} \qquad (2.7.6)$$

Equation (2.7.4) is the zero-order equation of perturbation theory, (2.7.5) is the first-order equation, etc., so the insoluble Schrödinger equation has been replaced by a set of coupled, but simpler, equations. Multiplying (2.7.5) through by $\psi^{(0)*}$ and integrating ($\langle \psi^{(0)}, \psi^{(1)} \rangle = 0$) we find

$$\varepsilon^{(1)} = \langle \psi^{(0)}, \hat{V}\psi^{(0)} \rangle \qquad (2.7.7)$$

The energy is determined to first order by the zero-order wavefunction, making it seem like we are pulling ourselves up by our own bootstraps.[29]

EXERCISE Given the zero-order problem $\hat{H}_0\psi^{(0)} = \varepsilon^{(0)}\psi^{(0)}$ as solved, and the full Hamiltonian $\hat{H} = \hat{H}_0 + \hat{V}/Z$ (where Z is the atomic number), suggest an expansion for ϕ and E in the insoluble atomic Schrödinger equation $\hat{H}\phi = E\phi$. Set up the first- and second-order perturbation equations for the problem and identify λ. Find $\varepsilon^{(1)}$. Do you think the expansions will converge?

The traditional method of solving (2.7.5) for $\psi^{(1)}$ is by use of the expansion theorem; $\psi^{(1)}$ is expanded in the complete set of known eigenfunctions to the Hermitian operator $\hat{H}_0$. Let $\psi_0^{(0)}$ be the particular unperturbed nondegenerate state of interest.[30]

$$\hat{H}_0\psi_k^{(0)} = \varepsilon_k^{(0)}\psi_k^{(0)} \qquad \psi_0^{(1)} = \sum_{k=1}^{\infty} c_k \psi_k^{(0)}$$

---

[29] It can be shown that $\psi^{(1)}$ determines the energy to third order, and, in general, the nth-order wavefunction determines the energy to order $2n + 1$.

[30] Degeneracy ($\varepsilon_0^{(0)} = \varepsilon_k^{(0)}$) is treated in the more advanced texts.

2.7 The Perturbation Approach

Equation (2.7.5) becomes

$$(\hat{H}_0 - \varepsilon_0^{(0)})\sum_{k=1}^{\infty} c_k \psi_k^{(0)} = (\varepsilon_0^{(1)} - \hat{V})\psi_0^{(0)}$$

$$\sum_{k=1}^{\infty} (\varepsilon_k^{(0)} - \varepsilon_0^{(0)}) c_k \psi_k^{(0)} = (\varepsilon_0^{(1)} - \hat{V})\psi_0^{(0)}$$

Multiplying the last equation from the left by $\psi_n^{(0)*}$ and integrating ($\langle \psi_n^{(0)}, \psi_k^{(0)} \rangle = \delta_{nk}$)

$$c_n = \frac{\langle \psi_n^{(0)}, \hat{V}\psi_0^{(0)}\rangle}{(\varepsilon_0^{(0)} - \varepsilon_n^{(0)})} \tag{2.7.8}$$

$$\psi_0^{(1)} = \sum_{k=1}^{\infty} \frac{\langle \psi_k^{(0)}, \hat{V}\psi_0^{(0)}\rangle}{(\varepsilon_0^{(0)} - \varepsilon_k^{(0)})} \psi_k^{(0)} \tag{2.7.9}$$

The second-order energy (multiply (2.7.6) by $\psi_0^{(0)*}$ and integrate) is

$$\varepsilon_0^{(2)} = \langle \psi_0^{(0)}, \hat{V}\psi_0^{(1)}\rangle = \sum_{k=1}^{\infty} \frac{|\langle \psi_k^{(0)}, \hat{V}\psi_0^{(0)}\rangle|^2}{(\varepsilon_0^{(0)} - \varepsilon_k^{(0)})} \tag{2.7.10}$$

This gives the first-order wavefunction and the energy to second-order in terms of known integrals over the perturbation $\hat{V}$. In order for a given term in the sum to make an appreciable contribution, the energy denominator $(\varepsilon_0^{(0)} - \varepsilon_k^{(0)})$ should be small. This means that the unperturbed states that are near to $\psi_0^{(0)}$ in energy make the largest contributions to $\psi_0^{(1)}$ and $\varepsilon_0^{(2)}$.

REFERENCES

Anderson, J. M., *Mathematics for Quantum Chemistry*, Benjamin, New York 1966.
Hameka, H., *Introduction to Quantum Theory*, Harper & Row, New York, 1967.
Pauling, L., and Wilson, E. B., *Introduction to Quantum Mechanics*, McGraw-Hill, New York, 1935.

PROBLEMS

1. Most chemical reactions have activation energies in the range of 15 to 90 kcal/mole. What is the corresponding range in (1) electron volts, (2) wavenumbers, (3) angstroms?
2. In a photoelectric experiment electrons are not ejected from the surface of tungsten metal by light of wavelength longer than 2700 A. What is the excess kinetic energy of electrons ejected by light of wavelength 1800 A? What is the velocity of these electrons? What is their de Broglie wavelength?
3. In a photoelectric experiment on a surface of sodium using light of wavelengths 3615 and 3125 A the photocurrent was reduced to zero by potentials of -0.950 volt and -0.382 volt, respectively. Find the threshold frequency and determine Planck's constant from the data.

4. Calculate and compare, by making a table, the velocity, radius, period, and frequency of electronic motion in the Bohr orbits $n = 1, 2, 3$, etc., if $Z = 1$ and $Z = 2$. Then calculate and compare the de Broglie wavelengths for electrons with these velocities.

5. In the Balmer series of H-atom spectral lines the final state (in emission) has $n = 2$. Calculate the difference in wavenumbers between the Balmer lines of hydrogen and deuterium. Compare this difference to the separation of lines in the series. This is the means by which the isotope was first detected.

6. Show that when an excited hydrogen atom undergoes a transition from a state having quantum number n to one having quantum number $n - 1$, the frequency of light radiated is approximately the same as the frequency of orbital motion of the electron predicted by the Bohr theory, *if n is very large in comparison with unity*.

7. An electron is confined to a one-dimensional box of length (1) 0.1 A, (2) 1 A, (3) 10 A. What is the minimum uncertainty in its momentum?

8. In order to "see" the details of molecular structure it is necessary to resolve distances of the order of a bond length 1–2 A. If a microscope cannot resolve distances much smaller than a wavelength, what energy (in eV) is required for electrons in an electron microscope in order to theoretically "see" molecular structure by this means?

9. Show that $xe^{-x^2/2}$ is an eigenfunction of the operator $-d^2/dx^2 + x^2$. What is the eigenvalue?

10. Find an entire class of operators which commutes with d/dx.

11. Show that $\sin kx$ is an eigenfunction of the operator d^2/dx^2. Find the eigenvalue.

12. The classical Hamiltonian for a hypothetical one-dimensional system of N particles of mass m and charge e interacting through Coulomb potentials is

$$H = \tfrac{1}{2} \sum_{i=1}^{N} p_i^2/m + \sum_{i>j}^{N} e^2/|x_i - x_j|$$

Derive the quantum mechanical Hamiltonian operator $\hat{H}$ for this system.

13. Show that in the region $-L \leq x \leq L$ the functions $f_0 = 1/\sqrt{2L}$, $f_n = \sqrt{L} \cos(n\pi/L)x$ and $g_n = 1/\sqrt{L} \sin(n\pi/L)x$ form an orthogonal and normalized (*orthonormal*) set.

14. Expand the function $\psi(x) = x^2$ in a Fourier series on the interval $0 \leq x \leq \pi$. Plot the first, second, and third approximations to the function.

15. Consider the physical system having the Hamiltonian

$$\hat{H} = -\frac{\hbar^2}{2m}\frac{d^2}{dx^2} + 2x$$

Given that the equation $\hat{H}\psi(x) = E\psi(x)$ has not been solved, take the trial wavefunction $\psi(x) = \exp(-kx)$ and find the value of k which gives the best wavefunction for the system. In what sense is this the "best" wavefunction?

3
THE PRINCIPLES OF QUANTUM MECHANICS WITH APPLICATIONS

**Actually we need not speak
of particles at all.**

W. HEISENBERG
Physics and Philosophy, 1958

THE WAVE–PARTICLE duality of matter—its statement, experimental verification, and justification in the Schrödinger equation—is a familiar and exciting part of recent scientific history. Agreement on the *interpretation* to be placed on features of the new theory was achieved only after long discussion and correspondence among the eminent physicists of the day. In the present chapter a concise statement of the principles and interpretation of quantum mechanics is given, with examples drawn from the problem of a particle in a box.

3.1 PROBABILITY AND THE WAVEFUNCTION

The Heisenberg uncertainty principle expresses an idea deeply embedded in quantum mechanics. While in classical mechanics both the position and the momentum of a body can be specified to any *observable* accuracy, simultaneously, this is just not possible for quantum mechanical particles because the very act of observation changes these variables. As the uncertainty in the coordinate is reduced, the uncertainty in the conjugate momentum must increase and vice versa, as $\Delta x\, \Delta p_x \geq \hbar/2$. We are thus constrained to surrender the trajectories of classical mechanics, or the orbits of the Bohr theory, in favor of wavefunctions $\Psi(x, t)$, functions of space and time. What interpretation can we place on these wavefunctions, and how are the properties of the system obtainable from them?

We will concern ourselves with conservative systems (energy is a constant of the motion) in which the energy is an eigenvalue of $\hat{H}$, so that the *stationary-state wavefunction* of the system has the form (Section 2.4) $\Psi(x, t) = \exp(-iEt/\hbar)\psi(x)$. The magnitude squared of the stationary-state wavefunction is independent of the time but varies from point to point as a function of x, $\Psi^*\Psi = |\Psi|^2 = |\psi(x)|^2$. This quantity is so important that it has been given several descriptive names.

$$\rho(x) = |\psi(x)|^2 \quad \text{THE PROBABILITY DISTRIBUTION FUNCTION OR THE PROBABILITY DENSITY} \quad (3.1.1)$$

The names arise from the following interpretation[1] of ρ:

$\rho(x)\,dx$ The probability of finding the particle between x and $x + dx$

$\rho(x, y, z)\,dV$ The probability of finding the particle in the volume element dV about the point x, y, z

The probabilistic interpretation of the relation between ρ and the location of a particle is an important aspect of quantum mechanics. In effect we *never* stop talking about particles, we just carefully (or tacitly) qualify statements about "particles" with this probabilistic interpretation, even though, as Heisenberg pointed out, we need not speak of particles at all. For example, chemists, knowing it is improper to speak of electrons as point particles, will instead want to know how the electron is "*distributed*" throughout the molecule (meaning the probability of finding the electron from point to point); the distribution is given by $\rho = |\psi|^2$, as in the *First Principle of quantum mechanics*.

I. The possible stationary states of the system have wavefunctions that are the solutions of the Schrödinger equation, $\hat{H}\psi = E\psi$; ψ completely specifies the dynamical properties of the system. The probability of finding a particle in the volume element dV is $|\psi|^2\,dV$.

COROLLARY ψ must be a well-behaved[2] function which is square-integrable so that it can be normalized to unity.

$$\int_{\substack{\text{all}\\\text{space}}} \rho\,dV = \langle \psi, \psi \rangle = 1 \tag{3.1.2}$$

Normalization to unity assures us that the total probability of finding the particle in space is 1. Finally, in order for $|\psi|^2$ to be physically interpretable as a probability density it is essential that ψ be well behaved (single-valued, continuous, and square-integrable).

EXERCISE   From the physical point of view what is wrong with a wavefunction which is not well behaved?

3.2   PARTICLE IN A ONE-DIMENSIONAL BOX

In order to display examples of wavefunctions, probability densities, normalization, and other quantum mechanical quantities, we choose the following simple physical problem having a solvable Schrödinger equation.

---

[1] M. Born, *Z. Phys.* **37**, 863 (1926); **38**, 803 (1926).
[2] See Appendix A.

3.2 Particle in a One-Dimensional Box

Consider a particle of mass m which moves in only one dimension (x) in a region of zero potential energy which extends from $x = 0$ to $x = L$. In order to confine the particle to this region the potential energy goes to infinity at the end points. The first step in solving a quantum mechanical problem is to write down the Hamiltonian.

$$\hat{H} = -\frac{\hbar^2}{2m}\frac{d^2}{dx^2} + \hat{V}(x) \tag{3.2.1}$$

Since $\hat{V}(x) = 0$ inside the box and the particle is confined to the box we have the Schrödinger equation

$$-\frac{\hbar^2}{2m}\frac{d^2}{dx^2}\psi(x) = E\psi(x) \quad 0 \leqslant x \leqslant L \tag{3.2.2}$$

The general solution of (3.2.2) is

$$\psi(x) = A\cos\left(\frac{2mE}{\hbar^2}\right)^{1/2} x + B\sin\left(\frac{2mE}{\hbar^2}\right)^{1/2} x$$

Since this solution extends beyond the end points we impose the *boundary conditions* $\psi(x = 0) = 0$ and $\psi(x = L) = 0$ on the wavefunction.[3] The first condition implies that $A = 0$, while the condition $\psi(x = L) = 0$, implies that

$$\psi(x = L) = B\sin\left(\frac{2mE}{\hbar^2}\right)^{1/2} L = 0$$

or

$$\left(\frac{2mE}{\hbar^2}\right)^{1/2} L = n\pi \quad \text{with} \quad n = 1, 2, 3, 4, \ldots.$$

$$E_n = \frac{n^2\hbar^2\pi^2}{2mL^2} = \frac{n^2h^2}{8mL^2} \tag{3.2.3}$$

$$\psi_n(x) = B\sin\frac{n\pi}{L}x \tag{3.2.4}$$

The $\psi_n(x)$ are Sturm-Liouville eigenfunctions.[4]

EXERCISE Show that (3.2.2) is a simple Sturm-Liouville equation and that the boundary conditions are S-L boundary conditions.

The $\psi_n(x)$ ($n = 1, 2, 3, 4, \ldots$) have all the properties of S-L eigenfunctions discussed in the last chapter. They are *orthogonal, normalizable*, form a

---

[3] The probability of finding the particle at the "wall" is now zero. In the region outside the box the infinite potential energy makes $\psi(x) = 0$ the only solution which is reasonable.
[4] See the example in Section 2.5.

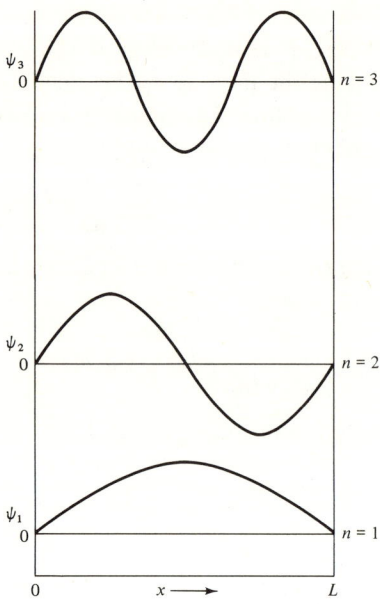

FIG. 3.2.1 Unnormalized wavefunctions $\psi_n = \sin(n\pi x/L)$ of particle in a box plotted at heights corresponding to relative energies.

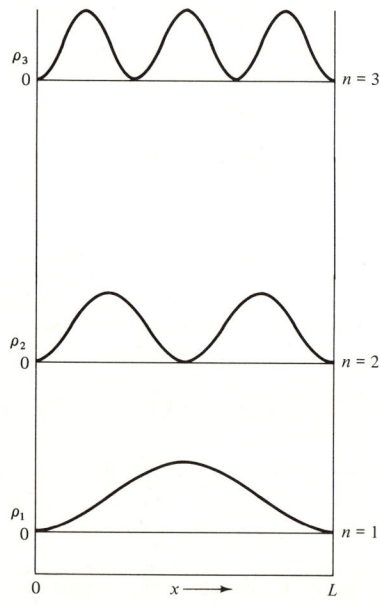

FIG. 3.2.2 Unnormalized probability densities ρ_n for particle in a box.

complete set, and have *discrete, real eigenvalues*, as seen in (3.2.3). We will call the $\psi_n(x)$ the *orbitals* of the particle in a box in order to introduce the following definition:

ORBITAL = ONE-PARTICLE WAVEFUNCTION

EXAMPLE Determine B such that the orbitals are normalized over the interval $0 \leq x \leq L$; $\langle \psi_n(x), \psi_n(x) \rangle = 1$.

$$\int_0^L |\psi(x)|^2 \, dx = \int_0^L B^2 \sin^2 \frac{n\pi}{L} x \, dx$$

$$= \frac{LB^2}{n\pi} \int_0^{n\pi} \sin^2 y \, dy$$

$$= \frac{LB^2}{n\pi} \left[-\tfrac{1}{2} \sin y \cos y + \tfrac{1}{2} y \right]_0^{n\pi}$$

$$= 1$$

or

$$B = \left(\frac{2}{L}\right)^{1/2} \quad \text{and} \quad \psi_n(x) = \left(\frac{2}{L}\right)^{1/2} \sin \frac{n\pi}{L} x$$

EXERCISE Show that $\psi_n(x)$ and $\psi_m(x)$ ($n \neq m$) are orthogonal, that is, $\langle \psi_n, \psi_m \rangle = 0$.

According to the first principle of quantum mechanics the probability of finding the particle in the element dx is $|\psi_n(x)|^2 \, dx$. The distribution functions $\rho_n = |\psi_n(x)|^2$ and the wavefunctions for several of the lower energies are plotted in Figs. 3.2.1 and 3.2.2. The probability of finding the particle in dx is not uniform throughout the box, but has maxima and minima depending on the value of n. In Section 3.7 we will correlate these maxima with the bonding in certain conjugated molecules.

3.3 EIGENVALUES AND OPERATORS

Having obtained the wavefunction, the probability distribution function, and the energy by solving the Schrödinger equation, we next ask how to calculate other observable properties from ψ. Principle II addresses itself to this question.

> II. With every observable property of a system there is associated a linear Hermitian operator $\hat{O}$. A measurement of the observable property yields an eigenvalue k, satisfying $\hat{O}\psi = k\psi$.

Examples of Hermitian operators associated with observables are $\hat{p}$, the momentum operator, and $\hat{H}$. To find other operators take their classical formulas and replace

$$p_x \text{ with } \frac{\hbar}{i}\frac{\partial}{\partial x}, \quad p_y \text{ with } \frac{\hbar}{i}\frac{\partial}{\partial y}, \quad \text{and } p_z \text{ with } \frac{\hbar}{i}\frac{\partial}{\partial z}$$

The coordinates x, y, and z are unchanged when used as operators.

EXAMPLE The particle-in-a-box equation $\hat{H}\psi = E\psi$ is an example of an eigenvalue equation. The Hamiltonian, the eigenfunctions, and the eigenvalues are known from (3.2.1) to (3.2.4). For $n = 1$ we have

$$\hat{H}\psi_1 = -\frac{\hbar^2}{2m}\frac{d^2}{dx^2}\left(B\sin\frac{\pi}{L}x\right)$$

$$= -\frac{\hbar^2}{2m}\frac{\pi}{L}\frac{d}{dx}\left(B\cos\frac{\pi}{L}x\right)$$

$$= \frac{h^2}{8mL^2}B\sin\frac{\pi}{L}x$$

$$= E_1\psi_1$$

EXERCISE Show that the particle-in-a-box wavefunctions are *not* eigenfunctions of the momentum $\hat{p}_x$ or the coordinate operators.

From the last exercise it follows that not *all* observables are associated with operators which have eigenvalues in a particular system. We might well ask what the conditions are under which two observables have simultaneous eigenvalues. In particular, since $\hat{H}\psi = E\psi$, what is the condition under which ψ is also an eigenfunction of $\hat{O}$? The answer follows from property J in Section 2.5. If ψ is simultaneously an eigenfunction of $\hat{O}$ and $\hat{H}$, then $\hat{O}\psi = k\psi$ and $\hat{H}\psi = E\psi$, so $\hat{H}\hat{O}\psi = \hat{O}\hat{H}\psi = kE\psi$, or $\hat{H}\hat{O} = \hat{O}\hat{H}$. That is, $\hat{O}$ *must commute with the Hamiltonian*. If two operators do not commute there is no experiment in which the corresponding observables can have simultaneous eigenfunctions.

EXERCISE Examine the operators for coordinate and conjugate momentum. Can they have simultaneous eigenfunctions? Discuss the connection with the uncertainty principle.

SYMMETRY

If a quantum mechanical system has a certain symmetry we should recognize that this symmetry is also an important observable property in the following sense. First of all, by symmetry we mean that the Hamiltonian is invariant to certain symmetry operations such as rotation about an

3.3 Eigenvalues and Operators

axis by 180° or inversion of coordinates through the origin ($x \to -x$, $y \to -y$, $z \to -z$). We can represent these operations by operators, that is, $\hat{C}_n$ is rotation about an axis by an angle $2\pi/n$ and $\hat{I}$ is inversion. Secondly, the statement that the physical situation is invariant to the symmetry operation means that the observable $|\psi|^2$ must be the same before and after the symmetry operation, that is, $|\psi|^2 = |\hat{O}\psi|^2$, which implies $\hat{O}\psi = \pm\psi$. Thus, in general, the existence of symmetry in the physical system implies that the *wavefunction is an eigenfunction* to the appropriate symmetry operator.[5] This important condition is usually easy to impose on approximate wavefunctions.

EXAMPLE   The use of a symmetry operator can be illustrated with particle-in-a-box wavefunctions. Note that the box is unchanged by rotation about the point $x = L/2$ by 180°. This is an elementary symmetry operation $\hat{C}_2$. The effect of $\hat{C}_2$ is to replace x with $(L - x)$, that is, $x = 0$ goes to $x = L$ and $x = \frac{1}{4}$ goes to $x = 3/4$, etc. Since the box is the same before and after the operation all the observables, for example, $|\psi_n|^2$ must be the same before and after.

$$\hat{C}_2 \psi_n(x) = \psi_n(L - x)$$

$$\hat{C}_2 B \sin\frac{n\pi x}{L} = B \sin\left(n\pi - \frac{n\pi x}{L}\right)$$

Using $\sin(a - b) = \sin a \cos b - \cos a \sin b$

$$\hat{C}_2 \psi_n(x) = \sin n\pi \cos\frac{n\pi x}{L} - \cos n\pi \sin\frac{n\pi x}{L}$$

$$= \pm \sin\frac{n\pi x}{L} \begin{cases} + \text{ if } n \text{ is odd} \\ - \text{ if } n \text{ is even} \end{cases}$$

ψ_n is an eigenfunction to $\hat{C}_2$ with eigenvalues ± 1 depending on whether n is odd or even. A simple symmetry operation serves to divide the particle-in-a-box wavefunctions into two classes of different symmetry: those functions which are symmetric ($+$) and those which are antisymmetric ($-$) with respect to $\hat{C}_2$. Figure 3.2.1 clearly illustrates the difference between the symmetric and antisymmetric particle-in-a-box wavefunctions and why they change sign under $\hat{C}_2$. Of course, $|\psi_n|^2 = |\pm\psi_n|^2 = |\hat{C}_2\psi_n|^2$. The *observable* probability density is unchanged (see Fig. 3.2.2).

EXERCISE   If $\hat{O}$ is a useful symmetry operator it must commute with the Hamiltonian. Then, $\hat{O}\hat{H} = \hat{H}\hat{O}$ and ψ is an eigenfunction to *both* $\hat{O}$ and $\hat{H}$. In the previous example the symmetry operator $\hat{C}_2$ was found to be useful in classifying the particle-in-a-box eigenfunctions into symmetric and antisymmetric functions with respect to $\hat{C}_2$ ($\hat{C}_2\psi_n = \pm\psi_n$). Since ψ_n is an eigenfunction to both $\hat{H}$ and $\hat{C}_2$, these operators must commute: Show it. Hint: Remember that $\hat{C}_2 f(x) = f(L - x)$ and show that $\hat{C}_2 \hat{H} f(x) = \hat{H} \hat{C}_2 f(x)$.

---

[5] In the absence of degeneracy; see Sections 6.3–6.6. The general relation is $\hat{O}\psi = \exp(i\theta)\psi$ so that the wavefunction is changed, at most, by a constant phase factor and $|\hat{O}\psi|^2 = \exp(-i\theta)\exp(i\theta)\psi^*\psi = |\psi|^2$.

3.4 AVERAGE VALUES

It may happen, and often does, that the operator $\hat{O}$ associated with a certain observable does not commute with the Hamiltonian. Then ψ is not an eigenfunction of $\hat{O}$; what then is the value of the observable? To answer, first examine the concept of probability. Consider a series of n situations having probabilities P_1/N for the first situation, P_2/N for the second, and so on up to P_n/N for the last. $N = \sum_{i=1}^{n} P_i$ so that the total probability is normalized to unity, $\sum_{i=1}^{n} P_i/N = 1$. With each situation is associated a certain value of a property Q_i. What do you expect the value of Q to be if a large number of measurements are made to determine its value? That is, what is the *expectation value* or *average value*, $\langle Q \rangle$, of Q? This is quite easy; weight each possible value of Q according to its probability, P_i/N.

$$\langle Q \rangle = \sum_{i=1}^{n} Q_i P_i/N \qquad (3.4.1)$$

EXAMPLE   Toss a coin. Probabilities of heads and tails are both $\frac{1}{2}$ and the total probability is normalized, $P_h = P_t = \frac{1}{2}$, $P_h + P_t = 1$. Suppose someone will give you 2¢ every time heads come up and take away 1¢ every time tails come up. $Q_h = +2¢$ and $Q_t = -1¢$. What is the expectation value?

$$\langle Q \rangle = \tfrac{1}{2} 2¢ - \tfrac{1}{2} 1¢ = +\tfrac{1}{2}¢ = \text{average financial gain per toss}$$

In quantum mechanics the probabilities are continuous so the sums are replaced by integrals. The expectation value is a statistical average over values of the property having Hermitian operator $\hat{O}$ as in the following statement of Principle III.

> III. The average value or expectation value for a number of measurements of the observable associated with the operator $\hat{O}$, on systems having wavefunction ψ, is
>
> $$\langle \hat{O} \rangle = \frac{\int \psi^* \hat{O} \psi \, dV}{\int \psi^* \psi \, dV} = \frac{\langle \psi, \hat{O}\psi \rangle}{\langle \psi, \psi \rangle} \qquad (3.4.2)$$

If ψ is normalized to unity, $\langle \hat{O} \rangle = \langle \psi, \hat{O}\psi \rangle$.

EXERCISE   If ψ is not an eigenfunction to $\hat{O}$ the average value $\langle \hat{O} \rangle$ is found using (3.4.2). Show that if ψ *is* an eigenfunction to $\hat{O}$ then (3.4.2) yields the eigenvalue.

EXERCISE $\hat{O}\phi_a = a\phi_a$ and $\hat{O}\phi_b = b\phi_b$ are given together with $\langle \phi_a, \phi_b \rangle = 0$. If $\psi = \tfrac{1}{2}\phi_a + \tfrac{1}{2}\phi_b$, find $\langle \hat{O} \rangle$ and interpret the result.

3.4 Average Values

Just as Hermitian operators have real eigenvalues, the Hermitian character of the operator $\hat{O}$ assures us that the average value of the observable is a *real number*.

PROOF

$$\langle \hat{O} \rangle = \langle \psi, \hat{O}\psi \rangle = \int \psi^* \hat{O} \psi \, dV$$

$$\langle \hat{O} \rangle^* = \langle \hat{O}\psi, \psi \rangle = \int \psi \hat{O}^* \psi^* \, dV$$

If $\langle \hat{O} \rangle$ is real then $\langle \hat{O} \rangle^* = \langle \hat{O} \rangle$ or $\langle \psi, \hat{O}\psi \rangle = \langle \hat{O}\psi, \psi \rangle$, but this is the definition of a Hermitian operator (Section 2.5).

According to the third principle of quantum mechanics average values of the properties of the "particle" can be calculated from the $\psi_n(x)$, the wavefunctions (orbitals) for the states of the particle. For example, what is the average position of the particle in a box?

EXAMPLE The average value of x, denoted $\langle \hat{x} \rangle$, is found from $\langle \hat{x} \rangle = \langle \psi_n, \hat{x}\psi_n \rangle$, when normalized wavefunctions are used.

$$\langle \hat{x} \rangle = \frac{2}{L} \int_0^L x \sin^2 \frac{n\pi}{L} x \, dx$$

$$= \frac{2L}{n^2 \pi^2} \int_0^{n\pi} y \sin^2 y \, dy$$

$$= \frac{2L}{4n^2 \pi^2} \left[y^2 - y \sin 2y - \tfrac{1}{2} \cos 2y \right]_0^{n\pi}$$

$$\langle \hat{x} \rangle = \frac{L}{2} \tag{3.4.3}$$

Since $\langle \hat{x} \rangle$ has just been shown to be independent of n, all states of the particle have the same expectation value of $\hat{x}$, $L/2$.

EXERCISE Two questions usually arise at this point which are worth a moment of consideration because they clarify what the introduction of averages and waves has done. (1) How can $\psi_2(x)$ have $\langle \hat{x} \rangle = L/2$ when Fig. 3.2.2 shows that the particle has zero probability of being at $L/2$ in this state? (2) In $\psi_2(x)$, how does the "particle" get from one side of the box to the other through a point ($x = L/2$) where the probability is zero?

A sharp distinction exists in quantum mechanics between those observables of the system which have well-defined values, that is, eigenvalues, and those observables which have average values. The particle in a box offers examples of each in the eigenvalue E_n (sharp and well defined for each n) and the average value of x (the same for all n, and an average).

3.5 ELECTRON SPIN, SPIN-ORBITALS, AND THE PAULI PRINCIPLE

The Bohr theory of the H atom explains the gross features of the H-atom spectrum as expressed in the Rydberg-Ritz formula [compare (2.2.6) to (2.2.8)]. However, a careful examination of the spectral lines shows that each line of, say, the Balmer series, has a *fine structure*; each line is a *multiplet* of closely spaced lines. This is not predicted by the simple Bohr theory, nor is the multiplet structure which results from placing an atom in a magnetic field while taking its spectrum (Zeeman effect) correctly predicted.

In 1924 Wolfgang Pauli suggested that many of these difficulties in the interpretation of atomic spectra are removed if one assumes that the electron has a fourth degree of freedom (in addition to the three degrees of freedom corresponding to motion in the x, y, and z directions). Pauli also rather cautiously postulated a "classically nondescribable two-valuedness" to this new degree of freedom. This is essentially the assignment of an additional quantum number to the electron with only two possible values. But physicists found it difficult to understand this additional degree of freedom because Pauli gave it no meaning in terms of a model of the electron. Uhlenbeck and Goudsmit[6] (two young graduate students at the University of Leiden) then stepped into the gap. They proposed that Pauli's suggestion and the observed effects could be interpreted in terms of an electron having an intrinsic angular momentum (i.e., a "*spinning*" *electron*; however, because electron "spin" cannot be described classically, we use the word *spin* as just an abstract name).

Pauli demonstrated that the assignment of a fourth degree of freedom to the electron fits nicely with the theory of relativity, in which, indeed, three space coordinates and a time coordinate (the fourth dimension) are necessary. Later, the relativistic treatment of the electron was given by Dirac (1928). He derived electron spin by combining the Schrödinger wave mechanics with the theory of relativity.

The spin is a *quantized* intrinsic angular momentum having quantum numbers $m_s = \pm\frac{1}{2}$, that is, magnitude of spin, $s = \frac{1}{2}$. A spinning charge gives rise to a magnetic moment, as would any circular electric current. Thus the electron has two equally fundamental electrical properties, a charge e and a magnetic moment of approximate magnitude[7]

$$\mu = s\frac{e\hbar}{mc} = 9 \times 10^{-21} \text{ erg/gauss}$$

---

[6] G. E. Uhlenbeck and S. A. Goudsmit, *Nature*, **117**, 264 (1926).

[7] This is a convenient unit for the measurement of atomic magnetic moments; it is called the Bohr magneton, $e\hbar/2mc = 9.2732 \times 10^{-21}$ erg/gauss. (See Section 6.9.)

3.5 Electron Spin, Spin-Orbitals, and the Pauli Principle

Atoms and molecules possessing a magnetic moment are called *paramagnetic*; they tend to orient themselves in an external magnetic field as a compass needle orients itself in the earth's magnetic field. Atoms and molecules which do *not* possess a magnetic moment are called *diamagnetic*. The intrinsic magnetic moment (spin) of the electron is experimentally demonstrated, among other ways, in magnetic field experiments such as (1) beam experiments, in which a beam of paramagnetic atoms is deflected by inhomogeneous magnetic fields; (2) Zeeman effect experiments, in which spectral lines are split by external magnetic fields. To give a simplified example, a beam of H atoms is split into two beams upon traversing an inhomogeneous magnetic field. The beams are composed of atoms with two possible components of magnetic moment in the field direction

$$\mu_z = -m_s \frac{e\hbar}{mc} = \pm\tfrac{1}{2}\frac{e\hbar}{mc}$$

This means the magnetic field resolves the ground state of H atom into two components, that is, two stationary states which are degenerate in the absence of the field, corresponding to $m_s = \pm\tfrac{1}{2}$. Other elementary particles (proton, neutron) also show effects interpretable in terms of an intrinsic angular momentum[8] or spin (see Section 6.9).

SPIN-ORBITALS

In the description of the wave mechanics of the electron the additional degree of freedom is specified by the *spin function*; that is, $\alpha(m_s = \tfrac{1}{2})$ and $\beta(m_s = -\tfrac{1}{2})$. A complete one-electron wavefunction is a *spin-orbital*.

$$\begin{aligned}\psi(x,y,z,s) &= \phi(x,y,z)\alpha \quad m_s = \tfrac{1}{2}\\ \psi(x,y,z,s) &= \phi(x,y,z)\beta \quad m_s = -\tfrac{1}{2}\end{aligned} \quad \text{SPIN-ORBITALS} \quad (3.5.1)$$

The spin-orbital is the product of a space orbital $\phi(x, y, z)$ times a spin function α or β.

For example, in the hypothetical case of an electron confined to a one-one-dimensional box each electron space orbital $\psi_n(x)$ goes to two spin-orbitals, $\psi_{nm_s}(x)$.

$$\psi_{n\tfrac{1}{2}} = \psi_n(x)\alpha = \left(\frac{2}{L}\right)^{1/2} \sin\left(\frac{n\pi}{L} x\right)\alpha$$

$$\psi_{n-\tfrac{1}{2}} = \psi_n(x)\beta = \left(\frac{2}{L}\right)^{1/2} \sin\left(\frac{n\pi}{L} x\right)\beta$$

---

[8] For a general discussion of spin angular momentum see H. Hameka, *Introduction to Quantum Theory*, Harper & Row, New York, 1967. An advanced formal treatise is P. A. M. Dirac, *The Principles of Quantum Mechanics*, 4th ed., Oxford, London, 1958.

By definition the spin functions are orthogonal and normalized.

$$\langle \psi_{n\frac{1}{2}}, \psi_{n\frac{1}{2}} \rangle = \langle \psi_n \alpha, \psi_n \alpha \rangle = \langle \psi_n, \psi_n \rangle \langle \alpha, \alpha \rangle$$
$$= \langle \alpha, \alpha \rangle = 1$$
$$\langle \psi_{n\frac{1}{2}}, \psi_{n-\frac{1}{2}} \rangle = \langle \psi_n \alpha, \psi_n \beta \rangle = \langle \psi_n, \psi_n \rangle \langle \alpha, \beta \rangle$$
$$= \langle \alpha, \beta \rangle = 0$$

THE PAULI PRINCIPLE

With the postulate of a fourth degree of freedom for the electron, Pauli moved on to state a principle which is vital to the understanding of many-electron systems. From an examination of the difficulties in atomic spectroscopy (1924) Pauli concluded that no two electrons in an atom can have identical quantum numbers (including m_s). This "Pauli exclusion principle" extends far beyond its initial application to atomic spectra. In its most general form, the Pauli principle is a statement about the permutation symmetry of the many-electron wavefunction which follows from the fact that electrons are *identical* and *indistinguishable*. (While in classical mechanics the trajectories of identical particles can be followed, in principle, and the particles thereby remain distinguishable, this is simply not possible for *matter waves* which are subject to interference.)

We saw in the previous principles of quantum mechanics that the properties of a system are given by its wavefunction. The wavefunction is a mathematical expression in terms of the space and spin coordinates of the particles, $\Psi(x_1, y_1, z_1, s_1; x_2, y_2, z_2, s_2; \ldots; x_N, y_N, z_N, s_N)$ or $\Psi(1, 2, 3, \ldots, N)$ for short. Let the operator $\hat{P}_{ij}$ permute (or exchange) electrons i and j. Because $\hat{P}_{ij}$ commutes with $\hat{H}$ the wavefunction must be an eigenfunction of $\hat{P}_{ij}$, that is, the wavefunction has exchange symmetry. Furthermore, because the electrons are identical and indistinguishable, the observable $|\Psi|^2$ must be the same before and after the symmetry operation $\hat{P}_{ij}$. Therefore $\hat{P}_{ij}\Psi = \pm\Psi$.

If two particles interchange space-spin coordinates and the wavefunction is unaffected, the wavefunction is *symmetric* ($\hat{P}_{ij}\Psi = \Psi$); but if the wavefunction changes sign it is *antisymmetric* ($\hat{P}_{ij}\Psi = -\Psi$). The Pauli principle states that the electronic wavefunction is antisymmetric.

IV. The wavefunction of an N-electron system is antisymmetric in the space-spin coordinates of the electrons.

Furthermore, wavefunctions pertaining to *all* particles having half-integral spin are antisymmetric, but those pertaining to particles having integral or zero spin are symmetric.

3.6 Many Particles in a Box

EXAMPLE   Consider a two-electron wavefunction, $\Psi(1, 2)$. $\hat{P}_{12}$ is the *permutation operator*; it operates on a function to exchange electrons 1 and 2.

$$\hat{P}_{12}\Psi(1, 2) = \Psi(2, 1) \quad \text{and according to Principle IV}$$

$$= -\Psi(1, 2) \qquad (3.5.2)$$

However, this is quite formal; in the next section we see how antisymmetry affects the system in practice by learning how to write a many-electron wavefunction that is antisymmetric.

3.6   MANY PARTICLES IN A BOX

If there are N particles present in the one-dimensional box, the simplest Hamiltonian is that for noninteracting particles.

$$\hat{H}_0 = -\frac{\hbar^2}{2m}\left[\frac{\partial^2}{\partial x_1^2} + \frac{\partial^2}{\partial x_2^2} + \cdots + \frac{\partial^2}{\partial x_N^2}\right] \qquad (3.6.1)$$

$x_1, x_2, \ldots, x_N$, are the coordinates of the N particles. Clearly, $\hat{H}_0$ is just a sum of N one-particle Hamiltonians.

$$\hat{H}_0 = -\frac{\hbar^2}{2m}\sum_{i=1}^{N}\frac{\partial^2}{\partial x_i^2} = \sum_{i=1}^{N}\hat{h}_0(i)$$

The Hamiltonian $\hat{H}_0$ consists of a sum of kinetic energy operators. From this point onward we turn our attention to the particular case of *electrons* in a one-dimensional box, in which case the Hamiltonian should really contain the interelectronic potential energy operators for electronic repulsion, that is, the total Hamiltonian is $\hat{H} = \hat{H}_0 + \hat{V}$, where $\hat{V}$ is the sum of one-dimensional interelectronic Coulomb repulsions.

$$\hat{V} = \sum_{i>j}^{N}\frac{e^2}{|x_i - x_j|} \qquad (3.6.2)$$

e is the electronic charge. The Coulomb repulsion is, of course, inversely proportional to the distance between the electrons, $|x_i - x_j|$. In contrast to $\hat{H}_0$, $\hat{V}$ is called a "two-electron operator," because it depends on the electronic coordinates taken two at a time.

    There is no exact solution[9] to a Schrödinger equation with $\hat{H} = \hat{H}_0 + \hat{V}$ and $N > 2$. For this reason we hold to the harmless fiction that $\hat{H}_0 \gg \hat{V}$, that is, that $\hat{V}$ is a mere *perturbation* of the physical situation characterized by $\hat{H}_0$.

---

[9] There is no exact solution to the classical problem of three interacting bodies, a fortiori there is no exact solution to the N-body problem. However, *approximation methods* are so highly developed in both classical and quantum mechanics that nonrelativistic solutions can be made as accurate as time and patience allow.

Then we can proceed immediately to find the eigenfunctions of $\hat{H}_0$.

$$\hat{H}_0 \Psi(x_1, x_2, x_3, \ldots, x_N) = E\Psi(x_1, x_2, x_3, \ldots, x_N) \quad (3.6.3)$$

where

$$\Psi(x_1, x_2, x_3, \ldots, x_N) = \psi_{n_1 m_{s_1}}(x_1) \psi_{n_2 m_{s_2}}(x_2) \cdots \psi_{n_N m_{s_N}}(x_N)$$

$$= \prod_{i=1}^{N} \psi_{n_i m_{s_i}}(x_i) \quad (3.6.4)$$

and

$$E = E_{n_1} + E_{n_2} + E_{n_3} + \cdots + E_{n_N} \quad (3.6.5)$$

The $\psi_{n_i m_{s_i}}$ are the particle-in-a-box spin-orbitals and the E_{n_i} are the corresponding energies, $E_{n_i} = n_i^2 h^2 / 8mL^2$.

PROOF By separation of variables, (3.6.3) with (3.6.1) and (3.6.4) can be written

$$\sum_{i=1}^{N} \hat{h}_0(i) \prod_{i=1}^{N} \psi_{n_i m_{s_i}}(x_i) = E \prod_{i=1}^{N} \psi_{n_i m_{s_i}}(x_i)$$

Divide both sides of this Schrödinger euqation by $\prod_{i=1}^{N} \psi_{n_i m_{s_i}}(x_i)$; the result is

$$\sum_{i=1}^{N} \frac{1}{\psi_{n_i m_{s_i}}(x_i)} \hat{h}_0(i) \psi_{n_i m_{s_i}}(x_i) = E$$

The last equation says that a sum of N functions of the N *independent variables*, $x_1, x_2, \ldots, x_N$, is equal to a constant for all values of the independent variables. This can only be true if each term in the sum is separately constant. Calling the constant E_{n_i}, we have

$$\frac{1}{\psi_{n_i m_{s_i}}(x_i)} \hat{h}_0(i) \psi_{n_i m_{s_i}}(x_i) = E_{n_i}$$

or

$$\hat{h}_0(i) \psi_{n_i m_{s_i}}(x_i) = E_{n_i} \psi_{n_i m_{s_i}}(x_i) \quad (3.6.6)$$

but (3.6.6) has been solved before. It is just the particle-in-a-box Schrödinger equation, (3.2.2). Thus $E = \sum_{i=1}^{N} E_{n_i}$, so (3.6.4) and (3.6.5) have been shown to satisfy the many-particle Schrödinger equation, (3.6.3).

EXERCISE Follow the proof through, step by step, for the two-electron case ($N = 2$) without using summation and product signs.

An N-electron wavefunction like (3.6.4) is called a *product wavefunction* for a *configuration* of N electrons occuping N spin-orbitals having quantum numbers $n_1 m_{s_1}, n_2 m_{s_2}, \ldots, n_N m_{s_N}$. Thus, the definition

CONFIGURATION = AN ARRANGEMENT OF N ELECTRONS OCCUPYING
N SPIN-ORBITALS

3.6 Many Particles in a Box

Although we assigned the electrons an ordered set of numbers, 1 through N, the electrons are completely indistinguishable from each other, that is, no electron can be tagged or followed with a microscope (see Sections 2.3 and 3.1) thereby preserving its identity. Because they *are* indistinguishable, any ordering of the electron numbers among the N spin-orbitals of $\Psi(x_1, x_2, \ldots, x_N)$ in (3.6.4) seems equally good. *However*, at this point the Pauli principle imposes a serious restriction on *how the electrons can be arranged among the spin-orbitals.*

The Pauli principle says that the electronic wavefunction,

$$\Psi(x_1, x_2, \ldots, x_N)$$

must be antisymmetric to exchange of the space-spin coordinates of the electrons. If $\hat{P}_{ij}$ is the operator which exchanges the space-spin coordinates x_i with x_j then

$$\hat{P}_{ij} \Psi(x_1, x_2, \ldots, x_i, x_j, \ldots, x_N) = \Psi(x_1, x_2, \ldots, x_j, x_i, \ldots, x_N)$$
$$= -\Psi(x_1, x_2, \ldots, x_i, x_j, \ldots, x_N)$$

Specifically, we start with two electrons and find the lowest-energy state (ground state) of the two-electron system which satisfies the Pauli principle. Choosing the lowest-energy spin-orbitals, we try,

$$\Psi'(x_1, x_2) = \psi_{1\frac{1}{2}}(x_1)\psi_{1\frac{1}{2}}(x_2)$$

but

$$\hat{P}_{12} \Psi'(x_1, x_2) = \Psi'(x_2, x_1) = \psi_{1\frac{1}{2}}(x_2)\psi_{1\frac{1}{2}}(x_1)$$

Clearly this function is symmetric, not antisymmetric. Under $\hat{P}_{12}$ the function $\psi_{1-\frac{1}{2}}(x_1)\psi_{1-\frac{1}{2}}(x_2)$, of the same energy, is also symmetric. But,

$$\hat{P}_{12} \psi_{1\frac{1}{2}}(x_1)\psi_{1-\frac{1}{2}}(x_2) = \psi_{1\frac{1}{2}}(x_2)\psi_{1-\frac{1}{2}}(x_1)$$

which suggests that

$$\Psi(x_1, x_2) = \frac{1}{\sqrt{2}} [\psi_{1\frac{1}{2}}(x_1)\psi_{1-\frac{1}{2}}(x_2) - \psi_{1\frac{1}{2}}(x_2)\psi_{1-\frac{1}{2}}(x_1)] \quad (3.6.7)$$

is the antisymmetric wavefunction of lowest energy. The $1/\sqrt{2}$ is for normalization.

EXAMPLE Normalization, $\langle \Psi(x_1, x_2), \Psi(x_1, x_2) \rangle = 1$

$$\int_0^L \int_0^L |\Psi(x_1, x_2)|^2 \, dx_1 \, dx_2 = \frac{1}{2} \left[\int_0^L |\psi_{1\frac{1}{2}}(x_1)|^2 \, dx_1 \int_0^L |\psi_{1-\frac{1}{2}}(x_2)|^2 \, dx_2 \right.$$

$$+ \int_0^L |\psi_{1\frac{1}{2}}(x_2)|^2 \, dx_2 \int_0^L |\psi_{1-\frac{1}{2}}(x_1)|^2 \, dx_1$$

$$\left. - 2 \int_0^L \psi_{1\frac{1}{2}}(x_1)\psi_{1-\frac{1}{2}}(x_1) \, dx_1 \int_0^L \psi_{1-\frac{1}{2}}(x_2)\psi_{1\frac{1}{2}}(x_2) \, dx_2 \right]$$

Since the "integrations" over the spin functions are understood, and spin functions of different m_s are orthogonal, $\langle\psi_{1\frac{1}{2}}(x_2), \psi_{1-\frac{1}{2}}(x_1)\rangle = 0$, we get $\langle\Psi(x_1, x_2), \Psi(x_1, x_2)\rangle = \frac{1}{2}(1+1) = 1$.

EXERCISE Show that $\Psi(x_1, x_2)$ has an average energy $\langle \hat{H}_0 \rangle$ equal to $2E_1$.
Hint: Do not omit normalization factors; use the eigenvalue equation $\hat{H}_0 \psi_{1\pm\frac{1}{2}} = E_1 \psi_{1\pm\frac{1}{2}}$.

From the definition of the determinant, equation (3.6.7) is written

$$\Psi(x_1, x_2) = \frac{1}{\sqrt{2}} \begin{vmatrix} \psi_{1\frac{1}{2}}(x_1) & \psi_{1-\frac{1}{2}}(x_1) \\ \psi_{1\frac{1}{2}}(x_2) & \psi_{1-\frac{1}{2}}(x_2) \end{vmatrix}$$

$$= \frac{1}{\sqrt{2}} \begin{vmatrix} \psi_1(x_1)\alpha(1) & \psi_1(x_1)\beta(1) \\ \psi_1(x_2)\alpha(2) & \psi_1(x_2)\beta(2) \end{vmatrix} \quad (3.6.8)$$

Such a determinantal wavefunction (*Slater determinant*) always satisfies the Pauli principle because $\hat{P}_{12}$ exchanges two rows of the determinant, a process which *changes the sign* of the determinant.

If one electron is present in the box, the ground state of the one-electron system has either $\psi_{1\frac{1}{2}}$ or $\psi_{1-\frac{1}{2}}$ occupied (it is immaterial which spin-orbital is occupied because they are degenerate), but the addition of a second electron to the box immediately brings the Pauli principle into play. Now, because the electrons are *identical* and all four occupations, $\psi_{1\frac{1}{2}}(1)\psi_{1\frac{1}{2}}(2)$, $\psi_{1-\frac{1}{2}}(1)\psi_{1-\frac{1}{2}}(2)$, $\psi_{1\frac{1}{2}}(1)\psi_{1-\frac{1}{2}}(2)$, and $\psi_{1\frac{1}{2}}(2)\psi_{1-\frac{1}{2}}(1)$, are *degenerate*, one might imagine all four occupations are equally satisfactory. However, the Pauli principle demands that the total wavefunction be antisymmetric and only the last two orbital products can be combined into an antisymmetric function [as seen in (3.6.7) or (3.6.8)]. The addition of a third electron to the box leads to a large number of degenerate product functions corresponding to permutations of the electrons among the two basic (and degenerate) occupations, $\psi_{1\frac{1}{2}}\psi_{1-\frac{1}{2}}\psi_{2\frac{1}{2}}$ and $\psi_{1\frac{1}{2}}\psi_{1-\frac{1}{2}}\psi_{2-\frac{1}{2}}$. In either case the antisymmetric wavefunctions are the Slater determinants of the spin-orbitals. So the *degenerate ground state* of three electrons in a box has the *degenerate, antisymmetric* wavefunctions

$$\begin{vmatrix} \psi_{1\frac{1}{2}}(x_1) & \psi_{1-\frac{1}{2}}(x_1) & \psi_{2\frac{1}{2}}(x_1) \\ \psi_{1\frac{1}{2}}(x_2) & \psi_{1-\frac{1}{2}}(x_2) & \psi_{2\frac{1}{2}}(x_2) \\ \psi_{1\frac{1}{2}}(x_3) & \psi_{1-\frac{1}{2}}(x_3) & \psi_{2\frac{1}{2}}(x_3) \end{vmatrix} \quad \text{and} \quad \begin{vmatrix} \psi_{1\frac{1}{2}}(x_1) & \psi_{1-\frac{1}{2}}(x_1) & \psi_{2-\frac{1}{2}}(x_1) \\ \psi_{1\frac{1}{2}}(x_2) & \psi_{1-\frac{1}{2}}(x_2) & \psi_{2-\frac{1}{2}}(x_2) \\ \psi_{1\frac{1}{2}}(x_3) & \psi_{1-\frac{1}{2}}(x_3) & \psi_{2-\frac{1}{2}}(x_3) \end{vmatrix}$$

both of which may be multiplied by $1/\sqrt{3!}$ for normalization.

Generalizing this result: *the N-electron wavefunction satisfying the Pauli principle is a Slater determinant of N spin-orbitals*. Now if two columns of a determinant are identical the determinant vanishes, which means that each spin-orbital can be used but *once* in the product wavefunction. So the lowest-energy state of the N-electron system has electrons in the N lowest-energy spin-orbitals, $\psi_1\alpha$, $\psi_1\beta$, $\psi_2\alpha$, $\psi_2\beta$, ..., $\psi_{N/2}\alpha$, $\psi_{N/2}\beta$. The Pauli principle is often stated as follows: *no two electrons can have the same set of quantum numbers*, that is, for the case of electrons in a one-dimensional box, n and/or m_s must differ for any two electrons. Consequently, the antisymmetry requirement causes an ordered occupation of the spin-orbitals by electrons in a box. Electrons in an atom show the same behavior as a result of the Pauli principle; as more electrons are added to the atomic system they do not all pack into the same spin-orbital, but rather, they stack up in an ordered array of occupied spin-orbitals, no spin-orbital more than singly occupied. The ordered occupation of the spin-orbitals is indicated by the *configuration* of the N-electron system.

A shorthand way of representing a configuration is

$$\text{Configuration} = (\psi_1)^2(\psi_2)^2(\psi_3)^2 \cdots (\psi_{N/2})^2$$

The superscript 2 indicates that two electrons of $m_s = \pm\tfrac{1}{2}$ occupy the same space orbital. A shorthand way of representing the normalized, antisymmetrized wavefunction is

$$\Psi(x_1, x_2, \ldots, x_N) = \hat{A}\psi_1\alpha(1)\psi_1\beta(2)\psi_2\alpha(3)\cdots\psi_{N/2}\beta(N) \qquad (3.6.9)$$

$\hat{A}$ is the "antisymmetrizer," an operator[10] which turns the product of spin-orbitals into a normalized determinant of spin-orbitals. Therefore (3.6.9), for $N = 2$, is entirely equivalent to either (3.6.8) or (3.6.7), but much easier to write down.

3.7 FREE-ELECTRON MOLECULAR ORBITAL METHOD FOR CONJUGATED MOLECULES

In the *molecular orbital* theory of conjugated molecules, particle-in-a-box wavefunctions are called "free-electron" orbitals to distinguish them from other molecular orbital approximations. The attraction of the positively charged nuclei for the electrons provides a potential well to which the electrons are confined, much like particles in a box. If the predominant dimension of the molecule is linear, then a one-dimensional box will rather nicely represent the electron-nuclear potential energy. We recall that in conjugated molecules each carbon atom donates one electron to be delocalized over the whole conjugated π-electron system. In the free-electron (FE) model, the electronic delocalization is represented by using particle-in-a-box spin-orbitals.

---

[10] Explicitly, $\hat{A} = 1/\sqrt{N!}\sum_P (-1)^p P$, that is, $\hat{A}$ is the sum over all permutations P of the electrons; p is the parity of the permutation.

Figure 3.7.1 presents the FE model of butadiene and ethylene. With four π electrons butadiene has a ground-state configuration, $(\psi_1)^2(\psi_2)^2$. Note that both orbitals are delocalized over the whole molecule, but $(\psi_2)^2$ does pile up electronic density where the bond formula says there are double bonds. Consider the electronic transition

$$\begin{array}{c}\text{Butadiene}\\ \text{Transition}\end{array} \quad \pi-\pi^* \ (\psi_1)^2(\psi_2)^2 \xrightarrow{h\nu} (\psi_1)^2(\psi_2)(\psi_3) \quad (3.7.1)$$

Calling the *excited* π orbital π^*, the $\pi-\pi^*$ transition of butadiene is $(\psi_2) \to (\psi_3)$, and the frequency at which absorption of light will occur can be predicted by calculating the energy difference between the FE molecular orbital configurations in (3.7.1), $\Delta E = h\nu$. For hexatriene, the ground-state configuration is $(\psi_1)^2(\psi_2)^2(\psi_3)^2$, and the $\pi-\pi^*$ transition is $(\psi_3) \to (\psi_4)$. Of course, the box size differs in different molecules, that is, it is propor-

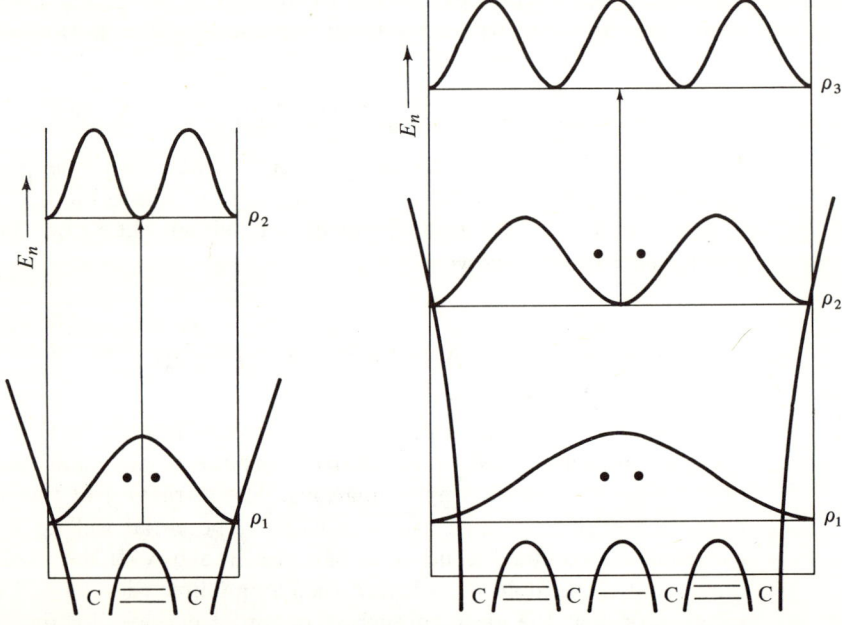

FIG. 3.7.1 Superposition of the one-dimensional box potential over the potential wells supplied by the nuclei in ethylene and butadiene. Note how the box potential approximates the nuclear well of the molecule. The probability densities of the particle-in-a-box orbitals used to describe the distribution of π electrons are given at their respective energies. Arrows indicate the $\pi \to \pi*$ transitions. (After Platt, J. R. "The Chemical Bond and the Distribution of Electrons in Molecules," in *Handbuch der Physik*, Bd. 37/2, pp. 173–281; Berlin-Göttingen-Heidelberg: Springen 1961. Used by permission.

3.8 Localized Orbitals

tional to the molecular length. Because the orbital energies go as $1/L^2$ they are more closely spaced in larger boxes, as illustrated in Fig. 3.7.1. As shown in Table 3.7.1, the free-electron molecular orbital spectral transitions predicted by this, admittedly crude, theory are in fair agreement with the observed spectrum.

TABLE 3.7.1

Molecule	$\Delta E_{calc}{}^a$ (eV)	ΔE_{obs} (eV)
Ethylene	7.85	7.65
Butadiene	5.95	5.9
Hexatriene	5.25	5.1
Octatetraene	4.55	4.1

[a] Semiempirical calculations of polyene spectra reported by S. R. La Paglia in *Theor. Chim. Acta* **11**, 89 (1968). Calculations include inter-electronic repulsions, equation (3.6.2). Used by permission.

3.8 LOCALIZED ORBITALS

Taking butadiene as an example, $(\psi_1)^2(\psi_2)^2$, the orbitals ψ_1 and ψ_2 are doubly occupied and delocalized over the whole molecule. We now want to show that there is a *localized representation*. We define *localized equivalent orbitals* χ_i from the *delocalized molecular orbitals* ψ_i.

$$\chi_1 = 1/\sqrt{2}(\psi_1 + \psi_2)$$
$$\chi_2 = 1/\sqrt{2}(\psi_1 - \psi_2)$$
(3.8.1)

The distribution functions of the localized orbitals are shown in Fig. 3.8.1 (compare to $|\psi_1|^2$ and $|\psi_2|^2$).

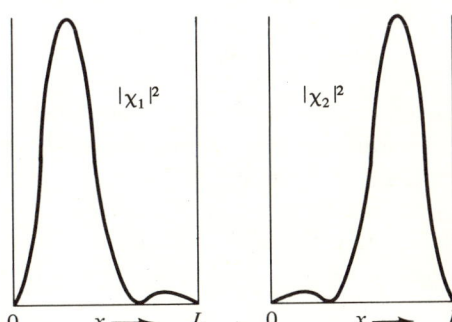

FIG. 3.8.1 Probability density distribution of the localized particle-in-a-box orbitals.

EXERCISE   Show that χ_1 and χ_2 are orthogonal and normalized. (Note that ψ_1 and ψ_2, being eigenfunctions of a Hermitian operator and having unlike eigenvalues, must be orthogonal.)

EXAMPLE   The expectation value of $\hat{x}$ should demonstrate the localization of the distributions in different ends of the molecule which is evident in the figure.

$$\langle \chi_1, \hat{x}\chi_1 \rangle = \tfrac{1}{2}[\langle \psi_1, \hat{x}\psi_1 \rangle + \langle \psi_2, \hat{x}\psi_2 \rangle + 2\langle \psi_1, \hat{x}\psi_2 \rangle]$$

$$= \frac{L}{2} + 2\langle \psi_1, \hat{x}\psi_2 \rangle = \frac{L}{2} - L\frac{32}{9\pi^2}$$

$$\langle \chi_2, \hat{x}\chi_2 \rangle = \frac{L}{2} - 2\langle \psi_1, \hat{x}\psi_2 \rangle = \frac{L}{2} + L\frac{32}{9\pi^2}$$

The average values of $\hat{x}$ show that the localized, equivalent orbitals localize the distributions on either side of the center of the box.

With results like these it is very tempting to take the ground state of butadiene to have a configuration $(\chi_1)^2(\chi_2)^2$ and a wavefunction

Localized Orbital Representation

$$X = \hat{A}\chi_1\alpha(1)\chi_1\beta(2)\chi_2\alpha(3)\chi_2\beta(4) \quad \text{compared to} \tag{3.8.2}$$

Molecular Orbital Representation

$$\Psi = \hat{A}\psi_1\alpha(1)\psi_1\beta(2)\psi_2\alpha(3)\psi_2\beta(4)$$

The piling up of charge density between the terminal carbon atoms in the localized representation is now due to *two π electrons paired in orbital χ_1 largely localized at one end of the molecule and two electrons paired in orbital χ_2 largely localized at the other end of the molecule*. This is much more like the Lewis structure for the shared pairs in the carbon chain, C :: C : C :: C.

EXERCISE   The final proof of the equivalence of Ψ and X, for now, is to show that the expectation value of the energy $\langle \hat{H}_0 \rangle$ is the same for both.

The χ representation of the electronic structure of butadiene gives two equivalent electron-pair π bonds which coincide with the double bonds of the classical chemical formula. However, χ_1 and χ_2 are *not* eigenfunctions to $\hat{H}_0$ even though both χ and ψ representations have the same *average energy*. In effect this means the *χ representation is useless for the prediction of spectroscopic transition energies or ionization potentials* (see following example).

EXERCISE   Show that $\chi_1 = 1/\sqrt{2}(\psi_1 + \psi_2)$ is *not* an eigenfunction to $\hat{H}_0$.

3.9 Particle in a Two-Dimensional Box: Degeneracy

EXAMPLE Calculate and compare the lowest spectroscopic transition energy of butadiene in molecular orbital (MO) and localized orbital (LO) representations.

MO $(\psi_1)^2(\psi_2)^2 \longrightarrow (\psi_1)^2(\psi_2)(\psi_3)$

$\Delta E = E_3 - E_2$, immediately, from the eigenvalues.

LO $(\chi_1)^2(\chi_2)^2 \longrightarrow (\chi_1)^2(\chi_2)(\chi_3)$

where $\chi_3 = 1/\sqrt{2}(\psi_3 + \psi_4)$

$\Delta E = \langle \chi_3, \hat{H}_0 \chi_3 \rangle - \langle \chi_2, \hat{H}_0 \chi_2 \rangle = \frac{1}{2}(E_3 - E_2 + E_4 - E_2)$

The transition energy in χ representation is seen to be an *average* transition energy for *two* of the transitions of the molecule.

3.9 PARTICLE IN A TWO-DIMENSIONAL BOX: DEGENERACY

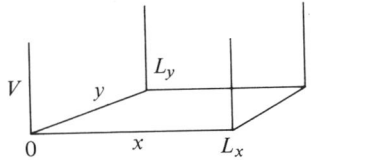

$V(x, y) = 0 \quad \begin{cases} 0 < x < L_x \\ 0 < y < L_y \end{cases}$

$V(x, y) = \infty \quad \begin{cases} 0 \geqslant x \geqslant L_x \\ 0 \geqslant y \geqslant L_y \end{cases}$

The potential energy of the particle is zero inside the box and infinity at and beyond the walls. The Schrödinger equation is *separated* in x and y; $\hat{H}$ is the sum of an operator in x and one in y, $\hat{H} = \hat{h}(x) + \hat{h}(y)$, where

$$\hat{h}(x) = -\frac{\hbar^2}{2m}\frac{\partial^2}{\partial x^2}$$

$$-\frac{\hbar^2}{2m}\left[\frac{\partial^2}{\partial x^2} + \frac{\partial^2}{\partial y^2}\right]\Psi(x, y) = E\Psi(x, y) \qquad (3.9.1)$$

$\Psi(x, y)$ is a two-dimensional, one-particle wavefunction (orbital). The $\Psi(x, y)$ which satisfies the boundary conditions $\Psi(x, y) = 0$ at $x, y = 0$ and $x, y = L_x, L_y$ is

$$\Psi_{n_x n_y}(x, y) = B_x B_y \sin\left(\frac{n_x \pi}{L_x}x\right)\sin\left(\frac{n_y \pi}{L_y}y\right) \qquad (3.9.2)$$

$$E_{n_x n_y} = \frac{h^2}{8m}\left(\frac{n_x^2}{L_x^2} + \frac{n_y^2}{L_y^2}\right) \qquad (3.9.3)$$

PROOF We prove that because $\hat{H} = \hat{h}(x) + \hat{h}(y)$ the substitution $\Psi_{n_x n_y}(x, y) = \psi_{n_x}(x)\psi_{n_y}(y)$, separates the Schrödinger equation, (3.9.1), into two equations, one in x, the other in y.

$$[\hat{h}(x) + \hat{h}(y)]\psi_{n_x}(x)\psi_{n_y}(y) = E\psi_{n_x}(x)\psi_{n_y}(y)$$

or

$$\psi_{n_y}(y)\hat{h}(x)\psi_{n_x}(x) + \psi_{n_x}(x)\hat{h}(y)\psi_{n_y}(y) = E\psi_{n_x}(x)\psi_{n_y}(y)$$

Divide both sides of the last equation by $\psi_{n_x}\psi_{n_y}$.

$$\frac{1}{\psi_{n_x}}\hat{h}(x)\psi_{n_x} + \frac{1}{\psi_{n_y}}\hat{h}(y)\psi_{n_y} = E$$

Now a sum of functions of the two independent variables x and y can only be equal to a constant E, if each of the terms is separately a constant; calling the constants E_{n_x} and E_{n_y}, we have

$$\frac{1}{\psi_{n_x}}\hat{h}(x)\psi_{n_x} = E_{n_x} \quad \text{and} \quad \frac{1}{\psi_{n_y}}\hat{h}(y)\psi_{n_y} = E_{n_y} \qquad (3.9.4)$$

The equations (3.9.4) have been solved before [they are just the one-dimensional particle-in-a-box Schrödinger equation (3.2.2)]; (3.9.3) and (3.9.2) then follow.

The particle in a two-dimensional box is characterized by two spatial quantum numbers, n_x and n_y. *Degeneracy* arises if the box is square, $L_x = L_y = L$, for then $E_{n_x n_y} = (h^2/8mL^2)(n_x^2 + n_y^2)$ and in general, $E_{ij} = E_{ji}$, for example, $E_{12} = E_{21} = (h^2/8mL^2)(5)$. So the wavefunctions Ψ_{12} and Ψ_{21} have the same energy even though they are obviously different

$$\Psi_{12} \neq \Psi_{21}$$

that is, $\sin(\pi x/L)\sin(2\pi y/L) \neq \sin(2\pi x/L)\sin(\pi y/L)$.

Wavefunctions which are different but have the same energy are said to be degenerate. Degeneracy is due to the *symmetry* of the physical situation. Looking down on the xy plane, the wavefunction is marked $+$ when it is above the $x = 0$, $y = 0$ plane, and $-$ when it is below this plane.

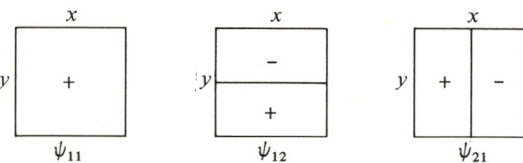

Ψ_{12} and Ψ_{21} are *degenerate by symmetry*, that is, a rotation of the square box by 90° brings them into each other.

EXERCISE   Show that χ_1 and χ_2, the localized orbitals of the free-electron model of butadiene, are brought into each other by the symmetry operation $\hat{C}_2$ of the one-dimensional box; also show that χ_1 and χ_2 are degenerate.

EXERCISE   Treat the problem of a particle in a three-dimensional box. Let the box have dimensions L_x, L_y, and L_z. Obtain the eigenfunctions and the eigenvalues. Discuss the conditions under which degeneracy will occur.

REFERENCES

Anderson, J. M., *Mathematics for Quantum Chemistry*, W. A. Benjamin, New York, 1966.

Eyring, H., Walter, J., and Kimball, G. E., *Quantum Chemistry*, Wiley, New York, 1944.

Feynman, R. P., Leighton, R. B., and Sands, M., *The Feynman Lectures on Physics. Quantum Mechanics*, Addison-Wesley, Reading, Mass., 1965.

Kauzmann, W., *Quantum Chemistry*, Academic Press, New York, 1957.

Margenau, H., and Murphy, G. M., *The Mathematics of Physics and Chemistry*, Vol. I, Van Nostrand, New York, 1943.

Pilar, F. L., *Elementary Quantum Chemistry*, McGraw-Hill, New York, 1968.

Platt, J. R., et al., *Free Electron Theory of Conjugated Molecules, Papers of the Chicago Group 1949–1961*, Wiley, New York, 1964.

PROBLEMS

1. An approximate wavefunction for a particle in a one-dimensional box of length L is $\tilde{\psi}(x) = N(b - x^2)x$, where N and b are disposable parameters. (1) Choose N and b so that the wavefunction is normalized and satisfies the particle-in-a-box boundary conditions. (2) Find the average value of the position of the particle from $\tilde{\psi}(x)$. (3) Find a power series expansion in x for the exact particle-in-a-box wavefunction; compare it to $\tilde{\psi}(x)$ and offer an improved approximation.

2. For the lowest energy level of the particle in a box calculate $\langle \hat{p} \rangle$, $\langle \hat{p}^2 \rangle$, $\langle \hat{x} \rangle$, and $\langle \hat{x}^2 \rangle$. Discuss the results in terms of the uncertainty principle.

3. Give a free-electron molecular orbital description of the π-electron configuration in ground and excited states of hexatriene.

4. Find the linear combinations of free-electron orbitals which represent the three localized π bonds in hexatriene.

5. In benzene the six π electrons are delocalized over the entire ring. To obtain a simple quantum mechanical description of this system assume that interactions between the electrons are negligible and that they move on a circle of radius 1.39 A. Derive an expression for the energy levels of the π electrons. (The kinetic energy of a particle of mass m located at angle θ on a circle of radius R is $p_\theta^2/2mR^2$.)

4
ATOMS AND THE PERIODIC SYSTEM

I rather expect that we shall someday find
a mathematico-mechanical explanation for what
we now call atoms which will render an account
of their properties.

A. KEKULÉ, 1867

IN THIS CHAPTER we tackle the problem of atomic structure. The best means of understanding how electrons behave in atoms is to appreciate the solutions of the one-electron atom (ion) Schrödinger equation. These solutions (wavefunctions) are called hydrogenic orbitals. From the hydrogenic orbitals and the Pauli principle we can proceed to build up the atoms of the periodic table and discuss atomic spectroscopy.

4.1 THE HYDROGEN ATOM

The hydrogenic atom is an atom (ion) with nuclear charge Ze and *one* electron. Being a two-particle problem (electron + nucleus) it has an exactly solvable Schrödinger equation. Just as the two-dimensional particle-in-a-box Schrödinger equation (3.9.1) separates into an equation in x and another in y, the hydrogenic Schrödinger equation separates, but *not* in Cartesian coordinates. The hydrogenic Schrödinger equation separates in spherical polar coordinates, r, θ, and ϕ.

The problem of the hydrogenic atom is analogous to the classical problem of two-body central field motion. In this problem, if m and M are the masses of the two bodies, one defines a reduced mass $\mu = Mm/(m + M)$ in terms of which the problem reduces to the motion of a particle of mass μ relative to the center of mass of the system. The nuclear mass M is at least ($Z = 1$) 1836 times greater than the electronic mass m, so $\mu = 1836m/1837 \approx m$. Thus in the hydrogenic atom the nucleus is so massive in comparison with the electron that we take the nucleus to be at the center of mass of the system, and the electron of mass m and charge $-e$ moves relative to the nucleus of charge $+Ze$ at a radial distance r.

Because electrons and (some) nuclei have spin (intrinsic angular momentum), a variety of magnetic interactions should be also included in the Hamiltonian. We neglect these interactions here because they are very small in light atoms and can always be included via perturbation theory (see Section 6.10). Neglecting magnetic interactions, the Hamiltonian of the hydrogenic atoms is simply

$$\hat{h} = \frac{-\hbar^2}{2m} \nabla^2 - \frac{Ze^2}{r} \qquad (4.1.1)$$

where

$$\frac{-\hbar^2}{2m}\nabla^2 = \frac{-\hbar^2}{2m}\left[\frac{\partial^2}{\partial x^2} + \frac{\partial^2}{\partial y^2} + \frac{\partial^2}{\partial z^2}\right]$$

is the kinetic energy operator in Cartesian coordinates x, y, and z. Converting to spherical polar coordinates,[1] r, θ, and ϕ, the Schrödinger equation, $(\hat{h} - E)\psi = 0$, is

$$\frac{-\hbar^2}{2m}\left[\frac{1}{r^2}\frac{\partial}{\partial r}\left(r^2\frac{\partial\psi}{\partial r}\right) + \frac{1}{r^2\sin\theta}\frac{\partial}{\partial\theta}\left(\sin\theta\frac{\partial\psi}{\partial\theta}\right)\right.$$
$$\left. + \frac{1}{r^2\sin^2\theta}\frac{\partial^2\psi}{\partial\phi^2}\right] - \left(\frac{Ze^2}{r} + E\right)\psi = 0 \quad (4.1.2)$$

The equation is separated by the substitution

$$\psi(r, \theta, \phi) = R(r)\Theta(\theta)\Phi(\phi) \quad (4.1.3)$$

Substitute (4.1.3) into (4.1.2) and divide through by $-\hbar^2 R\Theta\Phi/2mr^2\sin^2\theta$.

$$\frac{\sin^2\theta}{R}\frac{\partial}{\partial r}\left(r^2\frac{\partial R}{\partial r}\right) + \frac{\sin\theta}{\Theta}\frac{\partial}{\partial\theta}\left(\sin\theta\frac{\partial\Theta}{\partial\theta}\right) + \frac{1}{\Phi}\frac{\partial^2\Phi}{\partial\phi^2}$$
$$+ \frac{2mr^2\sin^2\theta}{\hbar^2}\left(\frac{Ze^2}{r} + E\right) = 0 \quad (4.1.4)$$

Because eq. (4.1.4) is of the form $f(r, \theta) + f(\phi) = 0$, and the relation holds for all values of the independent variables r, θ, and ϕ. the equation can only be true if the terms are constant, that is, $f(\phi) = -m^2$.

$$\frac{1}{\Phi}\frac{\partial^2\Phi}{\partial\phi^2} = -m^2 \quad (4.1.5)$$

where m^2 is a constant.[2]

THE ϕ EQUATION

Equation (4.1.5) has the solution

$$\Phi = A\exp(im\phi) \quad (4.1.6)$$

where $i = \sqrt{-1}$ and A is a constant. The angular coordinate ϕ varies from 0 to 2π, to 4π, etc., but the variable is periodic, that is,

$$\phi_0 = \phi_0 + 2\pi = \phi_0 + 4\pi = \text{etc.}$$

---

[1] See Appendix A.
[2] If $f(\phi) = m^2$, instead of $-m^2$, then the solutions $\exp(\pm m\phi)$ are not periodic when $\phi \to \phi + 2\pi k$ for any real k.

4.1 The Hydrogen Atom

Therefore, in order that the wavefunction $\Phi(\phi)$ be *single-valued*[3] it must satisfy the condition

$$\Phi(\phi_0) = \Phi(\phi_0 + 2\pi) = \Phi(\phi_0 + 4\pi) = \text{etc.}$$

$\Phi(\phi)$ satisfies this condition if

$$m = 0, \pm 1, \pm 2, \pm 3, \ldots \quad (4.1.7)$$

EXERCISE   Show that (4.1.6) is the solution to (4.1.5), and that (4.1.7) is the condition resulting from the single-valuedness of Φ. Hint: Use $\exp(im\phi) = \cos m\phi + i \sin m\phi$.

EXERCISE   The constant A in (4.1.6) is determined by normalization on the interval $(0, 2\pi)$; show that $A = 1/\sqrt{2\pi}$.

THE θ EQUATION

Substituting $-m^2$ into (4.1.4) and dividing through by $\sin^2 \theta$ we get a function of r, $f(r)$, plus a function of θ.

$$f(r) + \frac{-m^2}{\sin^2 \theta} + \frac{1}{\Theta \sin \theta} \frac{\partial}{\partial \theta}\left[\sin \theta \frac{\partial \Theta}{\partial \theta}\right] = 0$$

But if this equation is to hold for all values of the independent variables r and θ, each term must be a constant, or,

$$\frac{-m^2}{\sin^2 \theta} + \frac{1}{\Theta \sin \theta} \frac{\partial}{\partial \theta}\left[\sin \theta \frac{\partial \Theta}{\partial \theta}\right] = -l(l+1) \quad (4.1.8a)$$

where $l(l+1)$ is a constant. Making a change of variable to $w = \cos \theta$, this equation becomes

$$\left[l(l+1) - \frac{m^2}{1-w^2}\right]\Theta - 2w \frac{\partial \Theta}{\partial w} + (1-w^2)\frac{\partial^2 \Theta}{\partial w^2} = 0 \quad (4.1.8b)$$

This equation is a well-known differential equation in mathematics, the *associated Legendre equation*, which has the solutions

$$\Theta_{lm}(\theta) = P_l^{|m|}(\cos \theta) \quad (4.1.9)$$

$$\begin{aligned} l &= 0, 1, 2, 3, 4, \ldots \\ |m| &\leq l \end{aligned} \quad (4.1.10)$$

The $P_l^{|m|}$ are the associated Legendre functions.[4] Normalization of Θ_{lm} on the interval $(0, \pi)$ in θ [i.e., $(1, -1)$ in w] yields the normalization factor $[(2l+1)(l-|m|)!/2(l+|m|)!]^{1/2}$. The angular wavefunctions $\Phi_m(\phi)$ and $\Theta_{lm}(\theta)$ are listed in Tables 4.1.1 and 4.1.2, together with normalization

---

[3] See Section 3.1.
[4] The interesting properties of these and other functions of importance in atomic physics are described in H. F. Hameka, *Introduction to Quantum Theory*, Harper & Row, 1967, New York.

TABLE 4.1.1 The Hydrogenic Orbitals

Orbital quantum numbers n l m	Orbital designation	$R_{nl}(r)^a$	$\Theta_{lm}(\theta)$	$\Phi_m(\phi)$	Normalization Factor	Energy (eV)
1 0 0	$1s$	$\exp(-Zr)$	1	1	$(Z^3/\pi)^{1/2}$	$-Z^2 \times 13.6$
2 0 0	$2s$	$(2-Zr)\exp(-Zr/2)$	1	1	$\frac{1}{4}(Z^3/2\pi)^{1/2}$	$-\frac{Z^2}{4} \times 13.6$
3 0 0	$3s$	$(27-18Zr+2Z^2r^2)\exp(-Zr/3)$	1	1	$\frac{1}{81}(Z^3/3\pi)^{1/2}$	$-\frac{Z^2}{9} \times 13.6$
2 1 0	$2p_0$	$Zr\exp(-Zr/2)$	$\cos\theta$	1	$\frac{1}{4}(Z^3/2\pi)^{1/2}$	$-\frac{Z^2}{4} \times 13.6$
2 1 1	$2p_1$	$Zr\exp(-Zr/2)$	$\sin\theta$	$\exp(i\phi)$	$\frac{1}{8}(Z^3/\pi)^{1/2}$	$-\frac{Z^2}{4} \times 13.6$
2 1 −1	$2p_{-1}$	$Zr\exp(-Zr/2)$	$\sin\theta$	$\exp(-i\phi)$	$\frac{1}{8}(Z^3/\pi)^{1/2}$	$-\frac{Z^2}{4} \times 13.6$

4.1 The Hydrogen Atom

n	l	m		R(r)	Θ(θ)	Φ(φ)	Norm	Energy
3	1	0	$3p_0$	$(6Zr - Z^2r^2)\exp(-Zr/3)$	$\cos\theta$	1	$\frac{1}{81}(2Z^3/\pi)^{1/2}$	$-\frac{Z^2}{9} \times 13.6$
3	1	1	$3p_1$	$(6Zr - Z^2r^2)\exp(-Zr/3)$	$\sin\theta$	$\exp(i\phi)$	$\frac{1}{81}(Z^3/\pi)^{1/2}$	$-\frac{Z^2}{9} \times 13.6$
3	1	-1	$3p_{-1}$	$(6Zr - Z^2r^2)\exp(-Zr/3)$	$\sin\theta$	$\exp(-i\phi)$	$\frac{1}{81}(Z^3/\pi)^{1/2}$	$-\frac{Z^2}{9} \times 13.6$
3	2	2	$3d_2$	$Z^2r^2\exp(-Zr/3)$	$\sin^2\theta$	$\exp(2i\phi)$	$\frac{1}{81}(Z^3/4\pi)^{1/2}$	$-\frac{Z^2}{9} \times 13.6$
3	2	1	$3d_1$	$Z^2r^2\exp(-Zr/3)$	$\sin\theta\cos\theta$	$\exp(i\phi)$	$\frac{1}{81}(Z^3/\pi)^{1/2}$	$-\frac{Z^2}{9} \times 13.6$
3	2	0	$3d_0$	$Z^2r^2\exp(-Zr/3)$	$3\cos^2\theta - 1$	1	$\frac{1}{81}(Z^3/6\pi)^{1/2}$	$-\frac{Z^2}{9} \times 13.6$
3	2	-1	$3d_{-1}$	$Z^2r^2\exp(-Zr/3)$	$\sin\theta\cos\theta$	$\exp(-i\phi)$	$\frac{1}{81}(Z^3/\pi)^{1/2}$	$-\frac{Z^2}{9} \times 13.6$
3	2	-2	$3d_{-2}$	$Z^2r^2\exp(-Zr/3)$	$\sin^2\theta$	$\exp(-2i\phi)$	$\frac{1}{81}(Z^3/4\pi)^{1/2}$	$-\frac{Z^2}{9} \times 13.6$

[a] r is in bohr.

TABLE **4.1.2** The Hydrogenic Orbitals[a]

Purely real orbitals

n	l	$\lvert m \rvert$	Designation	Spherical harmonic	Normalization factor
2	1	0	$2p_z$	$\cos\theta$	$\frac{1}{4}(Z^3/2\pi)^{1/2}$
2	1	1	$2p_x$	$\sin\theta\cos\phi$	$\frac{1}{4}(Z^3/2\pi)^{1/2}$
2	1	1	$2p_y$	$\sin\theta\sin\phi$	$\frac{1}{4}(Z^3/2\pi)^{1/2}$
3	2	2	$3d_{x^2-y^2}$	$\sin^2\theta\cos 2\phi$	$\frac{1}{81}(Z^3/2\pi)^{1/2}$
3	2	1	$3d_{xz}$	$\sin\theta\cos\theta\cos\phi$	$\frac{1}{81}(2Z^3/\pi)^{1/2}$
3	2	0	$3d_{z^2}$	$3\cos^2\theta - 1$	$\frac{1}{81}(Z^3/6\pi)^{1/2}$
3	2	1	$3d_{yz}$	$\sin\theta\cos\theta\sin\phi$	$\frac{1}{81}(2Z^3\pi)^{1/2}$
3	2	2	$3d_{xy}$	$\sin^2\theta\sin\phi\cos\phi$	$\frac{1}{81}(Z^3/2\pi)^{1/2}$

[a] Radial parts and orbital energies are as in Table 4.1. r is in bohr. Note that only the magnitude of m is defined for the purely real orbitals, because they are linear combinations of functions with $\pm m$ values.

factors. The total angular eigenfunctions $Y_{lm}(\theta, \phi) = \Phi_m(\phi)\Theta_{lm}(\theta)$ are often called *spherical harmonics*.

THE r EQUATION

Substituting $-l(l + 1)$ into (4.1.4), making the change of variable $r = \rho\hbar/2\sqrt{-2mE}$, and defining

$$n = \frac{Z\sqrt{m}\, e^2}{\hbar\sqrt{-2E}} \qquad (4.1.11)$$

the equation is cast into the form

$$\frac{\partial^2 R}{\partial \rho^2} + \frac{2}{\rho}\frac{\partial R}{\partial \rho} - \left[\frac{1}{4} + \frac{l(l+1)}{\rho^2} - \frac{n}{\rho}\right]R = 0 \qquad 0 \leqslant \rho \leqslant \infty \qquad (4.1.12)$$

This equation is also a known differential equation; it has for solutions the *associated Laguerre functions*.[5]

$$R_{nl}(\rho) = L_{n+l}^{2l+1}(\rho)\rho^l \exp(-\rho/2) \begin{cases} n = 1, 2, 3, \ldots \\ l = n-1, n-2, \ldots, 0 \end{cases} \qquad (4.1.13)$$

[5] A detailed description of the mathematical methods of solving the r and θ equations is not necessary to understanding the following sections. Details are given by L. Pauling and E. B. Wilson, Jr. *Introduction to Quantum Mechanics*, chap. V, McGraw-Hill, New York, 1935. Also see chap. 4 and 8 of H. Hameka, *Introduction to Quantum Theory*, Harper & Row, New York, 1967.

4.1 The Hydrogen Atom

The solution is valid if l is zero or a positive integer and $l \leq n - 1$. The L_{n+l}^{2l+1} are polynomials in ρ; the "radial wavefunctions" R_n are listed in Table 4.1.1.

QUANTUM NUMBERS

Three quantum numbers n, l, and m have been found from the solution of the Schrödinger equation: n is called the *principal quantum number* and the quantum numbers l and m will later be shown to be related to the orbital angular momentum of the electron. (Of course, there is a fourth quantum number, the spin quantum number, $m_s = \pm\frac{1}{2}$.)

$$n = 1, 2, 3, 4, 5, \ldots, \infty$$

and because

$$l \leq n - 1$$
$$l = 0, 1, 2, 3, \ldots, (n-1)$$

From (4.1.10)

$$m = 0, \pm 1, \pm 2, \ldots, \pm l$$

and these are the only permissible values of the quantum numbers.

THE ENERGY

From the definition of n, (4.1.11),

$$E_n = -\frac{Z^2 m e^4}{2\hbar^2 n^2} \qquad n = 1, 2, 3, \ldots \qquad (4.1.14)$$

This is precisely the energy found in the Bohr model of the H atom, equation (2.2.9), so it gives the Rydberg-Ritz formula, ionization potential, etc. (see Section 2.2).

Because the energy depends solely on the principal quantum number, states with different l and m but the same n are degenerate. (This degeneracy with respect to l exists only in one-electron atoms and ions.) In addition to the negative energy eigenvalues (4.1.14) the H-atom Schrödinger equation has an infinite set of *positive continuous* eigenvalues which refer to *unbound* motion of the electron, that is, free-particle solutions. The free-particle wavefunctions are of vital importance in scattering problems; however, our chief interest lies with the solutions having negative eigenvalues (bound).

The energy and Schrödinger equation may be put into a simple form by a change to *atomic units*. Define a unit of length, the *bohr*,

$$a_H = \frac{\hbar^2}{me^2} = 1 \text{ bohr} \qquad (4.1.15)$$

and a unit of energy, the *hartree*,

$$\frac{me^4}{\hbar^2} = 1 \text{ hartree} \tag{4.1.16}$$

$a_H = 0.52918 \times 10^{-8}$ cm and 1 hartree = 27.210 eV. The energy levels of the hydrogenic atom are, by direct substitution in (4.1.14),

$$E_n = -\frac{1}{2}\frac{Z^2}{n^2} \text{ hartree} = -13.6\frac{Z^2}{n^2} \text{ eV} \tag{4.1.17}$$

and the hydrogenic Hamiltonian takes a simple form in atomic units.

$$\hat{h} = -\tfrac{1}{2}\nabla^2 - Z/r \tag{4.1.18}$$

In equation (4.1.18) and in all following tables and equations r is in bohr units, that is, $r(\text{bohr}) = r(\text{cm})/a_H(\text{cm})$.

THE RADIAL WAVEFUNCTION

As a consequence of solving the hydrogenic Schrödinger equation we have obtained an infinite set of orthogonal, normalized, hydrogenic orbitals, ψ_{nlm}, characterized by three quantum numbers n, l, and m, and a discrete energy E_n. Adding the spin degree of freedom requires m_s and yields spin-orbitals, $\psi_{nlm\,m_s}$.

$$\psi_{nlm} = R_{nl}(r)\Theta_{lm}(\theta)\Phi_m(\phi) \tag{4.1.19}$$

is the eigenfunction $\hat{h}\psi_{nlm} = E_n\psi_{nlm}$

$$\langle \psi_{nlm}, \psi_{n'l'm'} \rangle = \delta_{nn'}\,\delta_{ll'}\,\delta_{mm'} \tag{4.1.20}$$

The last equation expresses the orthogonality and normalization of the hydrogenic orbitals. The orthogonality, completeness, and other properties of the set of hydrogenic orbitals arise from the fact that the differential equations of which they are solutions are Sturm-Liouville equations (see Chapter 2).

The hydrogenic orbitals have a familiar designation 1s, 2s, 2p, etc., resulting from the following illogical but historical code.[6]

l	0	1	2	3	4	5
Symbol	s	p	d	f	g	h etc.

An orbital with $n = 1$, $l = 0$ is designated 1s; $n = 2$, $l = 1$, $m = 0$ is designated $2p_0$; $n = 3$, $l = 1$, $m = -1$, is designated $3p_{-1}$, etc.

$R_{nl}(r)$ is the radial part of the hydrogenic wavefunction (expressed in atomic units); it is plotted versus r in Fig. 4.1.1 for 1s and in Fig. 4.1.2 for several important orbitals. The probability of finding an electron at a

---

[6] Historically, s, p, d, f are symbols for the spectroscopists' visual designations of the spectral lines as sharp, principal, diffuse, and fundamental.

4.1 The Hydrogen Atom

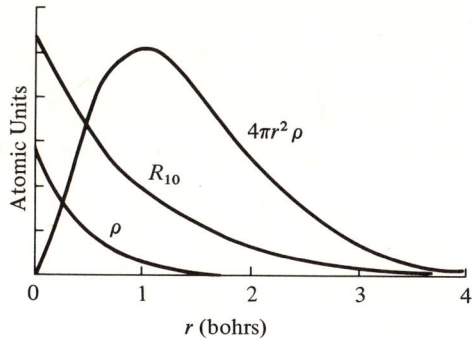

FIG. 4.1.1 The ground-state wavefunction of the hydrogen atom. $R_{10} = 2\exp(-r)$ is the normalized wavefunction; $\rho = |R_{10}|^2$ is the density function, and $4\pi r^2 \rho(r)\, dr$ is the probability of finding an electron between r and $r + dr$.

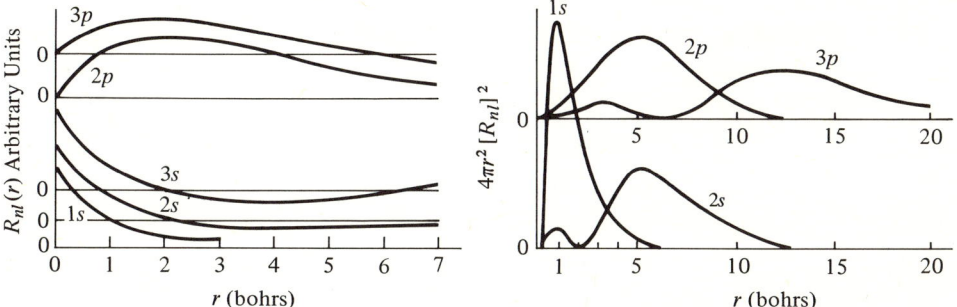

FIG. 4.1.2 Radial wavefunctions $R_{nl}(r)$, of some chemically important hydrogenic orbitals ($Z = 1$) and radial distribution functions $4\pi r^2 |R_{nl}|^2$. (The ordinate scale differs for different orbitals.)

distance r from the nucleus is found by multiplying the distribution $\rho(r) = |R_{nl}(r)|^2$ by the surface area of a sphere of radius r, $4\pi r^2$. The *radial distribution function* $4\pi r^2 \rho(r)\, dr$ is the probability of finding an electron between r and $r + dr$. As seen in Fig. 4.1.1 and Fig. 4.1.2, the radial distribution function is zero at the nucleus and depends on both n and l.

Turning our attention specifically to the 1s orbital, the ground state of the H atom, Fig. 4.1.1 shows that the maximum in the radial distribution function lies at the Bohr radius,[7] a_H. This is one resemblance to the Bohr

---

[7] In spite of the maximum in the radial distribution function at a_H, and its zero at the nucleus, the probability of finding the electron near the nucleus *per unit volume* is greater than elsewhere [see the maximum in $\rho(r)$].

theory of the H atom; however, as seen in Fig. 4.1.3, the 1s orbital is a spherical wavefunction with a probability density which is more like a "ball of charge" than an electron in a planar circular orbit. Figure 4.1.4 illustrates several alternative ways of depicting the probability density of an orbital.

The wavefunctions with $n = 2, 3, \ldots$ and $l = 0$, that is, 2s, 3s, etc., are

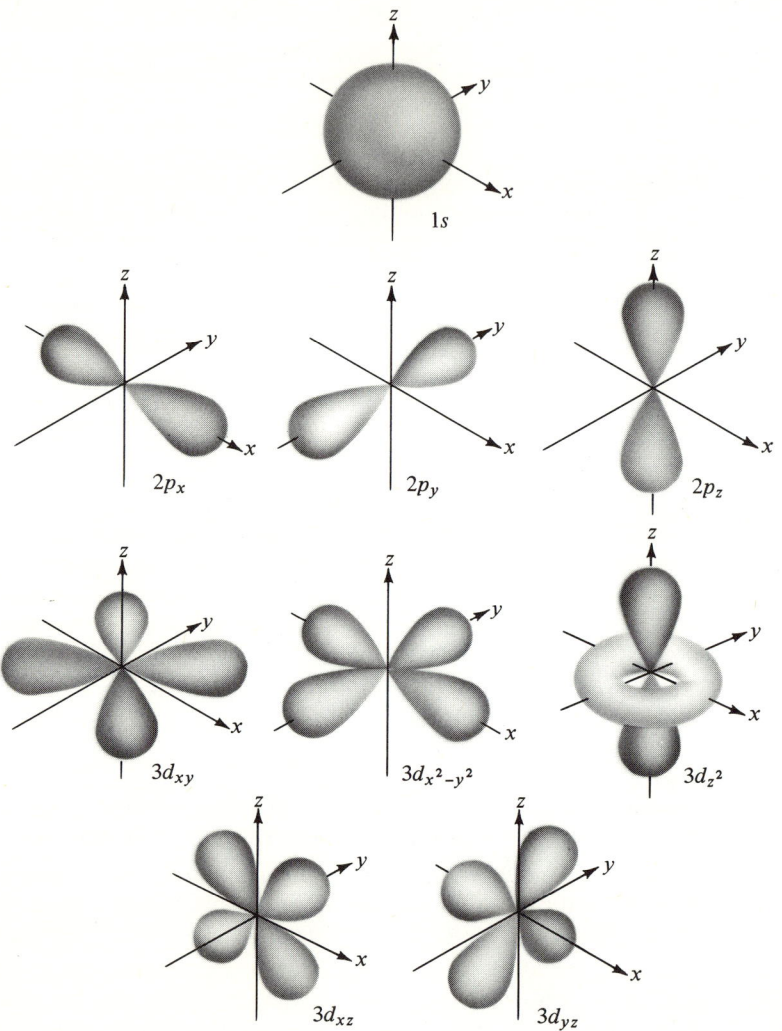

FIG. 4.1.3 Probability density contour surfaces of 1s, 2p, and 3d orbitals. These illustrations show by shading the actual shape of the outer contour of the orbitals.

4.1 The Hydrogen Atom

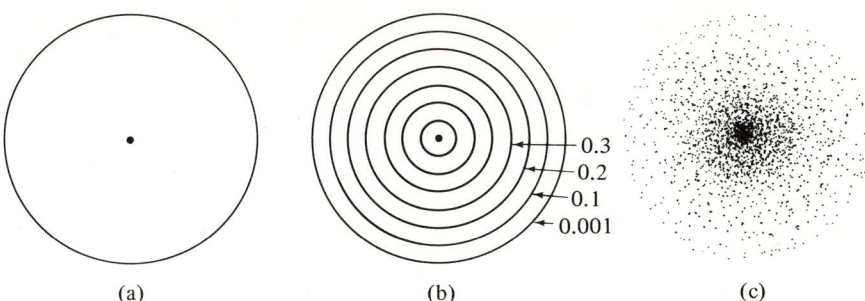

FIG. 4.1.4 Three popular methods of representing the probability density of an orbital, here illustrated on 1s. (a) Outer contour, contains 99% of the probability density within its boundary. (b) Successive contours of constant probability density (c) Probability density cloud (charge cloud).

also spherically symmetric, but show the presence of *radial nodes*.[8] At a radial node $R_{nl}(r) = 0$. The number of radial nodes in a hydrogenic orbital is $n - l - 1$; for example, a 3p function has one radial node (Fig. 4.1.2).

The average radial distance of the electron from the nucleus is the expectation value $\langle \hat{r} \rangle$; it has the value

$$\langle \hat{r} \rangle = \frac{1}{2Z} [3n^2 - l(l+1)] \text{ bohr} \qquad (4.1.21)$$

The "size" of an orbital increases as n^2 with constant l and Z, and decreases as $1/Z$ with constant l and n. Unfortunately, in comparing the sizes of atoms and ions there are several factors which change simultaneously (including the "screening," which we explain later) so (4.1.21) is valid only for one-electron atoms and ions, but it does help explain the relative positions of the radial distributions in Fig. 4.1.2.[9]

EXERCISE Show that the 1s, 2s, $2p_0$, etc., wavefunctions are normalized to unity as given in Tables 4.1.1 and 4.1.2.

EXERCISE Show that the 1s and 2s wavefunctions are orthogonal.

EXERCISE Calculate the average value of the potential energy of the hydrogenic atom ground state, $\langle 1s, (-Z/r) 1s \rangle$. Compare this energy to the total energy E_1 and deduce the average value of the kinetic energy. The result, called the *virial theorem*, is quite valid for all conservative systems with Coulomb potentials ($1/r$).

---

[8] The radial nodes serve the useful purpose of making orbitals of the same l but different n orthogonal to each other.
[9] The maximum in the radial distribution function of the 1s ($Z = 1$) orbital is at 1 bohr, but $\langle 1s, \hat{r} 1s \rangle = 3/2$ bohr. This is due to the long "tail" of the radial function.

THE ANGULAR WAVEFUNCTIONS

$Y_{lm}(\theta, \phi) = \Theta_{lm}(\theta)\Phi_m(\phi)$ is the angular part of the hydrogenic orbital (the spherical harmonic). The spherical harmonics give the orbitals their characteristic shapes. Because the $\Phi_m(\phi)$ ($m \neq 0$) are imaginary, it is much easier to work with their real versions. Taking linear combinations of the $+|m|$ and $-|m|$ Φ_m functions according to the identity $\cos|m|\phi = \frac{1}{2}[\exp(i|m|\phi) + \exp(-i|m|\phi)]$ and $\sin|m|\phi = (1/2i)[\exp(i|m|\phi) - \exp(-i|m|\phi)]$, purely real orbitals are obtained. The real orbitals are listed in Table 4.1.2 and are plotted in Figs. 4.1.3 and 4.1.5.

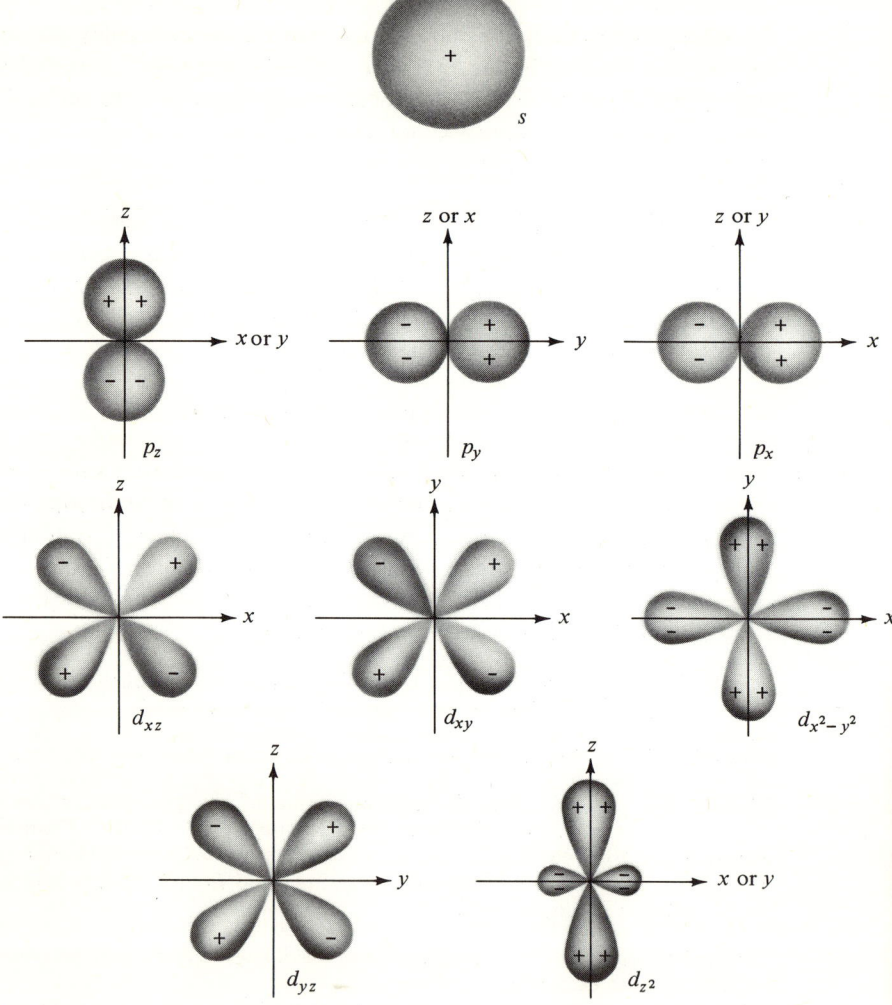

FIG. 4.1.5 Phase diagrams of the chemically important orbitals. Signs show the *phase* of the orbitals in different regions.

4.1 The Hydrogen Atom

EXERCISE The unnormalized spherical harmonics are $Y_{00} = 1$, $Y_{10} = \cos\theta$, $Y_{1\pm 1} = \sin\theta \exp(\pm i\phi)$, $Y_{20} = (3\cos^2\theta - 1)$, $Y_{2\pm 1} = \sin\theta\cos\theta \exp(\pm i\phi)$, and $Y_{2\pm 2} = \sin^2\theta \exp(\pm 2i\phi)$. Verify the real spherical harmonics listed in Table 4.1.2.

The real spherical harmonics are directly expressible in terms of the coordinates x, y, z, and r of Fig. A.1.1 (Appendix A).

EXAMPLE From the Cartesian–polar relations, $x = r\sin\theta\cos\phi$, $y = r\sin\theta\sin\phi$, and $z = r\cos\theta$, it follows that the real spherical harmonics are (Table 4.1.2) $p_x = x/r$, $p_y = y/r$, and $p_z = z/r$.

EXERCISE Show that the spherical harmonics for the d orbitals are

$$d_{x^2-y^2} = \frac{x^2 - y^2}{r^2}, \quad d_{z^2} = \frac{3z^2}{r^2} - 1, \quad d_{xy} = \frac{xy}{r^2}, \text{ etc.}$$

Figure 4.1.5 gives the phase diagrams of the chemically important orbitals. These diagrams do not indicate the *shape* of the orbitals as well as Fig. 4.1.3 does, but do show the sign of the *phase* of the wavefunction in each region. The probability density $\rho = |\psi_{nlm}|^2$, through its dependence on r, θ, and ϕ, gives the shape of the electronic distribution. However, the orbitals have another property which is also important for understanding the bonding in molecules: the phase. The phase of an orbital is determined by the angular part of the wavefunction. The s orbitals have positive phase for all θ and ϕ. But consider the plot of $2p_z$ in Fig. 4.1.6, $\psi_{210} = \frac{1}{4}(Z^3/2\pi)^{1/2} Zr \exp(-Zr/2) \cos\theta$. Clearly, the *phase* of $2p_z$ is positive above the xy plane and negative below. As the phase goes from $+$ to $-$, it passes through zero ($\cos\pi/2 = 0$); this introduces an *angular node* into the orbital in the

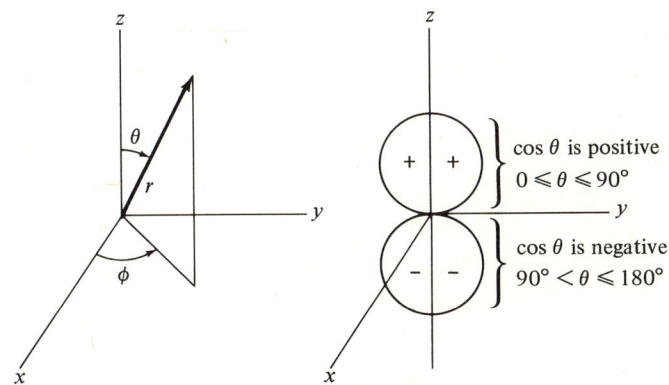

FIG. 4.1.6 Plot of the outer contour of $2p_z$ orbital showing how the phase of the function changes.

xy plane. The mathematical form of the angular part of the orbital determines the "shape" of the orbital in just this way. For $2p_x$ and $2p_y$, it is the combined zeros of $\sin\theta\cos\phi$ and $\sin\theta\sin\phi$, respectively, which give orbital *lobes* directed along x and y axes.

The phase determines the symmetry properties of the orbitals. Because ns orbitals have no angular nodes there is no rotation which can cause the function to change sign. However, a rotation of $2p_z$ by π about the x or y axis sends the positive lobe into the negative lobe, that is, the function changes sign. As an operator equation

$$C_x(180°)ns = ns \qquad [\text{also } C_x(90°)np_z = np_y$$
$$\text{but}\quad C_x(180°)np_z = -np_z \qquad \text{and } C_y(90°)np_z = np_x]$$

$C_x(180°)$ is the operator for rotation about x by $180°$. Since np_z, np_x, and np_y go into each other under symmetry operations (rotations of $\pi/2$ to which $\hat{h}$ is invariant) they are, of course, degenerate.

If $|\psi|^2$ is the observable and is independent of the sign of the phase, of what importance is this sign? As for any wave phenomenon, constructive and destructive interference is to be expected between overlapping waves. Whether the interference is constructive or destructive depends on the relative phase of the waves (wavefunctions) in the overlap region, that is, $(+)(+)$ or $(-)(-)$ yield constructive interference. This is extremely important when we later describe chemical bonding by constructively superimposing the orbitals of different atoms.

In the following sections we will go on to examine the wavefunctions and energies of two-electron and many-electron atoms. These systems have no exact solutions. Fortunately, we have approximations which are good enough to give quantitative agreement with experiment. Our starting point is to assume that each electron moves in a central, or spherically symmetric force field due to the nucleus and the other electrons (the *central field approximation*). This approximation accounts for the periodic system of the elements and the general size and energy of the atoms. It also gives the following result. Because atoms are spherically symmetric, the spherical harmonics are the angular wavefunctions for electrons in *all* atoms. Thus interelectronic interactions do not change the angular functions, whose properties we can learn once and for all. This permits us to speak of s, p, d, etc., orbitals for all atoms.

Unlike hydrogenic orbitals, the energy of other atomic orbitals depends on l as well as n. However, we can still use n as a quantum number, which is quite important for understanding the periodic table. For example, instead of Rydberg-Ritz terms (2.2.6) R/n^2, the spectra of the alkali atoms (Li, Na, K, etc.) can be organized by terms $R/(n - \Delta_l)^2$ where Δ_l is a small number, typical of l and the atom, called the *quantum defect*.

In the central field approximation, although the spherical harmonics are still the angular wavefunctions, the hydrogenic radial functions are no longer good wavefunctions for the many-electron atom. However, in the treatment of these atoms we will *start* with hydrogenic radial functions (Sections 4.2 and 4.3) and progress toward better approximations (Sections 4.9, 4.10, and 4.11).

4.2 THE HELIUM ATOM

There is no *exact* solution to the Schrödinger equation of three or more interacting particles; the helium atom, being the simplest of such insoluble problems, is a good introduction to the atomic many-electron problem which will concern us for the remainder of this chapter. We will describe the application of the perturbation method and the variational method to helium atom in some detail.

The helium atom is obtained by adding a second electron to the one-electron hydrogenic ion with $Z = 2$. The helium Hamiltonian displays the difference between helium and the hydrogenic ion (see Fig. 4.2.1).

$$\hat{H}(\text{helium}) = \left(-\tfrac{1}{2}\nabla_1^2 - \frac{2}{r_1}\right) + \left(-\tfrac{1}{2}\nabla_2^2 - \frac{2}{r_2}\right) + \frac{1}{|r_1 - r_2|} \quad (4.2.1)$$

or

$$\hat{H} = \hat{h}(1) + \hat{h}(2) + 1/r_{12} \quad (4.2.2)$$

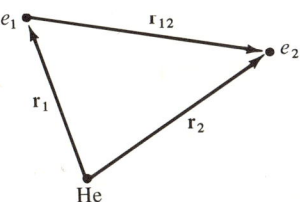

FIG. 4.2.1 Electronic coordinates in helium atom.

$\hat{h}(1)$ and $\hat{h}(2)$ are hydrogenic Hamiltonians, given in (4.1.18), for electrons 1 and 2, respectively, with $Z = 2$; $1/r_{12} = 1/|r_1 - r_2|$ is the Coulomb repulsion between the electrons. Since there is no exact solution to the Schrödinger equation with Hamiltonian (4.2.1) we take a perturbation approach to the problem (Section 2.7) by supposing that $1/r_{12}$ is merely a *perturbation*, that is,

$$\hat{H} = \hat{H}_0 + 1/r_{12} \quad (4.2.3)$$

$$\hat{H}_0 = \hat{h}(1) + \hat{h}(2) \quad (4.2.4)$$

$\hat{H}_0$ is the zero-order Hamiltonian of helium; the zero-order wavefunction is the eigenfunction to $\hat{H}_0$, $\Psi^{(0)}$.

$$\hat{H}_0 \Psi^{(0)} = \varepsilon^{(0)} \Psi^{(0)}. \tag{4.2.5}$$

$\Psi^{(0)}$ is a two-electron wavefunction, and it is easy to show by separation of variables that $\Psi^{(0)}(1,2) = \psi(1)\psi(2)$ satisfies (4.2.5) with $\varepsilon^{(0)} = 2E_1$ ($Z = 2$) and $\psi = 1s = (Z^3/\pi)^{1/2} \exp(-Zr)$ ($Z = 2$).

PROOF

$$\hat{H}_0 \Psi^{(0)}(1,2) = [\hat{h}(1) + \hat{h}(2)]\psi(1)\psi(2) = \varepsilon^{(0)} \Psi^{(0)}$$

$$\hat{h}(1)\psi(1)\psi(2) + \psi(1)\hat{h}(2)\psi(2) = \varepsilon^{(0)}\psi(1)\psi(2)$$

Divide through on both sides by $\psi(1)\psi(2)$

$$\frac{\hat{h}(1)\psi(1)}{\psi(1)} + \frac{\hat{h}(2)\psi(2)}{\psi(2)} = \varepsilon^{(0)} \tag{4.2.6}$$

Since the coordinates of electrons 1 and 2 are independent and yet (4.2.6) says that functions of these coordinates always sum to a constant, each function must be separately constant.

$$\hat{h}(1)\psi(1) = k_1 \psi(1) \tag{4.2.7a}$$

$$\hat{h}(2)\psi(2) = k_2 \psi(2) \tag{4.2.7b}$$

but (4.2.7a) and (4.2.7b) are just hydrogenic atom Schrödinger equations which may be solved to give exact ground-state solutions, $k_1 = k_2 = E_1(Z=2)$ and $\psi = 1s = (Z^3/\pi)^{1/2} \exp(-Zr)$ ($Z=2$). Since $\varepsilon^{(0)} = k_1 + k_2$, from (4.1.17),

$$\varepsilon^{(0)} = 2E_1(Z=2) = 2(-\tfrac{1}{2} \tfrac{4}{1}) \tag{4.2.8}$$

$$= -4 \text{ hartrees}$$

However, the zero-order wavefunction $1s(1)1s(2)$ is symmetric and will not satisfy the Pauli principle (Section 3.5), which requires that the wavefunction be antisymmetric in the space-spin coordinates of the electrons, If $\hat{P}_{12}$ is the permutation operator, the Pauli principle requires that

$$\hat{P}_{12} \Psi^{(0)}(1,2) = -\Psi^{(0)}(1,2)$$

The lowest-energy-antisymmetric wavefunction is built from spin-orbitals $1s\alpha$ and $1s\beta$. As in the case of two electrons in a box, the antisymmetric wavefunction is the *Slater determinant* of occupied spin-orbitals. (See Section 3.6.)

$$\Psi^{(0)}(1,2) = \hat{A} 1s\alpha(1) 1s\beta(2)$$

$$= 1/\sqrt{2} \begin{vmatrix} 1s\alpha(1) & 1s\beta(1) \\ 1s\alpha(2) & 1s\beta(2) \end{vmatrix}$$

$$= 1/\sqrt{2} [1s\alpha(1)1s\beta(2) - 1s\alpha(2)1s\beta(1)] \tag{4.2.9}$$

4.2 The Helium Atom

EXERCISE Prove that the energy of $\Psi^{(0)}(1, 2)$ is the same before and after antisymmetrization. Hint: find $\langle \Psi^{(0)}, \hat{H}_0 \Psi^{(0)} \rangle$ and use spin function orthogonality.

Continuing with the perturbation approach, the energy to first-order is easily obtainable from $\Psi^{(0)}$ [eq. (2.7.7)]

$$\varepsilon^{(1)} = \langle \Psi^{(0)}, 1/r_{12} \Psi^{(0)} \rangle \qquad (4.2.10)$$

Substituting (4.2.9) into (4.2.10) and integrating out the spin functions we obtain

$$\varepsilon^{(1)} = \langle [1s(1)]^2, (1/r_{12})[1s(2)]^2 \rangle = J_{1s1s} \qquad (4.2.11)$$

The integral

$$J_{1s1s} = \iint [1s(1)]^2 \frac{1}{r_{12}} [1s(2)]^2 \, dV_1 \, dV_2$$

is familiar in classical electrostatics as the *repulsive Coulomb interaction energy between the two charge distributions* $[1s(1)]^2$ and $[1s(2)]^2$. Thus the integral is called a *Coulomb integral*.

In Appendix B the integral J_{1s1s} is evaluated and shown to have the value $\frac{5}{8}Z$. The energy to first-order is now

$$\varepsilon^{(0)} + \varepsilon^{(1)} = -4 + \frac{10}{8} = -2.750 \text{ hartrees} \qquad (4.2.12)$$

This energy is still rather far from the exact energy of helium, $E_{\text{exact}} = -2.904$ hartrees, so $\varepsilon^{(2)}$ is important. Unfortunately, the evaluation of $\varepsilon^{(2)}$ requires the first-order wavefunction Section (2.7), and we might well ask if the perturbation approach starting from the hydrogenic model is the fastest or most physical means of getting a better result than (4.2.12). Is it really physically reasonable that each electron in helium see the full nuclear charge $Z = 2$? No, it is more reasonable to suppose that if electron 1 is near the nucleus, electron 2 will tend to be far from the nucleus, and that electron 1 therefore partially "screens" the nucleus from electron 2 and vice versa. The net result is to reduce Z from 2 to some "effective nuclear charge" Z_{eff}, in which case the Z that appears in the energy and in the wavefunction should be Z_{eff} not Z.

$$1s = (Z_{\text{eff}}^3/\pi)^{1/2} \exp(-Z_{\text{eff}} r) \qquad (4.2.13)$$

How does one find the "best" value for Z_{eff}? Regarding Z_{eff} as a nonlinear parameter, as in Section 2.6, we can use the variational principle to determine the best Z_{eff}.

$$W(Z_{\text{eff}}) = \frac{\langle \Psi^{(0)}, \hat{H} \Psi^{(0)} \rangle}{\langle \Psi^{(0)}, \Psi^{(0)} \rangle} \geq E_{\text{exact}}$$

where $\Psi^{(0)}$ contains $1s$ with Z_{eff} as in (4.2.13), but $\hat{H}$ is the full Hamiltonian of (4.2.1).

$$W(Z_{\text{eff}}) = Z_{\text{eff}}^2 - \frac{27}{8} Z_{\text{eff}} \qquad (4.2.14)$$

PROOF

$$W(Z_{\text{eff}}) = \langle \Psi^{(0)}, [\hat{h}(1) + \hat{h}(2)]\Psi^{(0)} \rangle + \langle \Psi^{(0)}, (1/r_{12})\Psi^{(0)} \rangle$$

where use has been made of the fact that $\Psi^{(0)}$ is normalized.

$$W(Z_{\text{eff}}) = \langle 1s(1), \hat{h}(1)1s(1) \rangle + \langle 1s(2), \hat{h}(2)1s(2) \rangle + \tfrac{5}{8}Z_{\text{eff}}$$

$$\langle 1s(1), \hat{h}(1)1s(1) \rangle = \langle 1s(2), \hat{h}(2)1s(2) \rangle = \langle 1s, (-\tfrac{1}{2}\nabla^2 - 2/r)1s \rangle$$

$$= \langle 1s, (-\tfrac{1}{2}\nabla^2 - Z_{\text{eff}}/r)1s \rangle + \langle 1s, [(Z_{\text{eff}} - 2)/r]1s \rangle$$

$$= -Z_{\text{eff}}^2/2 + (Z_{\text{eff}} - 2)\langle 1s, (1/r)1s \rangle$$

but from a previous exercise (on p. 91) $\langle 1s, (1/r)1s \rangle = Z_{\text{eff}}$.

$$\langle 1s, \hat{h}1s \rangle = -Z_{\text{eff}}^2/2 + (Z_{\text{eff}} - 2)Z_{\text{eff}}$$

so

$$W(Z_{\text{eff}}) = -Z_{\text{eff}}^2 + 2(Z_{\text{eff}} - 2)Z_{\text{eff}} + \tfrac{5}{8}Z_{\text{eff}}$$

$$= Z_{\text{eff}}^2 - \frac{27}{8} Z_{\text{eff}}$$

Minimizing $W(Z_{\text{eff}})$ with respect to Z_{eff} yields $dW/dZ_{\text{eff}} = 0 = 2Z_{\text{eff}} - 27/8$ or $Z_{\text{eff}} = 27/16 = 1.6875$, whereupon

$$W(1.6875) = -2.848 \text{ hartrees} \qquad (4.2.15)$$

which is appreciably better than (4.2.12) ($Z = 2$), -2.750 hartrees. The net reduction in the nuclear charge due to screening of the nucleus is $Z - Z_{\text{eff}} = 0.31$.

Although the difference between the screened orbital energy -2.848 hartrees and the exact energy -2.904 hartrees may seem small, it is ~ 1.4 eV or 31 kcal/mole, which is quite a large energy. Furthermore this is *not yet* the *best* wavefunction of the form $\hat{A}1s\alpha(1)\,1s\beta(2)$, simply because the variation of only one parameter (Z_{eff}) is not a complete variation of the orbital. The best wavefunction of this form is called the Hartree-Fock wavefunction and its energy is slightly lower, -2.862 hartrees, than the screened orbital energy. The Hartree-Fock method and the problem of the residual energy difference between the Hartree-Fock energy and the exact energy will be dealt with in later sections.

4.3 MANY-ELECTRON ATOMS: HYDROGENIC MODEL

When there are N electrons present in an atom their motions are strongly influenced by the positively charged nucleus *and* by the interelectronic Coulomb repulsions. In order to simplify the problem we use the perturbation approach. The perturbation is the total interelectronic repulsion, which depends on the distance between electrons as $1/|r_i - r_j| = 1/r_{ij}$.

With N electrons present in the atom the Hamiltonian of the atom is[10]

$$\hat{H} = \sum_{i=1}^{N}(-\tfrac{1}{2}\nabla_i^2 - Z/r_i) + \sum_{i>j}^{N} 1/r_{ij} \qquad (4.3.1)$$

or

$$\hat{H} = \sum_{i=1}^{N} \hat{h}(i) + \hat{V}(r_{ij}) = \hat{H}_0 + \hat{V}(r_{ij}) \qquad (4.3.2)$$

$\hat{H}_0$ is a sum of N hydrogenic Hamiltonians ($\hat{h}$) whose eigenfunctions satisfy the hydrogenic Schrödinger equation

$$\hat{h}\psi_{nlm}(r_i, \theta_i, \phi_i) = E_n \psi_{nlm}(r_i, \theta_i, \phi_i) \qquad (4.3.3)$$

The remainder of $\hat{H}$ is $\hat{V}(r_{ij})$, the total of all the interelectronic Coulomb repulsions. Using the hydrogenic orbitals ψ_{nlm} the zero-order problem is immediately solved by separation of variables,

$$\hat{H}_0 \Psi^{(0)}(1, 2, 3, \ldots, N) = \varepsilon^{(0)} \Psi^{(0)}(1, 2, 3, \ldots, N) \qquad (4.3.4)$$

where

$$\Psi^{(0)}(1, 2, 3, \ldots, N) = \psi_{n_1 l_1 m_1}(1)\psi_{n_2 l_2 m_2}(2)\psi_{n_3 l_3 m_3}(3) \cdots \psi_{n_N l_N m_N}(N) \qquad (4.3.5)$$

separates (4.3.4) so that the zero-order energy is

$$\varepsilon^{(0)} = \sum_{i=1}^{N} E_{n_i} \qquad (4.3.6)$$

EXERCISE Show that (4.3.5) separates the zero-order Schrödinger equation so that (4.3.6) is the zero-order energy.

As a result of (4.3.5) we may say that an atom (or ion) with N electrons ($N = Z$ for the neutral atom) is characterized by assigning the N sets of quantum numbers $nlmm_s$, thereby identifying the occupied spin orbitals. The *configuration* of the atom identifies the number of electrons in subshells of the same n and l; the number of electrons in each subshell is indicated by a superscript. For example, the lowest-energy (ground-state) configurations of lithium, beryllium, and carbon atoms are Li $(1s)^2(2s)$, Be

---

[10] Still neglecting small magnetic interactions; see Sections 4.7 and 6.10.

$(1s)^2(2s)^2$, and C $(1s)^2(2s)^2(2p)^2$; configurations of their ions are Li$^+$ $(1s)^2$, Be^{+2}$(1s)^2$, C^{+2} $(1s)^2(2s)^2$, etc.

How are the ground-state configurations of atoms and ions determined? As shown in the last section, the ground-state configuration of helium is $(1s)^2$. However, in order to proceed further along the periodic table we need the Pauli principle, which says that $\Psi^{(0)}$ must be antisymmetric in the space-spin coordinates of the electrons; as usual, this is achieved by taking an antisymmetrized product of the occupied spin-orbitals, that is,

$$\Psi^{(0)}(1, 2, \ldots, N) = \hat{A}\psi_{n_1l_1m_1m_{s_1}}(1)\psi_{n_2l_2m_2m_{s_2}}(2)\cdots\psi_{n_Nl_Nm_Nm_{s_N}}(N) \quad (4.3.7)$$

where

$$\psi_{n_1l_1m_1\frac{1}{2}} = \psi_{n_1l_1m_1}\alpha$$

and

$$\psi_{n_1l_1m_1-\frac{1}{2}} = \psi_{n_1l_1m_1}\beta$$

The antisymmetrizer $\hat{A}$ turns the product of spin-orbitals into a normalized determinant (Slater determinant). *If two columns or rows of a determinant are identical the determinant will vanish, therefore each electron must occupy a different spin-orbital.* As a consequence lithium has the ground-state configuration $(1s)^2(2s)$ and degenerate wavefunctions

$$\Psi(1, 2, 3) = \begin{cases} \hat{A}1s\alpha(1)1s\beta(2)2s\alpha(3) \\ \hat{A}1s\alpha(1)1s\beta(2)2s\beta(3) \end{cases}$$

$\hat{H}_0$ gives the zero-order energy of the many-electron atom, which from equations (4.3.4)–(4.3.6) is $\hat{H}_0\Psi^{(0)} = \sum_i E_{ni}\Psi^{(0)}$, and $\hat{V}(r_{ij})$ gives the energy to first-order (Section 2.1) $\varepsilon^{(1)} = \langle\Psi^{(0)}, \hat{V}(r_{ij})\Psi^{(0)}\rangle$.

EXAMPLE Lithium: The zero-order and first-order energies are

$$\varepsilon^{(0)} = 2E_1 + E_2 \quad \text{with} \quad Z = 3$$

and

$$\varepsilon^{(1)} = \langle \Psi^{(0)}, \sum_{i>j}^{3}(1/r_{ij})\Psi^{(0)}\rangle$$

$$= \langle \det |1s\alpha(1)1s\beta(2)2s\alpha(3)| \sum_{i>j}^{3}(1/r_{ij})1s\alpha(1)1s\beta(2)2s\alpha(3)\rangle$$

(Note that it is permissible to use the main diagonal of one determinant, if we omit the normalization factor, in taking the expectation values between Slater determinants.)

$$= \langle[1s(1)]^2, (1/r_{12})[1s(2)]^2\rangle + \langle[1s(1)]^2, (1/r_{13})[2s(3)]^2\rangle$$
$$+ \langle[1s(2)]^2, (1/r_{23})[2s(3)]^2\rangle - \langle 1s(1)2s(1), (1/r_{13})1s(3)2s(3)\rangle$$
$$\varepsilon^{(1)} = J_{1s1s} + 2J_{1s2s} - K_{1s2s}$$

4.4 The Electronic Charge Density

Again the J integrals are called "Coulomb integrals" because they represent the classical Coulomb repulsion between charge distributions $(1s)^2 \leftrightarrow (1s)^2$ and $(1s)^2 \leftrightarrow (2s)^2$. However, the K integral *has no classical counterpart*; it is called an *exchange integral*. Exchange integrals arise between spin-orbitals of the same spin as a result of antisymmetrization of $\Psi^{(0)}$. Although antisymmetrization has no effect on the zero-order energy (see previous exercises) it does add this nonclassical *exchange energy* in first order. Needless to say the exchange energy is a very important contribution to the energy of the atom.

In summary, the hydrogenic model provides us with a zero-order and a first-order energy. The zero-order energy is a sum of hydrogenic orbital energies for the occupied spin-orbitals; the first-order energy is a sum of Coulomb and exchange integrals (the exchange integrals arising only between spin-orbitals of the same spin, e.g., $1s\alpha$, $2s\alpha$ in the case of lithium). From our experience with the helium atom we know that a minimum improvement to the hydrogenic model would be the inclusion of screening, but we discuss this improvement in a later section.

EXAMPLE Beryllium: $\Psi^{(0)} = \hat{A} 1s\alpha(1)1s\beta(2)2s\alpha(3)2s\beta(4)$

$$\varepsilon^{(0)} = 2E_1 + 2E_2 \quad \text{with } Z = 4$$

$$\varepsilon^{(1)} = \langle \det |1s\alpha(1)1s\beta(2)2s\alpha(3)2s\beta(4)|, \sum_{i>j}^{4}(1/r_{ij})1s\alpha(1)1s\beta(2)2s\alpha(3)2s\beta(4)\rangle$$

$$= J_{1s1s} + J_{2s2s} + 4J_{1s2s} - 2K_{1s2s}$$

EXERCISE Find the zero-order energy of the boron ground state, and express the first-order energy in terms of Coulomb and exchange integrals.

4.4 THE ELECTRONIC CHARGE DENSITY

In dealing with the many-electron wavefunction $\Psi(1, 2, 3, \ldots, N)$ it is no longer so evident that $|\Psi|^2$ is a distribution function as defined in Section 3.1 because it is a function of the coordinates of N particles. We define the "one-electron probability density" or "electronic charge density" as

$$\rho(x, y, z) = N \iint \cdots \int |\Psi(1, 2, 3, \ldots, N)|^2 \, dV_2 \, dV_3 \cdots dV_N \quad (4.4.1)$$

where $\Psi(1, 2, \ldots, N)$ is an antisymmetrized and normalized wavefunction. Note that the dependencies on coordinates of electrons 2 to N are integrated out, so ρ is a function of the coordinates of one particle, here chosen to be particle 1. (Since Ψ is antisymmetrized it is immaterial which particle is singled out.) As implied by the name "charge" density, ρ is the real distribution of electronic charge in space and we visualize it as a continuous, three-dimensional classical charge distribution.[11]

$$\rho = \sum_{i=1}^{N} |\psi_{n_i l_i m_i m_{s_i}}|^2 \qquad (4.4.2)$$

EXAMPLE   Charge density in helium.

$$\rho = 2\int |\Psi(1,2)|^2 \, dV_2 = 2(1/\sqrt{2})^2 \int \begin{vmatrix} 1s\alpha(1) & 1s\beta(1) \\ 1s\alpha(2) & 1s\beta(2) \end{vmatrix}^2 dV_2$$

$$\rho = 1s\alpha(1)1s\alpha(1) \int 1s\beta(2)1s\beta(2) \, dV_2 + 1s\beta(1)1s\beta(1) \int 1s\alpha(2)1s\alpha(2) \, dV_2$$

$$\rho = 1s\alpha(1)1s\alpha(1) + 1s\beta(1)1s\beta(1) = |1s\alpha|^2 + |1s\beta|^2$$

or, considering just the spatial distribution,

$$\rho = 2|1s|^2$$

For neon, which has a configuration $(1s)^2(2s)^2(2p)^6$, we find

$$\rho = 2|1s|^2 + 2|2s|^2 + 2|2p_x|^2 + 2|2p_y|^2 + 2|2p_z|^2$$

Closed-shell charge densities, such as those of Ne, Be, and Ca, have spherically symmetric ρ. *The charge density ρ is a very important concept in molecules since the distribution of electronic charge throughout the molecule will determine the size, shape, and reactivity of the molecule.*

EXERCISE   Use the free-electron model of butadiene and prove that Ψ (molecular orbital) and χ (localized orbital) representations give the same ρ.

4.5 THE PERIODIC SYSTEM OF THE ELEMENTS

The Schrödinger equation $\hat{H}\Psi = E\Psi$, for the many-electron atom has no exact solution. Even approximate solutions are obtained with difficulty. We chose to work with the solvable equation $\hat{H}_0 \Psi^{(0)} = \varepsilon^{(0)} \Psi^{(0)}$ and to treat the interelectronic Coulomb repulsions $\hat{V}(r_{ij})$ as a perturbation. Each electron occupies a different spin-orbital, thus making it possible to antisymmetrize the wavefunction and satisfy the Pauli principle; the spin-orbitals are those of the hydrogenic atom (ion). The spin-orbitals are filled from orbitals of lowest energy on up to higher energy, but since the hydrogenic orbital energies depend on n, and not on l, we might be led to believe that $2s$ and $2p$ have the same energy even in the many-electron atom, that is, they are degenerate. However, $\hat{V}(r_{ij})$ serves to remove this degeneracy in first order so that a configuration $(1s)^2(2s)$ has a lower

---

[11] Strictly, $\rho \, dV$ is the probability of finding an electron in the volume element dV when all other electrons have arbitrary positions. Furthermore, ρ should be multiplied by the electronic charge e to have units of charge density.

energy than the configuration $(1s)^2(2p)$. This energy difference is not so large that Be $(1s)^2(2s)^2$ is an inert gas.[12]

Continuing to fill the spin-orbitals we reach Ne $(1s)^2(2s)^2(2p)^6$. Such completed shells are often denoted K, L, M, etc., for $n = 1$, 2, 3, etc., so neon has the configuration KL, The next spin-orbital has $n = 3$ and is considerably less tightly bound even in the hydrogenic model. Thus the configuration LK is highly stable and neon *is* an inert gas. With continued filling of the spin-orbitals we arrive at argon KL $(3s)^2(3p)^6$. The perturbation $\hat{V}(r_{ij})$ so splits the energy of configurations containing the $3p$ and $3d$ orbitals that they are occupied in the order $3p$, $4s$, then $3d$. As a result argon is an inert gas, and atomic numbers 19 and 20 (K and Ca) have $(4s)$ and $(4s)^2$ occupations.

The transition elements follow at values of Z beginning with $Z = 21$, and a long row ensues in which the $3d$ orbitals are filled before the next inert gas configuration is achieved; krypton KLM $(4s)^2(4p)^6$. Again the interelectronic repulsions so split the energy of configurations containing $4p$ and $4d$ orbitals that the orbitals are generally occupied in the order $4p$, $5s$, then $4d$. Elements $Z = 24$, 29, 41, 42, etc., offer exceptions to these orbital occupations orderings in the ground state.[13]

Another long row ensues containing the remainder of the transition metals ending with the inert gas xenon KLM $(5s)^2(4d)^{10}(5p)^6$. The transition metals are characterized by incomplete d shells, which are responsible for the interesting chemical and spectroscopic properties of these elements (see Chapter 9).

Beyond xenon the $6s$ shell is occupied, and then the $4f$ shell begins to fill, giving the lanthanide series of rare earths. In the next row the filling of $5f$ gives the actinide series which includes radioactive elements such as uranium. The heavier elements in the actinide series are manmade by bombardment in high-energy particle accelerators; they are very short-lived. However, the existence of stable[14] elements with atomic numbers in excess of 100 has been postulated[15], for example, $Z = 114$, 126. The ground state electronic configurations of the elements are listed in Table 4.5.1 together with spectroscopic information that will be explained in Sections 5.7–5.9.

---

[12] In fact Be $(1s)^2(2s)^2$ is so close in energy to Be $(1s)^2(2p)^2$ that a linear combination of these two configurations is a substantial improvement over the single configuration ground state; see Section 4.11.

[13] The ground state has the lowest total energy of the atom, which, as seen in the previous section, is not merely a sum of orbital energies, for example, see $\varepsilon^{(1)}$ of Li and Be. This makes it misleading at best to discuss orbital energies as if they gave the total energy of the many-electron system.

[14] By "stable" is meant having long half-lives.

[15] For example, T. E. Pierce and M. Blann, *Nucl. Phys.* **A106**, 14 (1967).

TABLE 4.5.1 The Periodic System of the Elements According to Atomic Number Z^a

Z	Atom	Orbital electronic configuration	Ground state	$IP_1(eV)$
1	H	$1s$	^2S	13.595
2	He	$1s^2$	^1S	24.580
3	Li	[He]$2s$	^2S	5.390
4	Be	[He]$2s^2$	^1S	9.320
5	B	[He]$2s^2 2p$	^2P	8.296
6	C	[He]$2s^2 2p^2$	^3P	11.264
7	N	[He]$2s^2 2p^3$	^4S	14.54
8	O	[He]$2s^2 2p^4$	^3P	13.614
9	F	[He]$2s^2 2p^5$	^2P	17.42
10	Ne	[He]$2s^2 2p^6$	^1S	21.559
11	Na	[Ne]$3s$	^2S	5.138
12	Mg	[Ne]$3s^2$	^1S	7.644
13	Al	[Ne]$3s^2 3p$	^2P	5.984
14	Si	[Ne]$3s^2 3p^2$	^3P	8.149
15	P	[Ne]$3s^2 3p^3$	^4S	11.0
16	S	[Ne]$3s^2 3p^4$	^3P	10.357
17	Cl	[Ne]$3s^2 3p^5$	^2P	13.01
18	Ar	[Ne]$3s^2 3p^6$	^1S	15.755
19	K	[Ar]$4s$	^2S	4.339
20	Ca	[Ar]$4s^2$	^1S	6.111
21	Sc	[Ar]$4s^2 3d$	^2D	6.56
22	Ti	[Ar]$4s^2 3d^2$	^3F	6.83
23	V	[Ar]$4s^2 3d^3$	^4F	6.74
24	Cr	[Ar]$4s\, 3d^5$	^7S	6.763
25	Mn	[Ar]$4s^2 3d^5$	^6S	7.432
26	Fe	[Ar]$4s^2 3d^6$	^5D	7.90
27	Co	[Ar]$4s^2 3d^7$	^4F	7.86
28	Ni	[Ar]$4s^2 3d^8$	^3F	7.633
29	Cu	[Ar]$4s\, 3d^{10}$	^2S	7.724
30	Zn	[Ar]$4s^2 3d^{10}$	^1S	9.391
31	Ga	[Ar]$4s^2 3d^{10} 4p$	^2P	6.00
32	Ge	[Ar]$4s^2 3d^{10} 4p^2$	^3P	7.88
33	As	[Ar]$4s^2 3d^{10} 4p^3$	^4S	9.81
34	Se	[Ar]$4s^2 3d^{10} 4p^4$	^3P	9.75
35	Br	[Ar]$4s^2 3d^{10} 4p^5$	^2P	11.84
36	Kr	[Ar]$4s^2 3d^{10} 4p^6$	^1S	13.996

4.5 The Periodic System of the Elements

TABLE 4.5.1—Continued

Z	Atom	Orbital electronic configuration	Ground state	IP_1 (eV)
37	Rb	[Kr]$5s$	^2S	4.176
38	Sr	[Kr]$5s^2$	^1S	5.692
39	Y	[Kr]$5s^24d$	^2D	6.5
40	Zr	[Kr]$5s^24d^2$	^3F	6.95
41	Nb	[Kr]$5s\,4d^4$	^6D	6.77
42	Mo	[Kr]$5s\,4d^5$	^7S	7.10
43	Tc	[Kr]$5s^24d^5$	^6S	7.28
44	Ru	[Kr]$5s\,4d^7$	^5F	7.364
45	Rh	[Kr]$5s\,4d^8$	^4F	7.46
46	Pd	[Kr]$4d^{10}$	^1S	8.33
47	Ag	[Kr]$5s\,4d^{10}$	^2S	7.574
48	Cd	[Kr]$5s^24d^{10}$	^1S	8.991
49	In	[Kr]$5s^24d^{10}5p$	^2P	5.785
50	Sn	[Kr]$5s^24d^{10}5p^2$	^3P	7.342
51	Sb	[Kr]$5s^24d^{10}5p^3$	^4S	8.639
52	Te	[Kr]$5s^24d^{10}5p^4$	^3P	9.01
53	I	[Kr]$5s^24d^{10}5p^5$	^2P	10.454
54	Xe	[Kr]$5s^24d^{10}5p^6$	^1S	12.127
55	Cs	[Xe]$6s$	^2S	3.893
56	Ba	[Xe]$6s^2$	^1S	5.210
57	La	[Xe]$6s^25d$	^2D	5.61
58	Ce	[Xe]$6s^24f5d$	^3H	6.91
59	Pr	[Xe]$6s^24f^3$	^4I	5.76
60	Nd	[Xe]$6s^24f^4$	^5I	6.31
61	Pm	[Xe]$6s^24f^5$	^6H	
62	Sm	[Xe]$6s^24f^6$	^7F	5.6
63	Eu	[Xe]$6s^24f^7$	^8S	5.6
64	Gd	[Xe]$6s^24f^75d$	^9D	6.16
65	Tb	[Xe]$6s^24f^9$	^6H	6.74
66	Dy	[Xe]$6s^24f^{10}$	^5I	6.82
67	Ho	[Xe]$6s^24f^{11}$	^4I	
68	Er	[Xe]$6s^24f^{12}$	^3H	6.08
69	Tm	[Xe]$6s^24f^{13}$	^2F	5.81
70	Yb	[Xe]$6s^24f^{14}$	^1S	6.2
71	Lu	[Xe]$6s^24f^{14}5d$	^2D	5.0
72	Hf	[Xe]$6s^24f^{14}5d^2$	^3F	
73	Ta	[Xe]$6s^24f^{14}5d^3$	^4F	7.88

TABLE 4.5.1.—Continued

Z	Atom	Orbital electronic configuration	Ground state	$IP_1\,(eV)$
74	W	$[Xe]6s^24f^{14}5d^4$	^5D	7.98
75	Re	$[Xe]6s^24f^{14}5d^5$	^6S	7.87
76	Os	$[Xe]6s^24f^{14}5d^6$	^5D	8.7
77	Ir	$[Xe]6s^24f^{14}5d^7$	^4F	9
78	Pt	$[Xe]6s^24f^{14}5d^9$	^3D	9.0
79	Au	$[Xe]6s\ 4f^{14}5d^{10}$	^2S	9.22
80	Hg	$[Xe]6s^24f^{14}5d^{10}$	^1S	10.43
81	Tl	$[Xe]6s^24f^{14}5d^{10}6p$	^2P	6.106
82	Pb	$[Xe]6s^24f^{14}5d^{10}6p^2$	^3P	7.415
83	Bi	$[Xe]6s^24f^{14}5d^{10}6p^3$	^4S	7.287
84	Po	$[Xe]6s^24f^{14}5d^{10}6p^4$	^3P	8.43
85	At	$[Xe]6s^24f^{14}5d^{10}6p^5$	^2P	
86	Rn	$[Xe]6s^24f^{14}5d^{10}5p^6$	^1S	10.746
87	Fr	$[Rn]7s$	^2S	
88	Ra	$[Rn]7s^2$	^1S	5.277
89	Ac	$[Rn]7s^26d$	^2D	
90	Th	$[Rn]7s^26d^2$	^3F	6.95
91	Pa	$[Rn]7s^25f^26d$	^4K	
92	U	$[Rn]7s^25f^36d$	^5L	6.1
93	Np	$[Rn]7s^25f^4\,6d$	^6L	
94	Pu	$[Rn]7s^25f^6$	^7F	5.1
95	Am	$[Rn]7s^25f^7$	^8S	6.0
96	Cm	$[Rn]7s^25f^76d$	^9D	
97	Bk	$[Rn]7s^25f^9$	^6H	
98	Cf	$[Rn]7s^25f^{10}$	^5I	
99	Es	$[Rn]7s^25f^{11}$	^4I	
100	Fm	$[Rn]7s^25f^{12}$	^3H	
101	Md	$[Rn]7s^25f^{13}$	^2F	
102	No	$[Rn]7s^25f^{14}$	^1S	
103	Lw	$[Rn]7s^25f^{14}6d$	^2D	

[a] IP_1 is the first ionization potential in electron volts. Data from Charlotte E. Moore, "Atomic Energy Levels," National Bureau of Standards Circular 457, vols. 1, 2, and 3.

In sum, the periodicity of the chemical elements and their properties seems to follow directly from the hydrogenic model and the Pauli principle, but many of the characteristic features of the periodic table arise from perturbation of the hydrogenic model by the interelectronic repulsions.

4.6 ANGULAR MOMENTUM

Thus $\hat{V}(r_{ij})$ is chemically very significant. It is also very significant for the spectra of the atom, because it often splits the configuration into states of different energy which have different angular momenta. We examine the atomic angular momenta in the next section and use the results in Section 4.7 on atomic spectra.

In Section 3.3 it was shown that if two observables have simultaneously well defined values (eigenvalues) then the respective operators for the observables commute. The converse is also true; if two operators commute they have simultaneous eigenfunctions. In order to understand the classification of atomic states and the meaning of the quantum numbers m and l we must find the set of operators which commute with $\hat{H}$.

If $\hat{O}\hat{P} = \hat{P}\hat{O}$ then $\hat{O}\psi = k\psi$ and $\hat{P}\psi = m\psi$; the condition for simultaneous eigenvalues is $[\hat{O}\hat{P} - \hat{P}\hat{O}] = 0$, $[\hat{O}\hat{P} - \hat{P}\hat{O}]$ is abbreviated $[\hat{O}, \hat{P}]$ and is called the *commutator bracket*. We must therefore look for operators such that $[\hat{H}, \hat{P}] = 0$, where $\hat{H}$ is the many-electron Hamiltonian equation (4.3.1). In classical mechanics the angular momentum of a particle about a center of mass is a constant of the motion. Therefore we examine the commutation properties of the quantum mechanical angular momentum operators. The classical orbital angular momentum is $\mathbf{L} = \mathbf{r} \times \mathbf{p}$ with components $L_x = yp_z - zp_y$, $L_y = zp_x - xp_z$, and $L_z = xp_y - yp_x$. Using the rule for translation to quantum mechanics (Section 3.3), we replace p_k with $\hbar/i\, \partial/\partial q_k$, where q_k is the coordinate.

$$\hat{L}_x = \frac{\hbar}{i}\left(y\frac{\partial}{\partial z} - z\frac{\partial}{\partial y}\right)$$

$$\hat{L}_y = \frac{\hbar}{i}\left(z\frac{\partial}{\partial x} - x\frac{\partial}{\partial z}\right) \quad (4.6.1)$$

$$\hat{L}_z = \frac{\hbar}{i}\left(x\frac{\partial}{\partial y} - y\frac{\partial}{\partial x}\right)$$

$$\hat{L} = \mathbf{i}\hat{L}_x + \mathbf{j}\hat{L}_y + \mathbf{k}\hat{L}_z \quad (4.6.2)$$

$$\hat{L}^2 = \hat{L}_x^2 + \hat{L}_y^2 + \hat{L}_z^2 \quad (4.6.3)$$

$$[\hat{L}_x, \hat{L}_y] = i\hbar\hat{L}_z \quad [\hat{L}_y, \hat{L}_z] = i\hbar\hat{L}_x \quad [\hat{L}_z, \hat{L}_x] = i\hbar\hat{L}_y \quad (4.6.4)$$

EXERCISE Establish the commutation properties of equation (4.6.4).

$$[\hat{L}^2, \hat{L}_x] = 0 \quad [\hat{L}^2, \hat{L}_y] = 0 \quad [\hat{L}^2, \hat{L}_z] = 0 \quad (4.6.5)$$

$$[\hat{L}^2, \hat{H}] = 0 \quad \text{and} \quad [\hat{L}_z, \hat{H}] = 0 \quad (4.6.6)$$

From (4.6.4) the components of the orbital angular momentum do not commute; therefore they are not simultaneous observables, and no experiment can be set up which precisely defines the value of more than *one* of these components simultaneously. However, $\hat{H}$ and $\hat{L}^2$ do commute with $\hat{L}_x$, $\hat{L}_y$, and $\hat{L}_z$. We may now choose *a commuting set of operators* $\hat{H}$, $\hat{L}^2$, and $\hat{L}_k$ (by convention $\hat{L}_k$ is the z component, $\hat{L}_z$) *which have simultaneous eigenfunctions and precisely defined observable values* (*eigenvalues*).

The eigenvalues of $\hat{L}^2$ and $\hat{L}_z$ are easily identified if we write the operators in spherical polar coordinates.

$$\hat{L}_z = -i\hbar \frac{\partial}{\partial \phi}$$

and

$$\hat{L}^2 = -\hbar^2 \left(\frac{1}{\sin\theta} \frac{\partial}{\partial\theta}\left(\sin\theta \frac{\partial}{\partial\theta}\right) + \frac{1}{\sin^2\theta}\frac{\partial^2}{\partial\phi^2}\right)$$

Now from (4.1.6)

$$\hat{L}_z \psi_{nlm}(r, \theta, \phi) = m\hbar \psi_{nlm}(r, \theta, \phi) \tag{4.6.7}$$

where $\psi_{nlm}(r, \theta, \phi)$ are the hydrogenic orbitals. From (4.1.8a)

$$\hat{L}^2 \psi_{nlm}(r, \theta, \phi) = l(l+1)\hbar^2 \psi_{nlm}(r, \theta, \phi) \tag{4.6.8}$$

EXERCISE   Prove (4.6.7) and (4.6.8).

The hydrogenic functions are eigenfunctions of $\hat{L}^2$ with eigenvalue $l(l+1)\hbar^2$, hence the states of the hydrogenic atom have a precisely defined orbital angular momentum of magnitude $\sqrt{l(l+1)\hbar^2}$. The s states have $l = 0$ and lack any orbital angular momentum. If a mental picture is necessary for states with $l \neq 0$, we might visualize the electron in orbit about the nucleus and thereby possessing an orbital angular momentum. However, as with any circulating charge, there is a magnetic moment associated with an electron having $l \neq 0$. As a result the atom is a tiny magnet and will orient itself in a magnetic field; the possible orientations are quantized (*space quantization*). The possible orientations are those such that the *component* $\hat{L}_z$ of the angular momentum in the field direction is $m\hbar$ (m is often called the "magnetic quantum number").

Turning to the spin angular momentum, there are the analogous operators $\hat{S}^2$ and $\hat{S}_z$ with the following properties:

$$\hat{S}_z \alpha = \tfrac{1}{2}\hbar\alpha \quad (m_s = \tfrac{1}{2}) \qquad \hat{S}_z \beta = -\tfrac{1}{2}\hbar\beta \quad (m_s = -\tfrac{1}{2}) \tag{4.6.9}$$

$$\hat{S}^2 \alpha = \tfrac{1}{2}(\tfrac{1}{2}+1)\hbar^2 \alpha \qquad \hat{S}^2 \beta = \tfrac{1}{2}(\tfrac{1}{2}+1)\hbar^2 \beta \tag{4.6.10}$$

where α and β are the spin functions discussed in Section 3.5 for the two possible spin states of the electron. What is important in the present context is that $\hat{L}^2$, $\hat{L}_z$, and $\hat{H}$ are all devoid of spin coordinates so they each commute with $\hat{S}^2$ and $\hat{S}_z$ ($\hat{S}^2$ and $\hat{S}_z$ commute with each other). We now have the commuting set of operators which specify the states of the one-electron atoms and ions, $\hat{H}$, $\hat{L}^2$, $\hat{L}_z$, $\hat{S}^2$, $\hat{S}_z$. The spin-orbitals ψ_{nlmm_s} are eigenfunctions to each operator of the commuting set and such a spin-orbital has precisely defined values of the corresponding observables (i.e., has quantum numbers n, l, m, and m_s).

4.7 THE SPECTROSCOPIC STATES OF ATOMS: VECTOR COUPLING

Spectroscopists observe an enormously larger number of spectral lines than can be accounted for as transitions between configurations. Since the hydrogenic model is certainly a correct starting point, the configurations must be split into several states of different energy by that part of $\hat{H}$ which the hydrogenic model ignores, $\hat{V}(r_{ij})$. To find out what characterizes these states (in addition to energies not easily calculated) we need only remember that the angular momentum is a constant of the motion. Each state corresponds to a certain coupling of the individual angular momenta of the electrons to give total atomic momenta. These couplings are predicted by a vector coupling rule.

RUSSELL-SAUNDERS COUPLING

The orbital angular momentum for the *i*th electron is a vector which couples by a vector addition rule with the angular momenta of the other electrons to give a *total orbital angular momentum L*. (Fig. 4.7.1). Russell and Saunders[16] prescribed how these vectors were to be coupled to give an empirical explanation of the calcium spectrum. While the scheme has been justified, we present it simply as a postulate, as they did. The spin

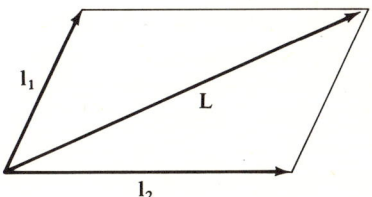

FIG. 4.7.1 Two angular momenta vectors couple to give the resultant.

[16] H. N. Russell and F. A. Saunders, *Astrophys. J.* **61**, 38 (1925).

angular momenta of the electrons must also be coupled to give a *total spin S*. The *total angular momentum* of the atom is then the resultant $S + L = J$. The operators for the total momenta are the sums of the individual operators for each electron, for example

$$\hat{L}_z = \sum_{i=1}^{N} \hat{l}_{z_i} \qquad \hat{L}_y = \sum_{i=1}^{N} \hat{l}_{y_i} \quad \text{etc.} \tag{4.7.1}$$

$$\hat{L}^2 = \hat{L}_x^2 + \hat{L}_y^2 + \hat{L}_z^2 \tag{4.7.2}$$

$$\hat{S}_z = \sum_{i=1}^{N} \hat{s}_{z_i} \quad \text{and} \quad \hat{S}^2 = \hat{S}_x^2 + \hat{S}_y^2 + \hat{S}_z^2 \tag{4.7.3}$$

$$\hat{J}_z = \sum_{i=1}^{N} (\hat{s}_{z_i} + \hat{l}_{z_i}) \qquad \hat{J}^2 = \hat{J}_x^2 + \hat{J}_y^2 + \hat{J}_z^2 \tag{4.7.4}$$

If the Hamiltonian is free of spin terms (i.e., spin-orbit interaction terms which we have not heretofore included in $\hat{H}$ or even discussed, but will shortly), then the operators $\hat{H}$, $\hat{L}^2$, $\hat{L}_z$, $\hat{S}^2$, $\hat{S}_z$, and $\hat{J}_z$ form a commuting set of operators with eigenvalues E, $L(L+1)\hbar^2$, $M_L\hbar$, $S(S+1)\hbar^2$, $M_S\hbar$, and $M_J\hbar$, respectively. Because the operators commute the wavefunction is simultaneously an eigenfunction to each of them. The many-electron wavefunction is then characterized by the quantum numbers L, M_L, S, M_S, and M_J; this is the so-called *Russell-Saunders coupling scheme*.[17] This set of commuting operators is commonly chosen to characterize the state of the many-electron atom. There are other possible choices, one of which, *j–j* coupling, will be dealt with shortly.

The many-electron wavefunction $\Psi_{LSM_L M_S M_J}$ in Russell-Saunders coupling satisfies the eigenvalue equations

$$\hat{J}_z \Psi = M_J \hbar \Psi \qquad \hat{L}^2 \Psi = L(L+1)\hbar^2 \Psi$$
$$\hat{L}_z \Psi = M_L \hbar \Psi \qquad \hat{S}^2 \Psi = S(S+1)\hbar^2 \Psi$$
$$\hat{S}_z \Psi = M_S \hbar \Psi \qquad \hat{J}^2 \Psi = J(J+1)\hbar^2 \Psi$$

The wavefunction has well-defined values (eigenvalues) of these total momenta and their components. It now appears that if we can find the possible values of L, M_L, M_S, etc., for a given arrangement of electrons in spin-orbitals we will know the possible states of the atomic system. This is correct, but how does one go about finding the values of these quantum numbers?

A configuration consists of a chosen set of n and l values for the electrons. *The possible couplings in such a configuration are obtained by writing down the possible occupations of m and m_s sublevels which obey the Pauli*

---

[17] Also called L-S Coupling.

4.7 The Spectroscopic States of Atoms: Vector Coupling

principle. For each such distribution of electrons we form the sum of $m_i(M_L)$ and the sum of $m_{s_i}(M_S)$, M_S, and M_L are the components of the total spin and orbital angular momentum, respectively. In order to find the total spin and orbital angular momenta we use a rule for the possible components, as follows:

$$M_L = \sum_i m_i \quad \text{and} \quad M_S = \sum_i m_{s_i}$$
$$M_L = L, L-1, L-2, \ldots, 0, \ldots, -L \tag{4.7.5}$$
$$M_S = S, S-1, S-2, \ldots, -S$$

The terminology for these states is summarized in Table 4.7.1.

TABLE **4.7.1**

L Symbol	0 S	1 P	2 D	3 F	4 G	5 H
S $2S+1$ Multiplicity	0 1 Singlet		$\frac{1}{2}$ 2 Doublet		1 3 Triplet	$\frac{3}{2}$ 4 Quartet	2 5 Quintet

The quantity $(2S+1)$ is called the *multiplicity*. The *spectroscopic term symbol* gives the values of S and L by placing the multiplicity as a left-hand superscript to the L value.

$$^{(2S+1)}L \qquad \text{THE SPECTROSCOPIC TERM SYMBOL}$$

Thus we have ^1S, ^2D, ^3P, etc.; one says, "singlet-ess," "doublet-dee," etc.

The total angular momentum of the atom takes the values

$$J = S+L, S+L-1, \ldots, 0, \ldots, (L-S) \quad \text{or} \quad (S-L) \tag{4.7.6}$$

whichever is positive. The J value of the state is written as a right-hand subscript to the term symbol, $^{(2S+1)}L_J$. This completely specifies the state of the atom except for the $(2J+1)$-fold degeneracy which arises from the possible values of M_J, $M_J = J, J-1, J-2, \ldots, -J$. This degeneracy can only be removed by an external field. States such as $^3P_2$, $^3P_1$, and $^3P_0$, which differ only in their total angular momentum, are called *multiplets*; they differ in energy because of spin-orbit coupling (see Fig. 4.7.3). The total degeneracy of a term is $(2S+1)(2L+1)$.

EXAMPLE   Boron has the ground-state term ^2P, which gives rise to the states $^2P_{1/2}$ and $^2P_{3/2}$. The degeneracy is

$$(2S+1)(2L+1) = (2 \times \tfrac{1}{2} + 1)(2 \times 1 + 1) = 6$$

The six-fold degeneracy of boron $(1s)^2(2s)^2(2p)$ corresponds to the two possible spins ($\pm \frac{1}{2}$) for each of three possible p orbitals. Alternatively, $^2P_{1/2}$ has $(2J+1) = 2$ possible M_J levels, and $^2P_{3/2}$ has $(2J+1) = 4$ possible M_J levels, again, a six-fold degeneracy. The degeneracy of the term ^2P is partially lifted by spin-orbit coupling which splits it into its component J states (see further).

As is evident from the previous example, an important simplification results from the fact that closed shells have total orbital and spin angular momenta of zero, that is, Ne $^1S_0$, Be $^1S_0$, etc., for we can *ignore all the electrons in closed shells in determining the spectroscopic terms.*

EXAMPLE   Helium $(1s)^2$, $m_1 = m_2 = 0$, therefore $M_L = 0$, so $L = 0$. $m_{s_1} = +\frac{1}{2}$, $m_{s_2} = -\frac{1}{2}$, therefore $M_S = 0$, so $S = 0$. From (4.7.6) $J = 0$ (and $M_J = 0$). The resulting state is He $(1s)^2$ $^1S_0$.

EXAMPLE   Boron $(1s)^2(2s)^2(2p)$. We can ignore the $^1S_0$ Be core, $(1s)^2(2s)^2$, and deal with the remaining electron, $(2p)$, with $m = 1, 0, -1$. Therefore $M_L = 1, 0, -1 = L, L-1, \ldots, 0, \ldots, -1$ from (4.7.5); hence $L = 1$ and this is a P state. $m_{s_1} = \pm \frac{1}{2}$, therefore $M_S = +\frac{1}{2}, -\frac{1}{2} = S, S-1, \ldots$ from (4.7.5), hence $S = \frac{1}{2}$ and $2S+1 = 2$, a doublet state. The possible J values are $1 \pm \frac{1}{2}$, so we have two states with symbols $^2P_{1/2}$ and $^2P_{3/2}$.

EXAMPLE   Find the spectroscopic states which arise from the excited configuration of Be $(1s)^2(2s)(2p)$. We may ignore the $^1S_0$ helium core. Note that although $n_1 = n_2$, $l_1 \neq l_2$, so the Pauli exclusion principle is already satisfied. The possible values of m_i and m_{s_i} are organized into a table.

m_1	m_2	M_L	m_{s_1}	m_{s_2}	M_S
0	0	0	$\pm\frac{1}{2}$	$\pm\frac{1}{2}$	1, −1
0	0	0	$\pm\frac{1}{2}$	$\mp\frac{1}{2}$	0, 0
0	1	1	$\pm\frac{1}{2}$	$\pm\frac{1}{2}$	1, −1
0	1	1	$\pm\frac{1}{2}$	$\mp\frac{1}{2}$	0, 0
0	−1	−1	$\pm\frac{1}{2}$	$\pm\frac{1}{2}$	1, −1
0	−1	−1	$\pm\frac{1}{2}$	$\mp\frac{1}{2}$	0, 0

Since $M_L = 1, 0, -1$, there are P states present. Furthermore, $M_S = 1, 0, -1$ ($M_L = 1, 0, -1$) and $M_S = 0$ ($M_L = 1, 0, -1$) account for all the combinations in the table. Hence we have ^3P and ^1P. The ^3P term is the multiplet $^3P_0$, $^3P_1$, $^3P_2$ and the other state is $^1P_1$.

EXERCISE   In the same way find the terms which arise from $(1s)^2(2s)^2(2p)(3p)$.

ANSWER ^3D, ^1D, ^3P, ^1P, ^3S, ^1S.

For other atoms, Table 4.7.2 is useful. It gives the terms which arise from configurations of unfilled subshells (all filled subhells are ^1S). The table distinguishes between the terms which arise from unfilled shells of equivalent electrons (electrons of the same nl subshell) and unfilled shells of nonequivalent electrons. In the former, certain states which are allowed

4.7 The Spectroscopic States of Atoms: Vector Coupling

among nonequivalent electrons are forbidden by the Pauli exclusion principle. The determination of the terms arising from unfilled shells of equivalent electrons, and nonequivalent electrons, is sufficiently tedious to recommend the use of Table 4.7.2, which lists these terms for the commonly occurring configurations.

EXAMPLE   As an example of a configuration of equivalent electrons consider carbon $(1s)^2(2s)^2(2p)^2$. The limitations imposed by the Pauli principle are expressed in the following table.

m_1	m_2	M_L	m_{s_1}	m_{s_2}	M_S
0	1	1	$\pm\tfrac{1}{2}$	$\pm\tfrac{1}{2}$	1, −1
0	1	1	$\pm\tfrac{1}{2}$	$\mp\tfrac{1}{2}$	0, 0
0	0	0	$\tfrac{1}{2}$	$-\tfrac{1}{2}$	0
0	−1	−1	$\pm\tfrac{1}{2}$	$\pm\tfrac{1}{2}$	1, −1
0	−1	−1	$\pm\tfrac{1}{2}$	$\mp\tfrac{1}{2}$	0, 0
1	1	2	$\tfrac{1}{2}$	$-\tfrac{1}{2}$	0
1	−1	0	$\pm\tfrac{1}{2}$	$\pm\tfrac{1}{2}$	1, −1
1	−1	0	$\pm\tfrac{1}{2}$	$\mp\tfrac{1}{2}$	0, 0
−1	−1	−2	$\tfrac{1}{2}$	$-\tfrac{1}{2}$	0

To find the resulting terms it is helpful to construct a table showing the number of times each M_L value occurs with each M_S value.

$$\begin{array}{c|ccc}
 & -1 & 0 & 1 \\ \hline
2 & & 1 & \\
1 & 1 & 2 & 1 \\
M_L\ 0 & 1 & 3 & 1 \\
-1 & 1 & 2 & 1 \\
-2 & & 1 & \\
\end{array}$$
$$M_S$$

From the maximum value of M_L we discern a D term having the array

$$\begin{array}{c|c}
2 & 1 \\
1 & 1 \\
M_L\ 0 & 1 \\
-1 & 1 \\
-2 & 1 \\
\end{array}$$
$L = 2$, $S = 0$, ^1D term

Subtracting out the ^1D array, the remaining array consists of the sum of

$$\begin{array}{c|ccc}
1 & 1 & 1 & 1 \\
M_L\ 0 & 1 & 1 & 1 \\
-1 & 1 & 1 & 1 \\
\end{array}$$
$L = 1$, $S = 1$, ^3P term

and

$$\begin{array}{c|c}
M_L\ 0 & 1 \\
\end{array}$$
$L = 0$, $S = 0$, ^1S term

Thus, the configuration (p^2) leads to the terms ^3P, ^1D, and ^1S. Figure 4.7.2 illustrates this result.

There is a useful equivalence between the terms arising from the number of vacancies in a subshell and the terms arising from the same number of electrons in the subshell. *The terms which arise from configurations $(nl)^{v-k}$, where v is the maximum number of electrons in the subshell of quantum number l, are identical to the terms which arise from the configuration $(nl)^k$.*

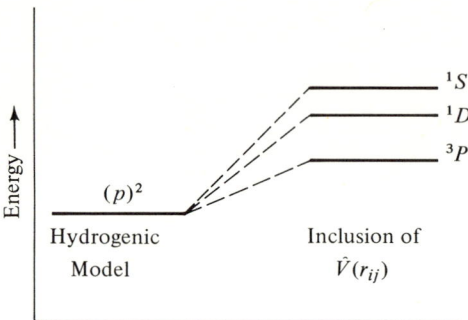

FIG. 4.7.2 Schematic of $(p)^2$ configuration splitting into spectroscopic states of different spin and orbital angular momentum.

EXAMPLE   Find the terms which arise from oxygen $(1s)^2(2s)^2(2p)^4$. Only $(2p)^4$ need be considered, $v = 6$ and $k = 4$, so the terms are the same as those of $(2p)^2$, ^3P, ^1D, and ^1S.

ENERGY

The lowest-energy term of a given configuration of equivalent electrons is given by Hund's rule.

Among terms of a given configuration the term of greatest multiplicity lies lowest. Among the terms of the same multiplicity that of greatest L lies lowest.

EXERCISE   The ground-state term symbols of the elements are given in Table 4.5.1. Compare these terms with the totality of terms of that configuration (Table 4.7.2) and verify Hund's rule.

The Russell-Saunders coupling scheme has now been described in sufficient detail. Russell-Saunders coupling is characterized by first forming L and S out of the individual momenta of the electrons and then adding these to form the total atomic momentum. With increasing atomic number there is a transition from Russell-Saunders to "jj coupling" brought about by the growing importance of the spin-orbit interaction terms in the

4.7 The Spectroscopic States of Atoms: Vector Coupling

TABLE **4.7.2**

Equivalent electrons		Nonequivalent electrons	
Configuration	Terms	Configuration	Terms
$(s)^1$	^2S	$(ns)(ks)$	$^1S, \, ^3S$
$(s)^2$	^1S	$(ns)(kp)$	$^1P, \, ^3P$
$(p)^1, (p)^5$	^2P	$(ns)(kd)$	$^1D, \, ^3D$
$(p)^2, (p)^4$	$^3P, \, ^1D, \, ^1S$	$(np)(kp)$	$^1P, \, ^3P, \, ^1D, \, ^3D, \, ^1S, \, ^3S$
$(p)^3$	$^2P, \, ^2D, \, ^4S$		
$(p)^6$	^1S		
$(d)^1, (d)^9$	^2D		
$(d)^2, (d)^8$	$^1S, \, ^1D, \, ^1G, \, ^3P, \, ^3F$		
$(d)^3, (d)^7$	$^2P, \, ^2D(2), \, ^2F, \, ^2G, \, ^2H, \, ^4P, \, ^4F$		
$(d)^4, (d)^6$	$\begin{cases}^1S(2), \, ^1D(2), \, ^1F, \, ^1G(2), \, ^1I, \, ^3P(2) \\ ^3D, \, ^3F(2), \, ^3G, \, ^3H, \, ^5D\end{cases}$		
$(d)^5$	$\begin{cases}^2S, \, ^2P, \, ^2D(3), \, ^2F(2), \, ^2G(2), \, ^2H \\ ^2I, \, ^4P, \, ^4D, \, ^4F, \, ^4G, \, ^6S\end{cases}$		
$(d)^{10}$	^1S		

Hamiltonian, which are strongly dependent on nuclear charge. In the jj coupling scheme the individual spin and orbital angular momenta of the electrons couple, forming a resultant vector j_i $(s_i + l_i = j_i)$; the j_i couple to form the resultant J. The spin-orbit interaction is a magnetic interaction between the spin magnetic moment of the electron and the magnetic field due to the orbital motion of the electron. The interaction is approximated by adding a term to the Hamiltonian, $\hat{H}' = \sum_{i=1}^{N} \zeta \hat{L}_i \cdot \hat{S}_i$, only $\hat{J}$ and $\hat{J}_z$ commute with $\hat{H} + \hat{H}'$ and as a result L, M_L, and M_S, as well as S, are no longer good quantum numbers. For our purpose, the net result of the spin-orbit interaction is to split the Russell-Saunders states of different J, as in Fig. 4.7.3. This leads to the *multiplet rule*: If a spectroscopic state arises from a configuration with a less than half-filled shell then the level of lowest J lies lowest in energy, but if the configuration has a more than half-filled shell this J energy ordering is *inverted*.

EIGENFUNCTIONS

The determination of the states which arise from a given configuration follows from the vector coupling rules (or from Table 4.7.2), but the construction of a wavefunction $\Psi_{LSM_LM_SM_J}$ which is an eigenfunction to the commuting set of operators is not as simple.[18] Most often the eigenfunction

---

[18] The construction of eigenfunctions to angular momentum operators is dealt with in more advanced texts.

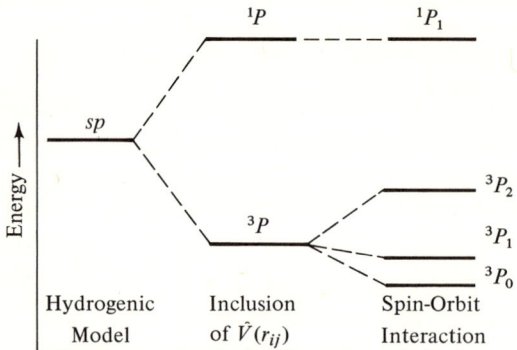

FIG. 4.7.3 Schematic splitting of the energy of the configuration (*sp*) first by inclusion of the interelectronic repulsions and again by the inclusion of spin-orbit magnetic interaction. (The splitting is exaggerated for clarity.)

is a *linear combination* of antisymmetrized products of spin-orbitals (non-closed shell configurations). In comparison, the closed-shell ($^1S_0$) eigenfunction is a single antisymmetrized product of spin-orbitals.

EXAMPLE Helium $(1s)(2s)$; ^1S and ^3S eigenfunctions are examples of non-closed-shell states. What is needed is an eigenfunction to $\hat{S}^2$ with eigenvalue $S(S+1)\hbar^2 = 0$, that is, $S = 0$, the singlet; also an eigenfunction to $\hat{S}^2$ with eigenvalue $2\hbar^2$, that is, $S = 1$, the triplet. The eigenfunctions are

$$^1S = (1/\sqrt{2})[\hat{A}1s\alpha(1)2s\beta(2) - \hat{A}1s\beta(1)2s\alpha(2)] \quad M_S = 0$$

$$^3S = \begin{cases} \hat{A}1s\alpha(1)2s\alpha(2) & M_S = 1 \\ \hat{A}1s\beta(1)2s\beta(2) & M_S = -1 \\ (1/\sqrt{2})[\hat{A}1s\alpha(1)2s\beta(2) + \hat{A}1s\beta(1)2s\alpha(2)] & M_S = 0 \end{cases}$$

In order to prove that the examples given are indeed eigenfunctions of the angular momentum, it is convenient to define *step-up* and *step-down* angular momentum *ladder operators* and to reexpress $\hat{S}^2$ and $\hat{L}^2$ in terms of these operators. Consider the spin operators: We define the step-up spin operator $\hat{S}_+$ and the step-down spin operator $\hat{S}_-$ such that

$$\hat{S}_+ \alpha = 0 \qquad \hat{S}_- \beta = 0$$
$$\hat{S}_+ \beta = \hbar\alpha \qquad \hat{S}_- \alpha = \hbar\beta$$

From these operators it can be shown that $\hat{S}^2 = \hat{S}_- \hat{S}_+ + \hat{S}_z^2 + \hbar \hat{S}_z$ or for the many-electron system

$$\hat{S}^2 = \sum_i \hat{s}_{-i} \sum_j \hat{s}_{+j} + \sum_i \hat{s}_{zi} \sum_j \hat{s}_{zj} + \hbar \sum_i \hat{s}_{zi} \quad (4.7.7)$$

4.8 *Energy Level Diagrams and Spectroscopic Transitions* **117**

EXERCISE Use the definition (4.7.7) of $\hat{S}^2$ to show that the eigenfunctions in the last example are, as stated, ^1S and ^3S, respectively (especially the $M_S = 0$ components of each state).

$\hat{L}^2$ may be expressed in terms of ladder operators[19] analogously to (4.7.7), however because we are concerned primarily with energy levels and transitions between energy levels, we will not need the eigenfunctions, explicitly, but only the information summarized in the term symbol.

4.8 ENERGY LEVEL DIAGRAMS AND SPECTROSCOPIC TRANSITIONS

An atomic energy level diagram gives the energies of the spectroscopic states of the atom. The zero of energy is usually chosen as the ground-state energy.[20] In the lithium atom energy levels of Fig. 4.8.1 we consider the

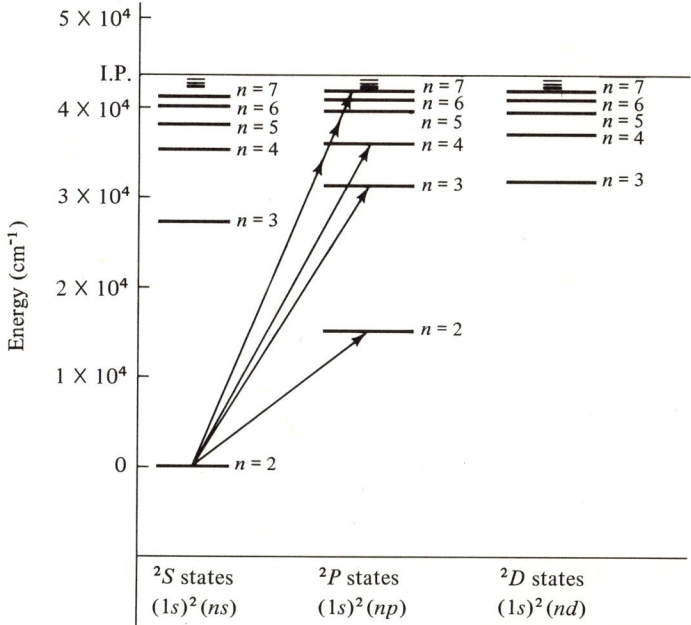

FIG. 4.8.1 Energy level diagram of lithium. The ground state is chosen as the arbitrary zero of energy. The arrows indicate the allowed optical transitions from the ground state.

---

[19] H. Hameka, *Introduction to Quantum Theory*, Harper & Row, New York, 1967.
[20] The zero of energy is, of course, entirely arbitrary. Heretofore we have chosen the zero of energy to be all the electrons separated from the nucleus and each other at infinity. On this scale the ground-state energy of the atom is a large negative number.

energy levels which results from exciting the 2s electron from the ground-state configuration $(1s)^2(2s)$ to the (ns) or (np) or (nd) orbitals, $n \geq 2$. The resulting $(1s)^2(ns)$ configurations have but one spectroscopic state $^2S_{1/2}$, the $(1s)^2(np)$ configurations have states $^2P_{3/2}$ and $^2P_{1/2}$, and so on, for other excited configurations. The energy splitting between states of different J is too small to be indicated in the figure.[21]

At the ionization limit the electron is removed from the atom altogether, leaving the ion $Li^+ = (1s)^2$. An infinity of states crowd in towards this ionization limit at 5.37 eV, blending into a continuum beyond the ionization energy. In Fig. 4.8.1 states are grouped according to the type of excited configuration (given at the bottom).

In general, a transition from the ground state to any of these states can be brought about by bombardment with electrons of sufficient energy, but for transition by absorption or emission of light there are *optical selection rules*.

$$\Delta l = \pm 1 \qquad \Delta L = \pm 1, 0$$
$$\Delta S = 0 \qquad \Delta J = \pm 1, 0*$$

OPTICAL SELECTION RULES

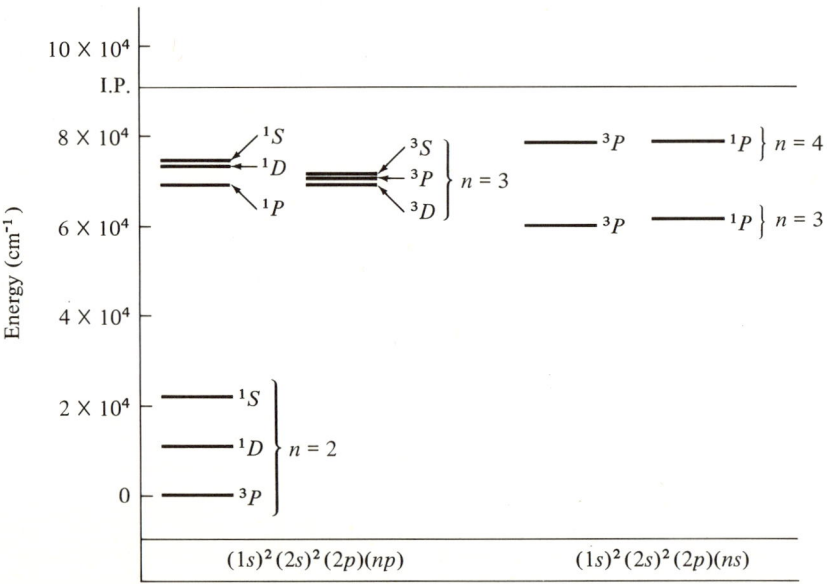

FIG. 4.8.2 A small portion of the energy levels of the carbon atom.

[21] Charlotte Moore, "Atomic Energy Levels," *National Bureau of Standard Circular* 457, Vol. I, 1949.

* Except $J = 0$ to $J' = 0$ is forbidden.

A typical way of observing the entire spectrum would be by bombarding lithium with very energetic electrons (i.e., an electrical discharge with lithium on the electrode) and observing the spectrum in emission as the atoms cascade down through the states emitting according to the selection rules.

As found in the previous section, in the case of carbon atom (Fig. 4.8.2) the configurations are split by $\hat{V}(r_{ij})$ into spectroscopic terms of different energy, for example, $(2p)^2$ into ^1S, ^1D, and ^3P, and $(2p)(ns)$ is split into ^1P and ^3P.

EXERCISE   Sketch an energy level diagram for beryllium. Indicate the allowed optical transitions from the ground state by arrows.

4.9 SCREENING, SLATER ORBITALS, IONIZATION POTENTIALS, AND ATOMIC RADIUS

In our dealings with configurations, states, and wavefunctions there has been an implicit assumption that the orbitals are hydrogenic spin-orbitals with Z equal to the atomic number of the atom. For example, Na $KL(3s)$, with $\psi_{3s} = (27 - 18\,Zr + 2Z^2r^2)\exp(-Zr/3)$ and $Z = 11$, the atomic number of Na. However, only if $\hat{V}(r_{ij}) = \sum_{i>j} 1/r_{ij} = 0$ is this choice of Z the *best choice* in the sense of the *variational theorem*; for only if the interelectronic repulsions were *zero* would each electron see solely the full nuclear charge Z.

As in the case of helium (Section 4.2) electrons near the nucleus screen the full nuclear charge from electrons further away. This is illustrated by Fig. 4.9.1, where a spherical shell of negative charge $-(Z-1)$ around the positive charge $+Z$ reduces the net potential outside the shell to just $+1/r$. In atoms with more than two electrons the screening effect on electrons outside a closed shell is very strong. The bulk of the contribution

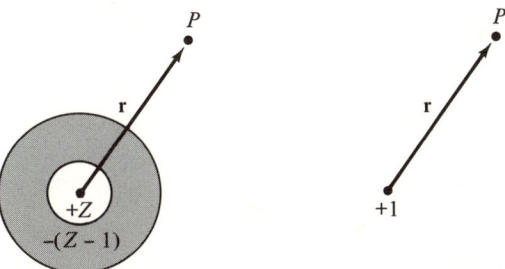

Physical Situation   Effective Result

FIG. 4.9.1   The screening effect. A positive charge of $+Z$ at the origin is screened from the point P by a spherical shell of charge $-(Z-1)$ centered at the origin. The net potential at P is $Z/r - (Z-1)/r = 1/r$.

of $\hat{V}(r_{ij})$ to the energy can be taken care of if Z is variationally determined for each orbital (nonlinear parameter variation). The variationally determined Z is called the *effective nuclear charge*, Z_{eff}.

$$Z_{\text{eff}} = Z - \mathfrak{S} \qquad \mathfrak{S} \text{ IS THE SCREENING CONSTANT} \qquad (4.9.1)$$

We digress for a moment to look more closely at the radial function. We saw that the angular nodes were quite important since they determine the shape and symmetry of the orbitals, but the radial nodes which are evident in plots of $2s$ and $3s$ are much less important. These nodes are the values of r at which $R_{nl}(r)$ is zero. They are necessary to assure the orthogonality of orbitals of different n and the same l, $\langle ns, ks \rangle = 0$. However, for most purposes of chemical interest, including the calculation of the energy, the nodes can be dispensed with; just take the highest power of r which appears in the polynomial under R_{nl} in Table 4.1.1 and modify the orbital as follows:

$$\psi_{nlm} = r^{n^*-1} \exp(-Z_{\text{eff}} r/n^*) \Theta_{lm}(\theta) \Phi_m(\phi) \qquad \text{SLATER TYPE ORBITAL} \quad (4.9.2)$$

Slater type orbitals (STO) are radially nodeless, easy to manipulate in chemical calculations, and very widely used.[22,23] Z_{eff} can be determined from the variational principle, but to avoid the complete variational calculation, an observed ionization potential can be used to find Z_{eff}. For a hydrogenic orbital the ionization potential is I.P. = $13.6 Z^2/n^2$ eV, which for Na would be $13.6(11)^2/(3)^2 = 182$ eV (observed I.P. = 5.15 eV!). So much for the hydrogenic model. Turning the relation around, use the observed I.P. to determine the Z_{eff}, $5.14 = 13.6 (Z_{\text{eff}})^2/(3)^2$, $Z_{\text{eff}} = 1.8$. Therefore, $\mathfrak{S} = 11 - 1.8 = 9.2$. So the $3s$ electron in Na has $(9.2/11) \times 100 = 83\%$ of the nuclear charge screened out. Each orbital will have its own screening constant depending on the nuclear charge Z and on the other occupied orbitals.

Slater[22,23] gave rules for determining Z_{eff} and n^* so that the I.P. was qualitatively predictable without doing the complete variational calculation. (Although variational calculations give values of Z_{eff} similar to Slater's values.) Roughly, one divides the orbitals into groups, $(1s)(2s2p)$ $(3s3p)(3d)(4s4p)(4d)(4f)(5s5p)$, etc. $\mathfrak{S}$ is found for an orbital of a group $(nl_1 nl_2)$ by adding the following contributions: (1) zero for any oribital further out; (2) 0.35 for each additional electron in the same group except for $1s$, where it is 0.30; (3) for s–p group add 0.85 for each electron with principal quantum number $n - 1$, plus 1.0 for each electron even further in, but for d or f groups add 1.0 for each additional electron further in regardless of n. Finally, the values of n^* are 1, 2, 3, 3.7, 4.0, and 4.2 for $n = 1, 2, 3, 4, 5$, and 6.

---

[22] J. C. Slater, *Phys. Rev.* **36**, 57 (1930).
[23] In the literature the symbols ζ_i (zeta) are usually used for the Z_{eff} of STO.

4.9 Screening, Slater Orbitals, Ionization Potentials, and Atomic Radius

EXAMPLE Going back to the Na case, we calculate Z_{eff} for $3s$ from Slater's rules to be $S = 2 \times 1$ for $(1s)^2 + 8 \times 0.85$ for $(2s2p)^8 = 8.8$, $Z_{eff} = 11 - 8.8 = 2.2$.

EXERCISE Calculate the first I.P. of (1) Li, K; (2) Mg, Ca; (3) C, Si: (4) O, S; compare to values in Table 4.5.1.

Slater's rules for S imply that orbitals in the same group inefficiently screen each other (i.e., because they have similar radial distributions), while an electron in an orbital closer in can completely screen a unit positive charge of the nucleus. Consequently, the inert gases have the maximum ionization potential in their rows, since the screening is quite incomplete in the last filled shell. Conversely, the alkali metals, with one electron outside of a closed core, have the minimum I.P. in the period. This is illustrated in Fig. 4.9.2.

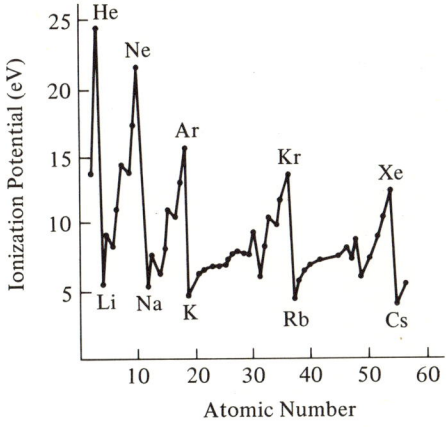

FIG. 4.9.2

To discuss the radius or "size" of the atoms we need a definition. There are many possible choices, but the use of ρ, the electronic density, is most satisfactory. We agree that the atomic radius is that value of r at which ρ has fallen to an arbitrarily small value, 0.01 or 0.0001, etc. On this basis we can compare atomic radii in a number of series. (1) Isoelectronic ions: Keep the number of electrons constant, but increase the nuclear charge; obviously the charge density will shrink around the nucleus (see discussion of radial wavefunction). (2) Move across a row of the periodic table: Because of incomplete screening in the outer shell, the net result is an increasing Z_{eff} and a considerable shrinkage as ρ increases near the nucleus. (Note that this means neon is a *smaller* atom than lithium.) (3) Move down a column of the periodic table: Note that $\langle \hat{r} \rangle \propto n^2$ so we can expect large increases in radii, but this is moderated by less efficient screening by the more numerous inner shells. (4) Move through a transition series: ρ, at larger r, is determined by the constant outer orbitals of larger n. As the inner d or f levels are filled they screen out the simultaneous increases in Z for a net constant Z_{eff}, and similar radii result.

4.10 THE HARTREE-FOCK METHOD

Up to this point we have not discussed how the atomic many-electron Schrödinger equation could be solved because there is no exact solution. Furthermore, the hydrogenic model, followed by the introduction of screening and the coupling of angular momenta to obtain the spectroscopic states, is a complete and correct *qualitative* description of atomic structure.

However, a fundamental understanding of what we do when we introduce screening and how it is a part of a process which converges toward a better wavefunction and energy for the atom is missing. This section and the next will try to give the student an appreciation of the convergence process by briefly outlining the *Hartree-Fock method* and the residual problem of *electron correlation*. The original Hartree-Fock procedure was developed by Hartree, Fock, and Slater about 1930, specifically for closed-shell systems. Although it has since been extended to open-shell configurations we will limit discussion to the closed-shell atom.

The key is again the variational principle, which says that if $\tilde{\Psi}$ is a trial many-electron ground-state wavefunction,[24] and if $W = \langle \tilde{\Psi}, \hat{H}\tilde{\Psi} \rangle / \langle \tilde{\Psi}, \tilde{\Psi} \rangle$, then $W \geq E$, where $\hat{H}$ is the complete Hamiltonian, $H = H_0 + \hat{V}(r_{ij})$, and E is the exact energy. Since W is always above E, $\tilde{\Psi}$ may be varied to minimize W. Let $\tilde{\psi}_i$ be trial one-electron wavefunctions (trial spin-orbitals); then $\tilde{\Psi}$, the trial N-electron closed-shell ground-state (^1S) wavefunction, is the single determinant

$$\tilde{\Psi}(1, 2, \ldots, N) = \hat{A}\tilde{\psi}_1(1)\tilde{\psi}_2(2) \cdots \tilde{\psi}_N(N) \qquad (4.10.1)$$

Substituting this trial wavefunction into the expression for W, and varying the $\tilde{\psi}_i$ so as to minimize[25] W, there results a wavefunction Ψ (the Hartree-Fock wavefunction), which is the *best* wavefunction of the single determinant form (4.10.1). The expectation value of the energy E_{HF} (the Hartree-Fock energy) is as close to E as possible with a single determinant wavefunction. The resulting spin-orbitals, $\psi_1, \psi_2, \ldots, \psi_N$, are the Hartree-Fock spin-orbitals; they are the *best set of one-electron wavefunctions* for the atom [and $\tilde{\Psi}$ of the form (4.10.1)].

---

[24] The variational principle may also be applied to excited trial wavefunctions with due care that the wavefunction is orthogonal to the ground state.

[25] The restriction on the variation is that $\tilde{\Psi}$ must remain a single determinantal ^1S state.

4.10 The Hartree-Fock Method

THE SELF-CONSISTENT FIELD

The complete variation of the trial orbitals in (4.10.1) reduces the many-electron problem to a set of coupled one-electron equations,[26] the Hartree-Fock equations.

$$\begin{aligned}\hat{H}_{\text{eff}}(1)\psi_1(1) &= E_1\psi_1(1) \\ \hat{H}_{\text{eff}}(2)\psi_2(2) &= E_2\psi_2(2) \\ &\vdots \qquad\qquad \vdots \end{aligned} \qquad (4.10.2)$$

The E_i are the *orbital energies* and again $-E_i$ are the *ionization potentials*[27] of the atom. $\hat{H}_{\text{eff}}$ is the effective one-electron Hamiltonian; it is the same for each electron. The form of $\hat{H}_{\text{eff}}$ illustrates the physical model which has resulted very naturally from orbital variation.

$$\hat{H}_{\text{eff}}(i) = \hat{h}(i) + \hat{V}_{\text{aver}}(i) \qquad (4.10.3)$$

$$\hat{V}_{\text{aver}}(i) = \sum_{j=1}^{N} \left\langle \psi_j(j), \frac{(1 - \hat{P}_{ij})}{r_{ij}} \psi_j(j) \right\rangle_j \qquad (4.10.4)$$

$\hat{H}_{\text{eff}}$ consists of the hydrogenic Hamiltonian $\hat{h}$ for a nucleus of charge Z = atomic number, plus an *average* (integral) *potential operator* $\hat{V}_{\text{aver}}(i)$, which is only a function of the coordinates of electron i (the dependence on the coordinates of electron j having been integrated out). The integral operator $\hat{V}_{\text{aver}}$ contains the permutation operator $\hat{P}_{ij}$, which is included to give the requisite exchange integrals.[28] To summarize, $\hat{H}_{\text{eff}}$ has electron i moving in a hydrogenic potential field plus the averaged potential field due to all the other occupied orbitals, and it includes exchange. The *spherical symmetry* of the Hamiltonian for *atoms* leads to a $\hat{V}_{\text{aver}}(i)$ which is only a function of r. The angular wavefunctions remain the spherical harmonics. This serves to make the Hartree-Fock method enormously easier for atoms than for molecules.

Antisymmetrization is responsible for the occurrence of the permutation operator in $\hat{V}_{\text{aver}}$ and for the appearance of exchange integrals in E_{HF}. The exchange integrals are $K_{ij} = \langle \psi_i(1)\psi_j(2), (1/r_{12})\psi_i(2)\psi_j(1) \rangle$. The Coulomb integrals are $J_{ij} = \langle \psi_i(1)\psi_j(2), (1/r_{12})\psi_i(1)\psi_j(2) \rangle$: note that $K_{ii} = J_{ii}$.

$$E_{\text{HF}} = \sum_{i=1}^{N} E_i - \sum_{i>j}^{N} (J_{ij} - K_{ij}) \qquad (4.10.5)$$

---

[26] The Hartree-Fock equation for the helium atom is derived in Appendix C.
[27] This is Koopmans' Theorem: Valid if ionization produces insignificant changes in the remaining orbitals.
[28] Note that the term in equation (4.10.4) having $j = i$ is formally zero because we have included $\hat{P}_{ij}$ and $\hat{P}_{ii} = 1$. The subscript j on the integral in (4.10.4) means integrate over the coordinates of the ith electron.

and from $E_i = \langle \psi_i, \hat{H}_{\text{eff}} \psi_i \rangle$ we also have

$$E_{\text{HF}} = \sum_{i=1}^{N} \langle \psi_i, \hat{h}_i \psi_i \rangle + \sum_{i>j}^{N} (J_{ij} - K_{ij}) \qquad (4.10.6)$$

Since all the orbitals appear in $\hat{V}_{\text{aver}}$, they occur in each equation of (4.10.2). Therefore the Hartree-Fock equations cannot be solved without making an initial guess at the ψ_i. One can then solve for the ψ_i and use these as input for another calculation (iteration procedure). Iteration is continued until the calculated set agrees with the input set as closely as desired. The calculation is then said to be self-consistent, that is, the orbitals calculated from (4.10.2) are consistent with the orbitals which supply the field $\hat{V}_{\text{aver}}$ for the calculation, thus a *self-consistent-field* (SCF) calculation. If the variation has been complete the final orbitals are not only self-consistent, but they are the Hartree-Fock orbitals of the atom.

The average potential $\hat{V}_{\text{aver}}$ largely takes care of the mutual interelectronic repulsion between electrons. Furthermore, this discussion of the Hartree-Fock method identifies $\hat{V}_{\text{aver}}$ as the source of the *screening* which reduces the nuclear charge to Z_{eff} and depends on the other occupied orbitals in the atom. The screening constant is a simple way of allowing for the average field of the other electrons. However, Z_{eff} is only one parameter per orbital; in general, several variational parameters should be introduced and varied to achieve an orbital of Hartree-Fock accuracy. A much-used procedure is to start with trial orbitals which are linear combinations of Slater-type orbitals. This is the *Hartree-Fock-Roothaan* procedure. The Hartree-Fock equations $\hat{H}_{\text{eff}} \psi_i = E_i \psi_i$ were originally solved by numerical methods which yields orbitals as tables of radial functions. For example, as $R_{nl}(r_i)$ versus r_i.[29] When highly refined this is probably the most accurate method, but it is seldom used. The Hartree-Fock-Roothaan procedure is easily adapted to rapid calculation on electronic computers and is very widely applied in atomic and molecular calculations. The following example outlines this procedure for the Hartree-Fock $1s$ orbital of the helium atom.

EXAMPLE He $(1s)^2$: The $1s$ function to be determined by the Hartree-Fock-Roothaan procedure is given a trial representation as a linear combination of STO.

$$1s(1) = \sum_{i=1}^{5} N_i c_i \exp(-Z_i r_1)$$

The N_i, c_i, and Z_i are normalization factor, coefficient, and nonlinear parameter, respectively. Of course, the $1s(2)$ trial function is identical to the above with r_2 substituted for r_1.

---

[29] The Hartree-Fock equations reduce to ordinary differential equations in r (rather than partial differential equations in r, θ, and ϕ) because $\hat{V}_{\text{aver}}$ is spherically symmetric. As a result the angular solutions must be the spherical functions discussed in Section 4.1 and the Hartree-Fock equations need only be solved for $R_{nl}(r)$.

4.10 The Hartree-Fock Method

PROCEDURE

The trial function is used to express the variational energy $W = \langle \Psi, \hat{H}\Psi \rangle / \langle \Psi, \Psi \rangle$. W is minimized with respect to the five coefficients, c_i, and the five parameters, Z_i. Explicitly, the c_i are determined from the five simultaneous equations $\partial W / \partial c_i = 0$ for an initial choice of the c_i and Z_i; the equations are then iterated to a self-consistent set of coefficients. The Z_i are varied by trial and error until the best choice of Z_i as well as self-consistent c_i are obtained.

RESULTS [30]

Z_i	c_i
1.4300	0.78503
2.4415	0.20284
4.0996	0.03693
6.4843	−0.00293
0.7978	0.00325

Orbital energy: −0.91796 hartree
Total energy calculated: −2.86168 hartrees
The calculated energy is to be compared with the hydrogenic model energy which we calculated in Section 4.2, −2.750 hartrees, and the exact energy, −2.904 hartrees.

EXAMPLE   As an additional example[30] of the Hartree-Fock-Roothaan procedure we present the table of coefficients and Z_i which define the orbitals of the nitrogen atom $(1s)^2(2s)^2(2p)^3$ ^4S. $c_i(1s)$ is the column of coefficients which defines the $1s$ Hartree-Fock-Roothaan orbital, $c_i(2s)$ defines the $2s$ orbital, etc.

STO Type	Z_i	$c_i(1s)$	$c_i(2s)$	$c_i(2p)$
1s	6.4595	0.92787	−0.21744	0
1s	10.8389	0.06535	−0.00843	0
2s	1.4699	−0.00067	0.27744	0
2s	1.9161	0.00188	0.54808	0
2s	3.1560	0.00017	0.33901	0
2s	5.0338	0.01665	−0.14212	0
2p	1.1937	0	0	0.29731
2p	1.7124	0	0	0.48388
2p	3.0112	0	0	0.28079
2p	7.1018	0	0	0.01352

Orbital energy: −15.6282   −0.94523   −0.56753 hartree
Total energy calculated: −54.4009 hartrees
The calculated energy is comparable to the experimental energy, −54.59 hartrees (found by adding up the ionization potentials of N, N^+, N^{2+} ... N^{6+}).

---

[30] Calculated by E. Clementi from a program due to C. C. J. Roothaan *et al.*, reported in "Tables of Atomic Functions," a supplement to the paper by E. Clementi, *IBM J. Res. Develop.* **9**, 2 (1965). Reprinted by permission.

Such calculations were impossible to carry out until the advent of really fast computers. Today the Hartree-Fock-Roothaan procedure has produced wavefunctions for the important atoms and ions of the periodic table and many small molecules. The results of these calculations are as follows:

1. Accurate *ionization potentials*,
2. *Electronic charge density*, close to physical reality.
3. Expectation values of one electron operators which are correct through first order, that is, certain electronic properties such as the *dipole moment* should be about as accurate as the energy.

Figure 4.10.1 gives a plot of the radial distribution function of argon. It clearly shows the three maxima corresponding to the K, L, and M

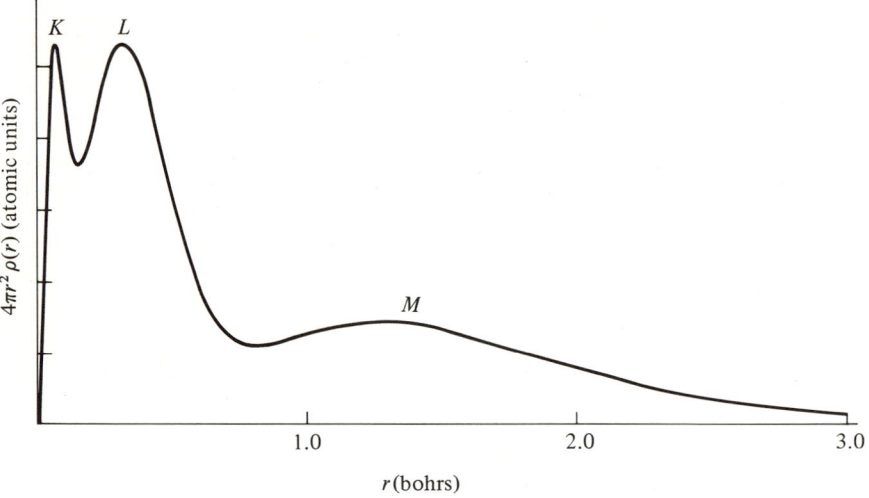

FIG. 4.10.1 The radial distribution function in argon, $(1s)^2(2s)^2(2p)^6(3s)^2(3p)^6$, from Hartree-Fock-Roothaan wavefunction. Experimental observations of electronic density distributions, for example, by electron or x-ray diffraction, do not give ρ to this precision, but are in essential agreement. Calculated by G. L. Malli. Used by permission. See G. L. Malli, *Can. J. phys.* **44**, 3121 (1966).

shells. Such plots of charge density inside the atom are obtainable experimentally from electron or x-ray diffraction experiments.[31] Experimental and purely theoretical plots are in substantial agreement; because of

[31] L. O. Brockway, *Phys. Rev.* **90**, 833 (1953).

experimental difficulties, the theoretical results are more precise (they may or may not, be more accurate). On the other hand Hartree-Fock calculations do *not* give reliable and consistent values of *transition energies*, molecular *dissociation energies*, and the *intensities* of *transitions*. The essential reason for this inadequacy is discussed next.

4.11 ELECTRON CORRELATION

In the Hartree-Fock method one obtains the best orbitals by minimizing the energy of Ψ, the antisymmetrized product wavefunction. The approximation which is inherent in a simple product wavefunction, $\Psi(1, 2) = \psi_1(1)\psi_2(2)$, is that the motion of an electron is *independent* of the position of all the other electrons (i.e., uncorrelated motion). This is evident from the fact that the two-electron probabilities distribution $|\Psi(1, 2)|^2$ is just the product of two one-electron probability distributions, $|\psi_1(1)|^2 |\psi_2(2)|^2$. This particle independence is moderated somewhat by the Pauli principle; the antisymmetrization of Ψ *correlates* the probability distribution in electrons of the same spin, so that there is zero probability of finding two electrons of the same spin close together. However, electrons of unlike spin remain uncorrelated in Ψ. Since the electrons repel each other strongly when found close together, there is an unwanted positive contribution to the energy in such a wavefunction, with the result $E_{HF} > E$, where E_{HF} is the Hartree-Fock energy. We define[32]

$$E - E_{HF} = E_{corr} \qquad \text{THE CORRELATION ENERGY} \qquad (4.11.1)$$

The magnitude of the correlation energy is small in comparison with E and E_{HF}; $E_{corr} \approx 1\%$ of E, so $E_{HF} \approx 99\%$ of E. Unfortunately, one is often only interested in energy differences, such as transition energies, or the difference in energy between reactants and products of chemical reactions; these energy differences are small differences between large energies. *If large energies of similar magnitude are in error by a percent or two they yield little if any information about their difference.*

The situation appeared bleak until systematic Hartree-Fock calculations and theoretical analysis showed that the correlation energy could be largely grouped into orbital-pair energies[33, 34]. In fact, the atomic correlation energy is found to be largely additive in pairs of electrons of unlike

---

[32] The E in this definition is not the strictly exact energy, but rather the "exact nonrelativistic energy." Because relativistic effects are only important at high velocity they make substantial contributions to the energy only in the atomic inner shells; relativistic energies are thus nearly constant in chemical changes and will not concern us further.

[33] L. C. Allen, E. Clementi, and H. M. Gladney, *Rev. Mod. Phys.* **35**, 465 (1963).

[34] O. Sinanoğlu, *J. Chem. Phys.* **36**, 706 (1962); *Advan. Chem. Phys.* **6**, 315 (1964).

spin, with the major contribution arising from the pairs occupying the same spatial orbital. For example, the correlation energy of a pair of 1s electrons, $(1s)^2$, is nearly independent of nuclear charge and equal to -1.2 eV.

A quantitative attack on the correlation problem uses the perturbation approach starting from the Hartree-Fock wavefunction as the zero-order wavefunction, and the Hartree-Fock energy as the energy to first order.[35] Then the perturbation expansion will converge if the perturbation $\hat{V}_F = \sum_{i>j} 1/r_{ij} - \sum_i \hat{V}_{aver}(i)$ is small. V_F has been called a *fluctuation potential*[36]; it is the difference between the *exact* interelectronic potential and the *mean* interelectronic potential of the atom. The chief feature of the fluctuation potential is its short range (in comparison with the Coulomb potential $1/r_{ij}$, which reaches to infinity). This means $\hat{V}_F$ has the most effect on electrons that get close to each other (see Fig. 4.11.1). In order to improve on the Hartree-Fock wavefunction it is necessary to include the effect of $\hat{V}_F$ in the wavefunction, that is, put a *correlation hole* into the wavefunction so that electrons have small probability of being found close together. At this

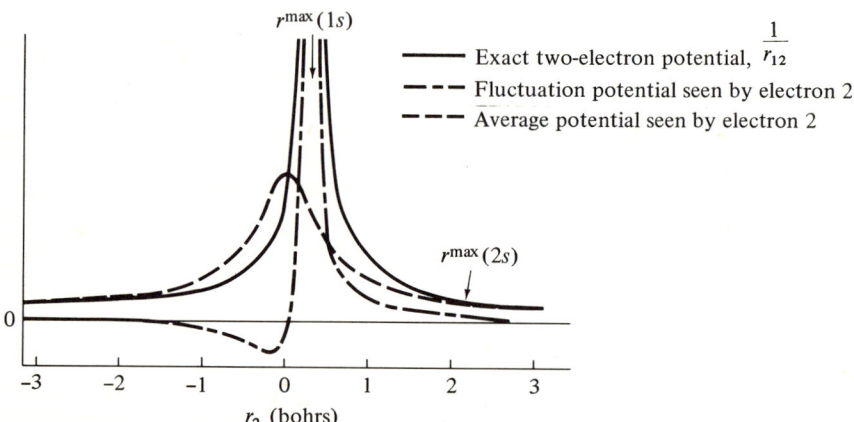

FIG. 4.11.1 The fluctuation potential for 1s electron in beryllium. The fluctuation potential is the difference between the *exact* interelectronic potential $1/r_{12}$, and the *average* interelectronic potential. The figure plots these potentials along the line joining the nucleus and electron 1. The potentials are functions of r_{12}, but the value of r_1 was chosen equal to its most probable value in the 1s Be orbital [$r^{max}(1s) = 0.27$ bohr]. Note that the fluctuation potential is small at the most probable radius of the 2s electron with this choice of r_1. After O. Sinanoğlu, *Proc. Nat. Acad. Sci.* **47**, 1217 (1961). Used by permission.

---

[35] E_{HF} is the energy to first order because $E_{HF} = \langle \Psi, \hat{H}\Psi \rangle = \langle \Psi(\hat{H}_0 + \hat{V}_F)\Psi \rangle$.
[36] O. Sinanoğlu, *Proc. Nat. Acad. Sci.* **47**, 1217 (1961).

4.11 Electron Correlation

point we emphasize that the Hamiltonian is not at fault, nor the Schrödinger equation; the problem is in the Hartree-Fock wavefunction, which failed to *correlate* the motions of electrons of unlike spin. We improve on the Hartree-Fock wavefunction by adding terms to it. These terms are determined by the fluctuation potential; they will put the correlation hole into the wavefunction and lower the energy towards its exact value.

The perturbation method proceeds by finding the first-order wavefunction. If Ψ' is the exact wavefunction and Ψ is the Hartree-Fock wavefunction then

$$\Psi' = \Psi + \chi = \Psi + \chi^{(1)} + \chi^{(2)} + \cdots \qquad (4.11.2)$$

$\chi^{(1)}$ will appreciably improve on the Hartree-Fock wavefunction but some nice features of the independent particle model are lost. Specifically, an often-used procedure is to represent χ by including *configuration interaction*[37] (CI). CI, as the name implies, represents χ by taking a linear combination of wavefunctions of N-electron configurations, each configuration having a different assignment of electrons among atomic orbitals of different n and l. Clearly, we have lost the nice concept of electrons occupying spin-orbitals with assignable n, l, and m values; all that remains are the total angular momenta, for example, $(1s)^2$ ^1S is mixed in CI only with other ^1S wavefunctions.

EXAMPLE   Helium atom. The principal configurations in the $\chi^{(1)}$ (the only ones in $\chi^{(1)}$) assign the electrons to such doubly excited configurations as $(2s)^2$, $(2s)(3s)$, $(3s)^2$ ...; $(2p)^2$, $(2p)(3p)$, $(3p)^2$ ...; etc., all ^1S wavefunctions,

$$\Psi'(He) = c_1\Psi + c_2\chi[(2s)^2] + \ldots + d_1\chi[(2p)^2] + \ldots$$

where the coefficients c_i, d_i, etc., are to be determined.[38, 39]

Physically, configuration interaction is not difficult to interpret. The configurations $(2s)^2$, $(3s)^2$, etc., correlate the motion of the electrons in He such that when one electron has high probability near the nucleus the other electron has greatest probability at larger r. The configurations $(2p)^2$, $(3p)^2$, etc., introduce angular correlation such that when one electron has a large probability in a particular direction the other electron has its largest probability in a different direction. Thus, slowly and painfully, the correlation hole is built in.

---

[37] E. A. Hylleraas, *Z. Phys.* **48**, 469 (1928); often called a superposition of configurations.
[38] According to a theorem (Brillouin's theorem) "singly excited" configurations such as $(1s)(ns)$ ^1S do not occur in $\chi^{(1)}$ but do occur in higher order.
[39] There exists a set of orbitals (different for each atom) that maximizes the rate of convergence of the CI series, the so-called *natural* spin-orbitals. [For He see E. R. Davidson, *J. Chem. Phys.* **39**, 875 (1963).]

It is much quicker to correlate the electrons via a wavefunction containing the interelectronic distance explicitly, for example, $1s(1)1s(2)(1 + kr_{12})$, where k is a variationally determined coefficient. This method is generalizable to N-electron systems if, for example, we adopt an *independent pair model*[40] for χ to augment the *independent particle model* of Ψ. Such a model recognizes that because of the Pauli principle and the short range of $\hat{V}_F$ the important effects in χ are due to binary encounters between electrons. Pair functions (sometimes called *geminals*) give $\chi^{(1)}$ in its entirety and in test calculations on small systems give the major portion of E_{corr}. This is the basis for the statement made earlier that the correlation energy is largely additive in electron pairs of unlike spin, with a major contribution arising from pairs occupying the same spatial orbitals.

REFERENCES

Berry, R. S. *J. Chem. Educ.* **43**, 283 (1966). An elementary but precise discussion of the Schrödinger equation, the Hartree-Fock method, and the problem of electron correlation.

Herzberg, G., *Atomic Spectra and Atomic Structure*, Dover, New York, 1944.

Kauzmann, W., *Quantum Chemistry*, Academic Press, New York, 1957.

Pilar, F. L., *Elementary Quantum Chemistry*, McGraw-Hill, New York, 1968.

Sinanoğlu, O., *Proc. Nat. Acad. Sci.* **47**, 1217 (1961) and *J. Phys. Chem.* **66**, 2283 (1962). A simple presentation and analysis of the role of electron correlation in the electronic structure of atoms, molecules, and their interactions.

Slater, J. C., *Quantum Theory of Atomic Structure*, 2 vols, McGraw-Hill, New York, 1960.

PROBLEMS

1. Test the orthonormality of $\Phi_m(\phi)$, $\Theta_{lm}(\theta)$, and $R_{nl}(r)$ on their respective intervals for several values of n, l, and m.
2. Show that the Φ_m, Φ_{lm}, and R_{nl} differential equations are of the Sturm-Liouville form. The latter two are the associated Legendre (4.1.8) and the associated Laguerre equations (4.1.12), respectively.
3. Evaluate $\langle \hat{r} \rangle$ for $1s$, $2s$, and $2p$ and show that equation (4.1.21) is correct.
4. Find the states which arise from the following orbital configurations. In each case designate the state of lowest energy. (1) $2s$, (2) $2p^3$, (3) $2p^22s$, (4) $2p3p$, (5) $2p3d$, (6) $3d^2$, (7) $3d^8$, (8) $3d^9$, (9) $2s4f$, (10) $2p^5$, (11) $3d^34s$.
5. Give the spectroscopic state designation for the ground states and a few excited states of (1) Si, (2) Mn, (3) Rb, (4) Ni. What are the *allowed* transitions from the ground state in each case?

[40] Pair models are an important part of modern quantum chemical theory (see previous footnotes).

4.11 Electron Correlation

6. How do the ^2P states arising from the configurations p^1 and p^5 split due to spin-orbit coupling?
7. Find and compare the probability distribution of the space coordinates of electrons in the excited states of helium $(1s)(2s)$, ^1S and ^3S, by integrating out the spin coordinates. Compare these results to the probability distribution which results from a simple product function (i.e., no antisymmetrization).

5
THE DIATOMIC MOLECULE

The underlying physical laws necessary
for the mathematical theory of a large part
of physics and the whole of chemistry
are completely known.

A remark attributed to **P. A. M. DIRAC,** 1929

T O A GREAT EXTENT the study of the chemical bond is the study of the diatomic molecule. We now have reached a stage of development which permits us to fully demonstrate the power of quantum mechanical principles applied to the study of the chemical bond. To begin with the simplest diatomic is to begin with the molecule-ion H_2^+.

Molecular Methods

5.1 H_2^+

The molecular ion H_2^+ is admittedly of little chemical interest, but its one-electron bond can be treated with so much ease and exactness that it occupies a positive relative to molecules analogous to that of the H atom towards atoms.

In Fig. 5.1.1 there are nuclei located at a and b a distance R apart. The electron is at a radial distance r_a from a, and r_b from b. The remaining coordinates are as for the H atom but are subscripted a or b depending on the coordinate origin, except for ϕ, which is the same for both.

While all these Cartesian and polar coordinates look very elaborate they do not help with the exact solution of the problem because the Schrödinger equation does not separate in these coordinates. It does separate in confocal elliptical coordinates. However, since all chemically interesting

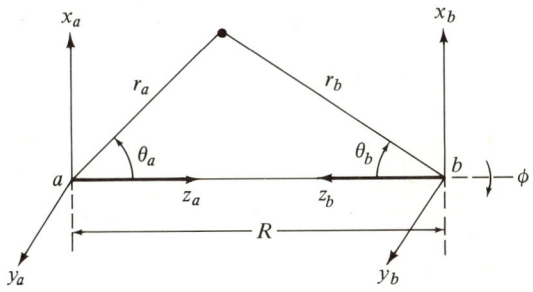

FIG. 5.1.1 Coordinates of an electron in a diatomic molecule.

diatomic molecules are many-electron systems with approximate (but good) solutions, there is no need to dwell on exact methods for H_2^+. Rather we proceed directly to methods of wide applicability to diatomic and even polyatomic molecules.

The Hamiltonian of H_2^+ should reflect the three-body nature of the problem (two protons and one electron in motion about a common center of mass). However, because the proton is about 1800 times more massive than the electron, to high approximation (the Born-Oppenheimer approximation[1]) we take the protons to be *stationary* at a distance R apart; thus the Hamiltonian of H_2^+ is that of an electron in motion about two stationary protons.

$$\hat{h} = -\tfrac{1}{2}\nabla^2 - \frac{1}{r_a} - \frac{1}{r_b} \tag{5.1.1}$$

$\hat{h}$ consists of the kinetic energy operator and the electron–proton potential energy [Coulomb attraction $= -(1/r_a + 1/r_b)$].

$$\hat{h}\psi_n = E_n\psi_n \tag{5.1.2}$$

In the Schrödinger equation all the quantities depend on the fixed internuclear distance R. For example, when $R = 0$ the protons have coalesced into an atom with nuclear charge $Z = 2$, so ψ_n is the hydrogenic wavefunction for $Z = 2$ and $E_n = -Z^2/2n^2$ hartree $= (-4/n^2) \times 13.6$ eV. However, at $R = \infty$ the protons have been separated and we have an H atom and a bare proton, so ψ_n is the hydrogenic wavefunction for $Z = 1$ and $E_n = -13.6\,n^2$ eV. At intermediate distances we need the exact solutions to (5.1.2) at each fixed value of R. In the ground state, $n = 1$, so the *electronic energy* is denoted E_1 even at intermediate distances. Finally, the *total molecular energy* $E(R)$ must include the internuclear repulsion $1/R$.

$$E(R) = E_1 + 1/R$$

Figure 5.1.2 shows all these quantities and their dependence on R; note the limits at $R = 0$ and ∞.

The very shallow minimum in the plot of $E(R) = E_1 + 1/R$ versus R shows that a *stable* species H_2^+ exists, with an *equilibrium internuclear distance* $R_e = 2$ bohrs $= 1.06$ Å. The two protons are in equilibrium (have zero net force exerted on them) only at a distance of 2 bohrs; away from 2 bohrs they will experience a force tending to restore them to R_e. Therefore in the absence of all external perturbations the average separation of two protons in H_2^+ is R_e. An amount of energy D_e, the *dissociation energy*, $D_e(H_2^+) = E(H \text{ atom}) - E(H_2^+ \text{ at } R_e)$, is indicated by arrows

---

[1] The Born-Oppenheimer approximation is discussed in Appendix D.

5.2 The MO as Simple LCAO

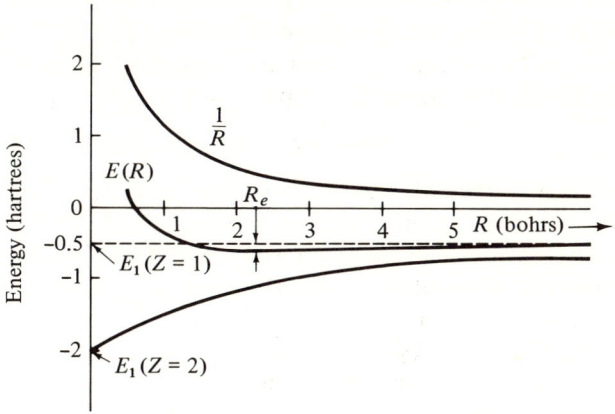

FIG. 5.1.2 The total molecular energy of H_2^+ $E(R)$, the electronic energy E_1, and the internuclear repulsion $1/R$, plotted as a function of the internuclear distance. Energies in hartree units, R in bohrs $E_1(Z)$ is the hydrogenic atom energy. After H. Wind, *J. Chem. Phys.* **42**, 2371 (1965). Used by permission.

in Fig. 5.1.2. D_e is the energy necessary to dissociate the ion H_2^+ into an H atom and a proton. Figure 5.1.2 is a rather graphic illustration of the primary chemical fact in the formation of the bond, *the chemical bond arises from a small energy difference (D_e) between the separated atoms and the molecule.*[2] D_e for H_2^+ ($D_e = 0.10$ hartree $= 2.8$ eV $= 64$ kcal/mole) is 1/5 the energy of the separated atoms and an order of magnitude smaller than the electronic energy at R_e. The magnitude of D_e for most other diatomic molecules is in the range 1 to 6 eV (25 to 150 kcal/mole) and, of course, the bond energies of polyatomic molecules are in the same range. Chemists can gain an appreciation of molecular fragility from this point of view.

5.2 THE MO AS SIMPLE LCAO

The ψ_n in equation (5.1.2) are molecular orbitals, MO. *A molecular orbital is a one-electron wavefunction which satisfies the molecular one-electron Schrödinger equation.* ψ_1 is the lowest-energy MO of H_2^+. An approximate representation of ψ_1 is easily obtained when we realize that in the vicinity of proton a ψ_1 must resemble $1s_a$ (the 1s hydrogen atomic orbital on

---

[2] Plots of $E(R)$ versus R are most often presented on expanded energy scale, so that the minimum in $E(R)$ looks quite deep but $1/R$ and E_n are off-scale. Such plots may lend a deceptive feeling of stability to the chemical bond.

proton a) because of the overpowering influence of the Coulomb potential of the proton a on the electronic motion, and in the vicinity of proton b ψ_1 must resemble $1s_b$. Thus the following linear combination suggests itself as a trial molecular orbital:

$$\tilde{\psi}_1 = c_1 1s_a + c_2 1s_b \qquad (5.2.1)$$

Molecular Orbitals which are written as *Linear Combination* of *Atomic Orbitals* are denoted by the acronym, LCAO–MO. Since the use of LCAO–MO is so widespread we take all MO to be LCAO unless specified otherwise, remembering that there *are* also free-electron MO as well as the exact MO of H_2^+.

Defining $W = \langle \tilde{\psi}_1, \hat{h}\tilde{\psi}_1 \rangle / \langle \tilde{\psi}_1, \tilde{\psi}_1 \rangle$, the variational principle requires that $W \geq E_1$, so W can be minimized with respect to the coefficients c_1 and c_2.

$$W = \frac{c_1^2 H_{aa} + c_2^2 H_{bb} + 2c_1 c_2 H_{ab}}{c_1^2 + c_2^2 + 2c_1 c_2 S_{ab}} \qquad (5.2.2)$$

EXERCISE In this notation the *overlap integral* is $S_{ab} = \langle 1s_a, 1s_b \rangle$ and the energy integrals are $H_{aa} = \langle 1s_a, \hat{h}1s_a \rangle$, $H_{ab} = \langle 1s_a, \hat{h}1s_b \rangle$. Explain why $H_{aa} = H_{bb}$ and $H_{ab} = H_{ba}$. Also compare equation (5.2.2) to equation (2.6.10).

At the variational minimum we have the conditions, $\partial W/\partial c_1 = 0$ and $\partial W/\partial c_2 = 0$; imposing these conditions on (5.2.2) we obtain

$$\frac{\partial W}{\partial c_1} = \frac{2c_1 H_{aa} + 2c_2 H_{ab}}{c_1^2 + c_2^2 + 2c_1 c_2 S_{ab}}$$

$$- \frac{(2c_1 + 2c_2 S_{ab})(c_1^2 H_{aa} + c_2^2 H_{bb} + 2c_1 c_2 H_{ab})}{(c_1^2 + c_2^2 + 2c_1 c_2 S_{ab})^2} = 0$$

or

$$\frac{\partial W}{\partial c_1} = \frac{2c_1 H_{aa} + 2c_2 H_{ab} - (2c_1 + 2c_2 S_{ab})W}{c_1^2 + c_2^2 + 2c_1 c_2 S_{ab}} = 0$$

or

$$c_1(H_{aa} - W) + c_2(H_{ab} - S_{ab} W) = 0 \qquad (5.2.3)$$

Similarly, from $\partial W/\partial c_2 = 0$ one gets

$$c_1(H_{ab} - S_{ab} W) + c_2(H_{bb} - W) = 0 \qquad (5.2.4)$$

Equations (5.2.3) and (5.2.4) can be written as one matrix equation.[3]

$$(\mathbf{H} - \mathbf{S}W)\mathbf{c} = 0$$

---

[3] See Appendix A.

5.2 The MO as Simple LCAO

that is

$$\begin{pmatrix} H_{aa} - W & H_{ab} - S_{ab}W \\ H_{ab} - S_{ab}W & H_{bb} - W \end{pmatrix} \begin{pmatrix} c_1 \\ c_2 \end{pmatrix} = 0 \qquad (5.2.5)$$

The condition that the matrix equation have nontrivial solutions ($c_i = 0$ is trivial) is that the *secular determinant* must vanish,[3] $\det |\mathbf{H} - \mathbf{S}W| = 0$, that is

$$\begin{vmatrix} H_{aa} - W & H_{ab} - S_{ab}W \\ H_{ab} - S_{ab}W & H_{bb} - W \end{vmatrix} = 0 \qquad (5.2.6)$$

Multiplying out the determinant yields a quadratic equation, *the secular equation*, with two roots, W_1 and W_2. Denoting the coefficients associated with W_1 in matrix equation (5.2.5) as c_{11} and c_{12}, and coefficients associated with W_2 as c_{21} and c_{22}, the results may be summarized as follows:

$$W_1 = \frac{H_{aa} + H_{ab}}{1 + S_{ab}} \qquad W_2 = \frac{H_{aa} - H_{ab}}{1 - S_{ab}}$$

$$\frac{c_{11}}{c_{12}} = +1 \qquad \frac{c_{21}}{c_{22}} = -1$$

$$\psi_1 = 1\sigma_g = N_1(1s_a + 1s_b) \qquad \psi_2 = 1\sigma_u = N_2(1s_a - 1s_b)$$

where N_1 and N_2 are normalization factors. Using symmetry from the beginning, ψ_1 is denoted $1\sigma_g$ because $(1s_a + 1s_b)$ is *symmetric* (*g* for the German gerade or even) under the symmetry operation of inversion,[4] and ψ_2 is denoted $1\sigma_u$ because $(1s_a - 1s_b)$ is antisymmetric (*u* for ungerade or odd) under inversion. The σ denotes the cylindrical symmetry of these functions about the bond; symmetry will be more fully explored in a later section. Since H_{aa} and H_{ab} are both negative, W_1 is the lowest root of the secular equation and $1\sigma_g$ is the ground-state molecular orbital.

PROOF   The secular equation for H_2^+ is $(H_{aa} - W)(H_{bb} - W) - (H_{ab} - S_{ab}W)^2 = 0$. Let $H_{aa} = \alpha = H_{bb}$, $H_{ab} = \beta$, and $S_{ab} = S$. So $(\alpha - W)^2 - (\beta - WS)^2 = 0$, and because $a^2 - b^2 = (a + b)(a - b)$ we can write $(\alpha - W + \beta - SW)(\alpha - W - \beta + SW) = 0$ or

$$W_{1,2} = \frac{\alpha \pm \beta}{1 \pm S} = \frac{H_{aa} \pm H_{ab}}{1 \pm S}$$

Then from $c_1(\alpha - SW) + c_2(\beta - W) = 0$ it follows that

$$\frac{c_2}{c_1} = \frac{SW - \alpha}{\beta - W}$$

and for $W = W_1$

$$\frac{c_{12}}{c_{11}} = \frac{S(\alpha + \beta) - \alpha(1 + S)}{\beta(1 + S) - (\alpha + \beta)} = \frac{\beta S - \alpha}{\beta S - \alpha} = 1$$

[4] Under inversion $x \to -x$, $y \to -y$ and $z \to -z$.

So $1\sigma_g = N_1(1s_a + 1s_b)$, where N_1 should be determined from the requirement that $1\sigma_g$ be normalized to unity. Using normalized atomic orbitals we have

$$\langle 1\sigma_g, 1\sigma_g \rangle = 1 = N_1^2 \langle (1s_a + 1s_b), (1s_a + 1s_b) \rangle = N_1^2(2 + 2S_{ab})$$

so

$$N_1 = (2 + 2S_{ab})^{-\frac{1}{2}}$$

EXERCISE Similarly, from W_2, find $1\sigma_u$ (normalized).

The roots of the secular equation depend on R because H_{aa}, H_{ab}, and S_{ab} all depend on R. Figure 5.2.1 shows the overlap of the two $1s$ functions as a shaded volume. The overlap integral (proportional to the shaded volume) is strongly R-dependent. H_{aa} and H_{ab} also contain terms with

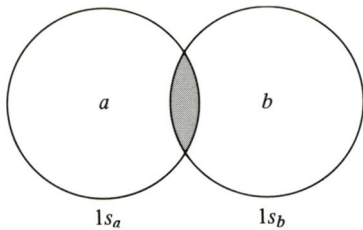

FIG. 5.2.1 The overlap of the two $1s$ functions is indicated by the shaded volume.

inverse exponential dependence on R, for example, $\langle 1s_a, (1/r_b)1s_a \rangle = (1/R)[1 - (1 + R)\exp(-2R)]$ and $\langle 1s_a, (1/r_b)1s_b \rangle = (1 + R)\exp(-R)$. A plot of the LCAO–MO energies corresponding to the two roots W_1 and W_2 displays the R dependence and shows that only the lowest root of the secular equation leads to a stable state, $1\sigma_g$ (Figs. 5.2.2 and 5.2.3). The figures show that the total molecular energy $E(R) = W_1 + 1/R$, has a minimum, but that the root W_2 (first excited state) leads to a repulsive interaction [no minimum in $E(R) = W_2 + 1/R$]. We say that $1\sigma_g$ is a *bonding orbital* but that $1\sigma_u$ is an *antibonding orbital*.

EXERCISE Evaluate the overlap integral S_{ab} to find the functional dependence on the internuclear distance R. Hints: Use the normalized atomic orbitals, $1s_a = \pi^{-1/2}\exp(-r_a)$, $1s_b = \pi^{-1/2}\exp(-r_b)$. Transform the integral to the coordinates $\mu = (r_a + r_b)/R$ and $\eta = (r_a - r_b)/R$ with volume element $dV = (R^3/8)(\mu^2 - \eta^2)\,d\mu\,d\eta\,d\phi$ and limits of integration $1 \leq \mu \leq \infty$, $-1 \leq \eta \leq +1$, and $0 \leq \phi \leq 2\pi$. Finally,

$$\int_1^\infty x^n \exp(-ax)\,dx = \frac{n!\exp(-a)}{a^{n+1}} \sum_{k=0}^n \frac{a^k}{k!}$$

5.2 The MO as Simple LCAO

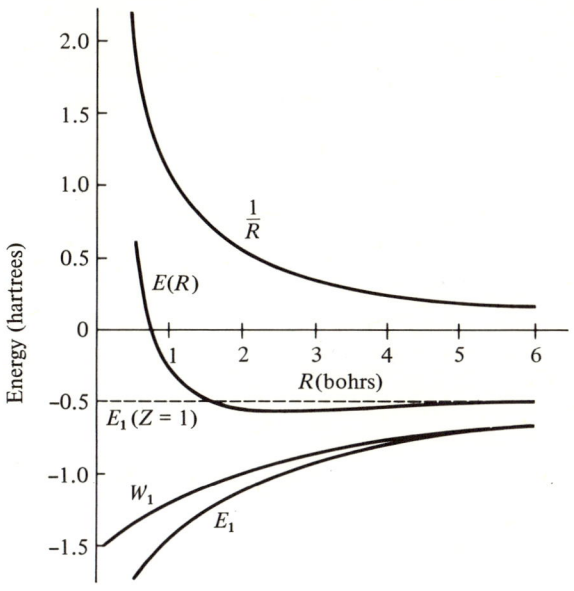

FIG. 5.2.2 Hydrogen molecule-ion energies as a function of R (ground state). W_1 is the electronic energy of the LCAO–MO $1s_a + 1s_b$ (compared to the exact electronic energy E_1). $E(R)$ is the total calculated energy, $E(R) = W_1 + 1/R$.

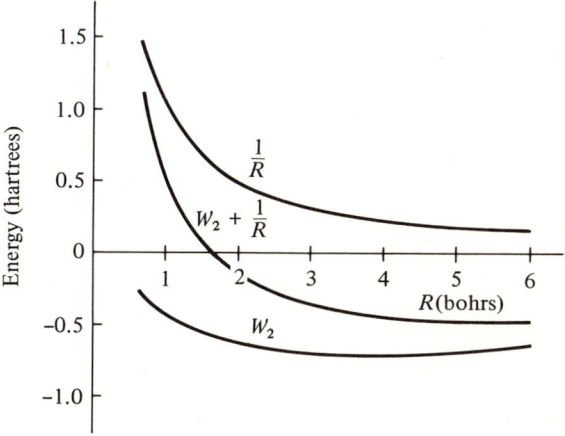

FIG. 5.2.3 Hydrogen molecule-ion energy as a function of R (excited state). W_2 is the electronic energy of the LCAO–MO $1s_a - 1s_b$.

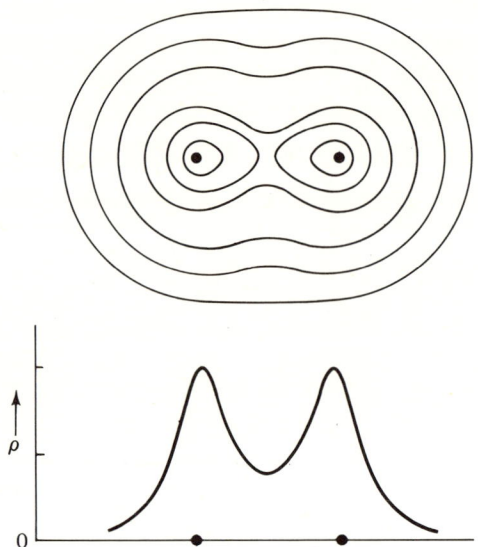

FIG. 5.2.4 *Top*: Contours of constant ρ, $\rho = |1\sigma_g|^2$. *Bottom*: Profile of ρ along the internuclear line. Both the contours and the profile of ρ show the piling up of electronic charge density between the nuclei.

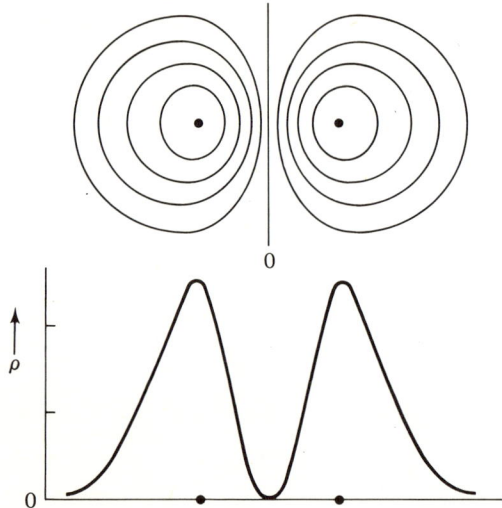

FIG. 5.2.5 *Top*: Contours of constant ρ, $\rho = |1\sigma_u|^2$. *Bottom*: Profile of ρ along the internuclear line. Both the contours and the profile of ρ show that the electronic density falls to zero between the nuclei. The nodal line ($\rho = 0$) is drawn in the contour map.

5.3 Observable Features of the Chemical Bond

ELECTRONIC CHARGE DENSITY

Reasons behind the bonding or antibonding character of MO are made intuitively clear by *electronic charge density contour maps* and *electronic charge density profiles*. Remember the definition $\rho = \sum_{i=1}^{N} |\psi_i|^2$, where ψ_i are the occupied spin-orbitals and ρ is the real three-dimensional distribution of electronic charge in space. A ρ *contour map* consists of a series of lines of constant ρ drawn in a plane containing the nuclei. A ρ *profile* is the plot of ρ along the line joining the nuclei. Figures 5.2.4 and 5.2.5 illustrate the nature of these ρ plots for $1\sigma_g$ and $1\sigma_u$.

The ρ contours and profiles of $1\sigma_g$ and $1\sigma_u$ graphically illustrate the difference between bonding and antibonding MO. The antibonding MO has a node ($\rho = 0$) between the nuclei. The existence of the node is evident in the form of the MO as LCAO, $1\sigma_u = N_2(1s_a - 1s_b)$. Because the wavefunction changes sign midway between the nuclei it must go through zero (Fig. 5.2.6).

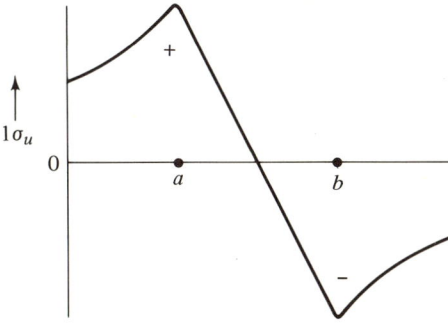

FIG. 5.2.6

5.3 OBSERVABLE FEATURES OF THE CHEMICAL BOND

Quantum mechanically, the result of an experiment measuring the observable property having linear, Hermitian operator $\hat{O}$ is $\langle \Psi, \hat{O}\Psi \rangle / \langle \Psi, \Psi \rangle$. When we say "observable" we mean observable by a physical experiment. Admittedly, very precise observations of atomic and molecular structure are often made by very indirect experiments. Usually a spectroscopic experiment is involved, but not always. For example, ρ and R_e are observable, in principle, from x-ray diffraction experiments. Spectroscopic experiments give precise determinations of the observables R_e and D_e, two important indexes of "bond strength." In general, a smaller R_e and a larger D_e indicate a stronger bond, and "bond strength" is a chemical observable. But R_e and D_e are just two characteristics of the total molecular energy curve, $E(R) = E(\text{electronic}) + Z_a Z_b / R$, where $Z_a Z_b / R$ is the Coulomb repulsion between nuclei of charge Z_a and Z_b at a distance R (see

Fig. 5.3.1). The curve $E(R)$, which is experimentally obtainable from spectroscopy, is most often called "*the potential energy curve*," because $E(R)$ is the potential energy of the nuclei as a function of R (Appendix D). With nuclear potential energy $E(R)$ classical mechanics gives the force acting between the nuclei from the slope dE/dR, Force $= -dE(R)/dR$. At the minimum of a stable $E(R)$ curve, $R = R_e$ and $(dE/dR)_{R_e} = 0$, and there is no force on the nuclei. A repulsive potential energy curve never has zero slope so the force on the nuclei is always repulsive (pushing them apart).

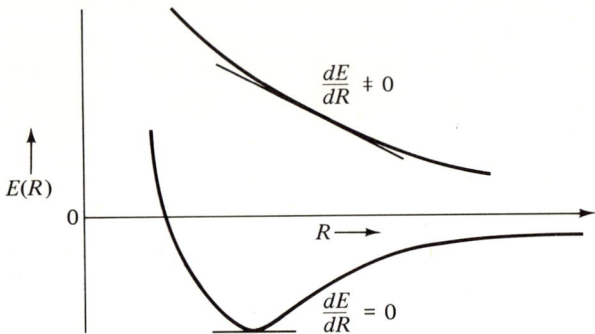

FIG. 5.3.1 Contrasting the total molecular energy curve (nuclear potential energy curve) $E(R)$ of the unstable, repulsive and dissociative state, $dE/dR \neq 0$, to that of the stable electronic state, $(dE/dR)_{R_e} = 0$.

There is a simple relation between the force on the nuclei and the electronic charge density.[5]

$$\text{Force} = -\frac{dE(R)}{dR} = \frac{Z_a Z_b}{R^2} - \left\langle \Psi, \frac{\partial \hat{H}}{\partial R} \Psi \right\rangle \quad \text{HELLMANN-FEYNMAN THEOREM}$$

(5.3.1)

$\hat{H}$ is the electronic Hamiltonian. Making use of the coordinates of Fig. 5.1.1, (5.3.1) becomes[5]

$$\text{Force} = \frac{Z_a Z_b}{R^2} - \tfrac{1}{2}\left[\int \rho \frac{Z_a}{r_a^2} \cos\theta_a \, dV + \int \rho \frac{Z_b}{r_b^2} \cos\theta_b \, dV \right]$$

ELECTROSTATIC THEOREM (5.3.2)

The electrostatic theorem has a classical interpretation; the first term is the repulsive force of the positively charged nuclei, the second term is the component along the bond of the attractive force between the nuclei and

---

[5] Derived in Appendix E.

5.3 Observable Features of the Chemical Bond

an electronic charge distribution ρ. Qualitatively speaking, if the electronic density is concentrated between the nuclei, there will be a net attractive force for some value of R and binding results: But if sufficient electronic charge density is located on the far sides of the nuclei for all R, then $Z_a Z_b / R^2$ is never neutralized and the $E(R)$ curve is repulsive.[6] Qualitatively, the difference between (a) and (b) of Fig. 5.3.2 is the difference between the $1\sigma_g$ and $1\sigma_u$ ρ contours and profiles. Equation (5.3.2) makes the relation between ρ and binding quantitatively precise.[7]

The formation of the chemical bond is thus related to very classical electrostatic ideas.[8] Unfortunately, in order to predict binding one needs

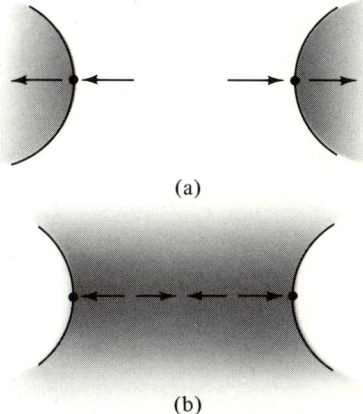

FIG. 5.3.2 (a) Arrows indicate forces arising from the repulsive charge density (shaded) and the repulsive nuclear interaction. (b) Arrows indicate forces arising from the attractive charge density (shaded) and the repulsive nuclear interaction.

---

[6] In discussing energy and nodal patterns of MO the terms bonding and antibonding are used, for example, $1\sigma_g$ and $1\sigma_u$. However, in the discussion of the forces which electronic charge densities exert on the nuclei, the terms binding and antibinding are preferable.

[7] T. Berlin, *J. Chem. Phys.* **19**, 208 (1951)

[8] In 1902 Kelvin proposed an electrostatic model of the atom in which a stable equilibrium was achieved between point charge electrons at rest and the positive charge in the atom by distribution of the positive charge uniformly throughout a sphere and embedding the electrons therein. Earnshaw's theorem (Chapter 1, footnote 16) does not apply because the distribution of positive charge is continuous. J. J. Thomson (*The Corpuscular Theory of Matter*, Scribners, New York, 1907) examined the stability of possible geometrical arrangements of stationary electrons in such a sphere. As a result the model is known as the Thomson model of the atom. The present electrostatic description of molecular structure is a curious negative image of the Kelvin-Thomson model in that *positive* point charges are embedded in a *non*uniform distribution of *negative* charge to achieve molecular stability.

a very accurate ρ and this is only obtainable from the quantum mechanical calculation of the nonobservable wavefunction!

Of the many physical and chemical observable properties of molecules we found that ρ, $E(R)$, D_e, and R_e are important observables for the characterization of bond formation. In chemistry we should try to *stay with observables for the interpretation of observables*. The nonobservable molecular wavefunction may get very, very complicated in order to give very accurate observables. But we should not be overwhelmed by the details of the wavefunction, or try to give chemical interpretations for everything that goes into it.

5.4 H$_2$ AND OTHER SMALL SPECIES

H$_2$ is the prototype of the covalent chemical bond. Although it has no exact solution, just highly accurate solutions, nothing essentially new is introduced in the treatment. We proceed in going from H$_2^+$ to H$_2$ as we did in going from H to He. Figure 5.4.1 gives the coordinates of the two electrons in H$_2$. The Hamiltonian is as follows:

$$\hat{H} = \left(-\tfrac{1}{2}\nabla_1^2 - \frac{1}{r_{1a}} - \frac{1}{r_{1b}}\right) + \left(-\tfrac{1}{2}\Delta_2^2 - \frac{1}{r_{2a}} - \frac{1}{r_{2b}}\right) + \frac{1}{r_{12}} \quad (5.4.1)$$

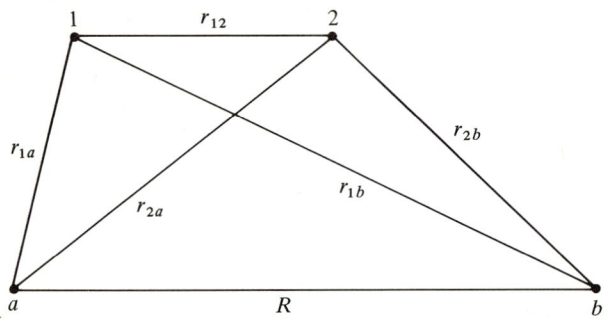

FIG. 5.4.1

Noting that in equation (5.4.1) the operators in parentheses are H$_2^+$ Hamiltonians (5.1.1) $\hat{h}(1)$ and $\hat{h}(2)$, respectively, we let $\hat{H}_0 = \hat{h}(1) + \hat{h}(2)$ be the zero-order Hamiltonian; then $\hat{V}(r_{12}) = 1/r_{12}$ is the perturbation.

$$\hat{H} = \hat{H}_0 + \hat{V}(r_{12}) \quad (5.4.2)$$

Taking $\hat{H}_0$ to be much greater than $\hat{V}(r_{12})$ we can temporarily ignore $\hat{V}(r_{12})$ and look for the solution to $\hat{H}_0\Psi = E\Psi$. $\hat{H}_0$ being a *sum* of H$_2^+$ Hamiltonians, Ψ is a *product* of H$_2^+$ molecular orbitals.

5.4 H_2 and Other Small Species

PROOF

$$\hat{H}_0\Psi = E\Psi \quad \text{or} \quad (\hat{h}(1) + \hat{h}(2))\psi(1)\psi(2) = E\psi(1)\psi(2)$$

or

$$\hat{h}(1)\psi(1)\psi(2) + \psi(1)\hat{h}(2)\psi(2) = E\psi(1)\psi(2)$$

dividing each side by $\psi(1)\psi(2)$ we obtain

$$\frac{1}{\psi(1)}\hat{h}(1)\psi(1) + \frac{1}{\psi(2)}\hat{h}(2)\psi(2) = E$$

By separation of variables we have shown that each term on the left-hand side must be a constant, or $\hat{h}(1)\psi(1) = \text{constant} \times \psi(1)$ and $\hat{h}(2)\psi(2) = \text{constant} \times \psi(2)$, which are just H_2^+ Schrödinger equations. As a result $\psi(1)$ and $\psi(2)$ are the H_2^+ MO, $1\sigma_g$ or $1\sigma_u$, and E is a sum of H_2^+ orbital energies.

The electron spin enters at this point through the Pauli principle, which requires the wavefunction to be antisymmetric in the space-spin coordinates. So the wavefunction of H_2 is written in terms of the two lowest-energy *molecular spin-orbitals*, $1\sigma_g\alpha$ and $1\sigma_g\beta$ as

$$\Psi(H_2) = \hat{A}1\sigma_g\alpha(1)1\sigma_g\beta(2) = \frac{1}{\sqrt{2}}\begin{vmatrix} 1\sigma_g\alpha(1) & 1\sigma_g\beta(1) \\ 1\sigma_g\alpha(2) & 1\sigma_g\beta(2) \end{vmatrix} \quad (5.4.3)$$

A molecular spin-orbital is just a molecular orbital times the appropriate spin function. From here on MO may mean either molecular spin-orbital or molecular orbital unless it is important to distinguish between them. The electronic *configuration* of the ground state of H_2 is $(1\sigma_g)^2$.

Because of the perturbation $\hat{V}(r_{12})$ the wavefunction (5.4.3) is very far from the best orbital product wavefunction when $1\sigma_g$ is the H_2^+ LCAO–MO. It could be improved, for example, by the introduction of screening (as in the case of helium). The best wavefunction of the form (5.4.3) is the Hartree-Fock wavefunction. The Hartree-Fock wavefunction is obtained by complete variation of $1\sigma_g$ in the application of the variational principle to the equation $[\hat{H}_0 + \hat{V}(r_{12})]\Psi = E\Psi$. Even this wavefunction leaves something to be desired. In Table 5.4.1 the dissociation energies predicted by successive approximations are compared to the experimental D_e. The most accurate D_e is found from the wavefunction of Kolos and Wolniewicz,[9] which is not of the form (5.4.3) but is an expansion in elliptical coordinates.[10] $E(R)$ calculated from this very accurate wavefunction is plotted in Fig. 5.4.2. The difference between the Hartree-Fock and the experimental energies is due to the residual effect called *electron*

---

[9] W. Kolos and L. Wolniewicz, *J. Chem. Phys.* **43**, 2429 (1965).
[10] A method first used in the accurate calculations of H. M. James and A. S. Coolidge, *J. Chem. Phys.* **1**, 825 (1933).

TABLE 5.4.1

	$D_e^{*\dagger}$ (eV)	R_e (bohrs)
$\Psi = (1\sigma_g)^2$ $1\sigma_g =$ LCAO-MO of H_2^+	2.65	1.57
$\Psi = (1\sigma_g)^2$ $1\sigma_g =$ Hartree-Fock[a]	3.636	1.375
$\Psi =$ Accurate wavefunction[b]	4.747†	1.40
Experiment[c]	4.74$_7$	1.400

[a] W. Kolos and C. C. J. Roothaan, *Rev. Mod. Phys.* **32**, 219 (1960).
[b] W. Kolos and L. Wolniewicz, *J. Chem. Phys.* **41**, 3663 (1964).
[c] G. Herzberg and A. Monfils, *J. Mol. Spectrosc.* **5**, 482 (1960).
* $D_e(H_2) = E(2 \text{ H atoms}) - E(H_2 \text{ at } R_e)$.
† Calculated values utilize the Born-Oppenheimer approximation.

correlation.[11] Principally, this effect is due to Coulomb repulsion of electrons of unlike spin confined to the same spatial orbitals, for example, the $1s\alpha \leftrightarrow 1s\beta$ pair of helium or the $1\sigma_g\alpha \leftrightarrow 1\sigma_g\beta$ pair of H_2. From Table 5.4.1 the correlation energy of H_2 at R_e is $E_{\text{corr}} = E - E_{\text{HF}} = -1.1$ eV (which is about the same as the E_{corr} of helium; see Sections 4.10 and 4.11).

Other diatomic species can be described by simple generalization of the procedure just applied to H_2. The general diatomic electronic Hamiltonian is

$$\hat{H} = \sum_{i=1}^{N} \left(-\tfrac{1}{2}\nabla_i^2 - \frac{Z_a}{r_{ia}} - \frac{Z_b}{r_{ib}} \right) + \sum_{i>j}^{N} \frac{1}{r_{ij}}$$

$$\hat{H} = \sum_{i=1}^{N} \hat{h}(i) + \hat{V}(r_{ij}) = \hat{H}_0 + \hat{V}(r_{ij})$$

(5.4.4)

The zero-order wavefunctions are the eigenfunctions in $\hat{H}_0 \Psi = E\Psi$; Ψ is an antisymmetrized product of H_2^+ MO. Thus, from just $1\sigma_g$ and $1\sigma_u$ we can build the configurations of species containing up to four electrons (see Table 5.4.2).

The *bond order* is defined as

$$\frac{\text{NUMBER OF ELECTRONS IN BONDING MO} - \text{NUMBER OF ELECTRONS IN ANTIBONDING MO}}{2}$$

Species of zero bond order have equal number of electrons in bonding and antibonding orbitals. From the electrostatic arguments presented previously

[11] Application of the Hartree-Fock-Roothaan method to the $1\sigma_g$ MO of H_2 is discussed in Section 5.7 and the residual problem of electron correlation in Section 5.15.

5.4 H_2 and Other Small Species

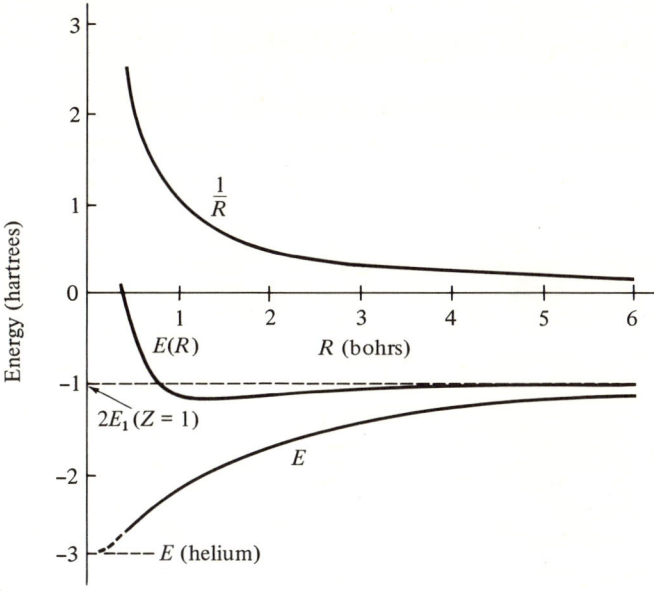

FIG. 5.4.2 Hydrogen molecule energy as a function of the internuclear distance. E is the accurate electronic energy of W. Kolos and L. Wolniewicz [*J. Chem. Phys.* **43**, 2429 (1965); used by permission]. $E(R) = E + 1/R$, and E_1 is the hydrogenic atom energy.

one expects *no* stable species to exist under such circumstances. In fact, He_2 does not exist. The correlation of D_e and R_e with bond order is indicative of the fact that all three are quantitative measures of the chemical idea of bond strength.

EXERCISE   Is the bond order, as defined, an observable? Why not?

TABLE **5.4.2**

Species	Configuration	Bond order	$D_e(eV)$	R_e(bohrs)	State symbol[a]
H_2^+	$(1\sigma_g)$	$\frac{1}{2}$	2.8	2.	$^2\Sigma_g$
H_2	$(1\sigma_g)^2$	1	4.7	1.4	$^1\Sigma_g$
He_2^+	$(1\sigma_g)^2(1\sigma_u)$	$\frac{1}{2}$	(3)	2.	$^2\Sigma_u$
He_2	$(1\sigma_g)^2(1\sigma_u)^2$	0	0	—	$^1\Sigma_g$

[a] State symbols are part of the spectroscopic terminology to be developed in a later section

The General LCAO-MO Method

5.5 OVERLAPPING ATOMIC ORBITALS: SYMMETRY

Figure 5.5.1 shows some atomic orbitals on nuclei a and b which have nonvanishing overlap $S_{ab} = \langle \phi_a, \phi_b \rangle \neq 0$. Linear combinations of these

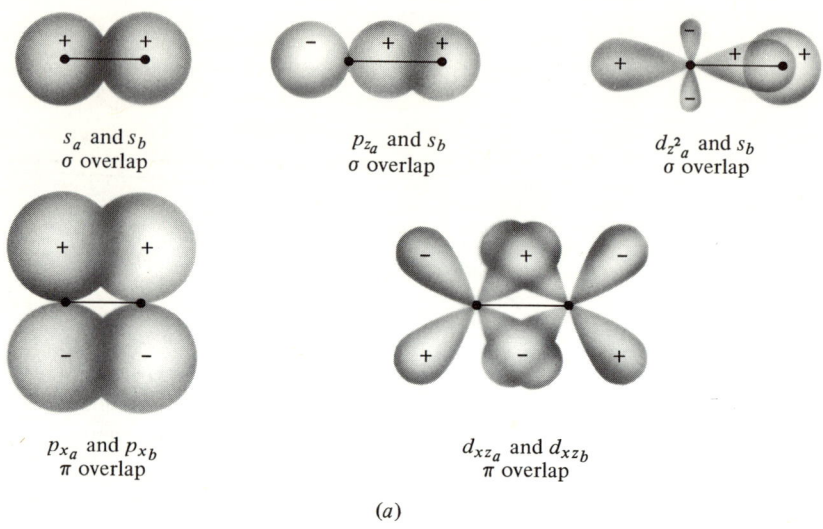

s_a and s_b
σ overlap

p_{z_a} and s_b
σ overlap

$d_{z^2_a}$ and s_b
σ overlap

p_{x_a} and p_{x_b}
π overlap

d_{xz_a} and d_{xz_b}
π overlap

(a)

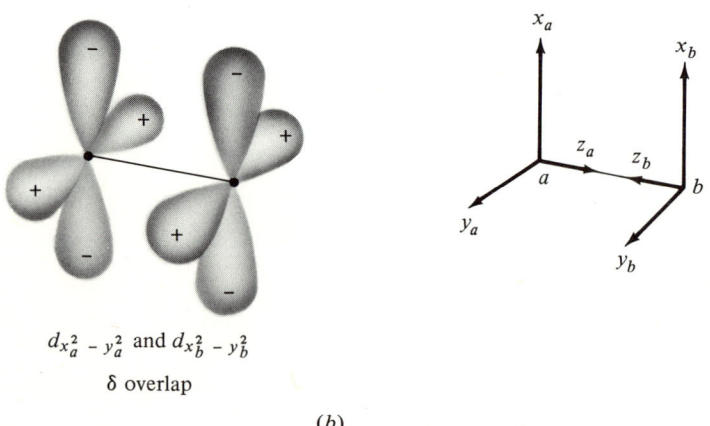

$d_{x^2_a - y^2_a}$ and $d_{x^2_b - y^2_b}$
δ overlap

(b)

FIG. 5.5.1 Overlapping atomic orbitals. Phases are indicated by the signs in lobes.

5.5 Overlapping Atomic Orbitals: Symmetry

atomic orbitals to form MO could be taken by inspection of the figure. However, the use of symmetry makes the classification of the LCAO–MO more systematic. Because symmetry operators commute with the Hamiltonian of the molecule we may demand that the molecular orbitals (to be worthy of the name) be eigenfunctions to the symmetry operators.[12] We are chiefly concerned with only a few of the symmetry operators which bring a homonuclear diatomic molecule into itself.

Figure 5.5.2 gives the explicit demonstration of some of the symmetry operations of the homonuclear diatomic molecule in the coordinate system shown. On the left is the molecule before the operation, on the right after the operation. Lines and a star have been attached to one atom in order to see what the operation accomplished. Of course, no real atom

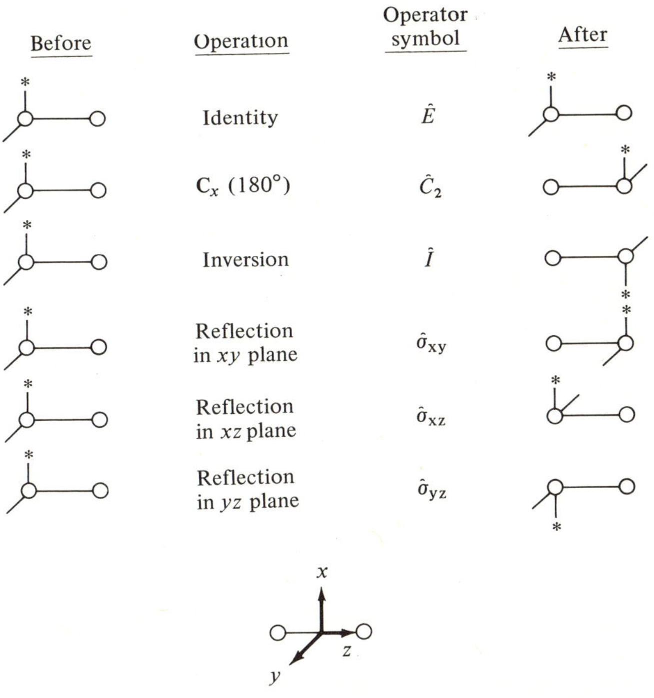

Before	Operation	Operator symbol	After
	Identity	$\hat{E}$	
	C_x (180°)	$\hat{C}_2$	
	Inversion	$\hat{I}$	
	Reflection in xy plane	$\hat{\sigma}_{xy}$	
	Reflection in xz plane	$\hat{\sigma}_{xz}$	
	Reflection in yz plane	$\hat{\sigma}_{yz}$	

FIG. 5.5.2

has marks on it, so a real molecule is the same before and after the operation. The invariance of the molecule to symmetry operations means that $|\psi|^2$ is unchanged. If $\hat{O}$ is one of the symmetry operators, what is the

---

[12] If the molecular orbitals are not degenerate. While degeneracy is largely ignored in the following, an extensive treatment of group theoretical methods is given in Chap. 6.

TABLE 5.5.1 MO Nomenclature and Eigenvalues to Some Symmetry Operators

MO	E	I	σ_{xz}	σ_{yz}
σ_g	1	1	1	1
σ_u	1	−1	1	1
π_{xg}	1	1	1	−1
π_{xu}	1	−1	1	−1
π_{yg}	1	1	−1	1
π_{yu}	1	−1	−1	1
δ_{xyg}	1	1	−1	−1
δ_{xyu}	1	−1	−1	−1

effect of $\hat{O}$ on ψ such that $|\psi|^2$ is not changed? There are only two possibilities,[13] $\hat{O}\psi = \pm\psi$.

$$|\hat{O}\psi|^2 = |\pm\psi|^2 = |\psi|^2 \tag{5.5.1}$$

Therefore ψ is either symmetric, $\hat{O}\psi = \psi$, or antisymmetric, $\hat{O}\psi = -\psi$, under the symmetry operation $\hat{O}$; that is, ψ has eigenvalues ± 1. The nomenclature for the homonuclear diatomic molecular orbitals is given in Table 5.5.1 from the eigenvalues in the operator relations $\hat{O}\psi = \pm\psi$.

The g and u MO are, respectively, symmetric and antisymmetric to inversion. The σ MO, being cylindrically symmetric about the bond, are symmetric to reflections in xz and yz planes, but the π MO are antisymmetric to one or another of these reflections. It is not difficult to find LCAO which have these symmetries. First note that linear combinations of AO of the same type have definite symmetry, for example, $\hat{I}\, 1s_a = 1s_b$ no symmetry, but $\hat{I}(1s_a + 1s_b) = \hat{I}\, 1s_a + \hat{I}\, 1s_b = 1s_b + 1s_a = (1s_a + 1s_b)$, therefore $(1s_a + 1s_b)$ is g.

The symmetry classifications of the LCAO explain the nonvanishing overlaps in Fig. 5.5.1. Overlap between AO having the same symmetry about the bond is nonzero, for example, $\langle px_a, d_{xz_b}\rangle = \pi$ overlap. If AO symmetry about the bond is different, the overlap is zero. Figure 5.5.3 illustrates the cancellation of positive and negative contributions to the overlap integral for a net zero result.

EXERCISE   Find the combinations of $2s$, $2p$, and $3d$ AO on different nuclei which have net zero overlap.

[13] Again, more generally, $\hat{O}\psi = \exp(i\theta)\psi$, so $|\hat{O}\psi|^2 = \exp(-i\theta)\exp(i\theta)\psi^*\psi = |\psi|^2$.

5.5 Overlapping Atomic Orbitals: Symmetry

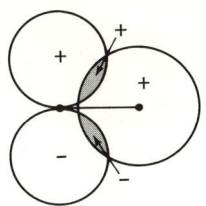

FIG. 5.5.3 Net zero overlap, $\langle s_a, p_{xa}\rangle = 0$.

EXERCISE Find the eigenvalues of the LCAO as given in Table 5.5.2 by the actual application of the symmetry operators to the LCAO (use orbital phases).

TABLE 5.5.2 LCAO Eigenfunctions to Some Symmetry Operators with Eigenvalues[a]

LCAO	E	I	σ_{xz}	σ_{yz}	Symmetry type	λ[b]	Bonding character[c]
$s_a + s_b$	1	1	1	1	σ_g	0	Bonding
$s_a - s_b$	1	−1	1	1	σ_u	0	Antibonding
$d_{z_a^2} + d_{z_b^2}$	1	1	1	1	σ_g	0	Bonding
$d_{z_a^2} - d_{z_b^2}$	1	−1	1	1	σ_u	0	Antibonding
$p_{z_a} + p_{z_b}$	1	1	1	1	σ_g	0	Bonding
$p_{z_a} - p_{z_b}$	1	−1	1	1	σ_u	0	Antibonding
$p_{x_a} + p_{x_b}$	1	−1	1	−1	π_{xu}	1	Bonding
$p_{x_a} - p_{x_b}$	1	1	1	−1	π_{xg}	1	Antibonding
$p_{y_a} + p_{y_b}$	1	−1	−1	1	π_{yu}	1	Bonding
$p_{y_a} - p_{y_b}$	1	1	−1	1	π_{yg}	1	Antibonding
$d_{xy_a} + d_{xy_b}$	1	1	−1	−1	δ_{xyg}	2	Bonding
$d_{xy_a} - d_{xy_b}$	1	−1	−1	−1	δ_{xyu}	2	Antibonding
$d_{xz_a} + d_{xz_b}$	1	−1	1	−1	π_{xu}	1	Bonding
$d_{xz_a} - d_{xz_b}$	1	1	1	−1	π_{xg}	1	Antibonding
$d_{yz_a} + d_{yz_b}$	1	−1	−1	1	π_{yu}	1	Bonding
$d_{yz_a} - d_{yz_b}$	1	1	−1	1	π_{yg}	1	Antibonding

[a] The table gives the symmetries of some *homonuclear* LCAO. For simplicity only enough symmetry operators are included to distinguish between the listed LCAO, for example, $d_{x^2-y^2}$ forms δ LCAO and has $\lambda = 2$. The coordinate system used is identical to that of Fig. 5.5.1.
[b] λ is the magnitude of the component of orbital angular momentum along the z axis.
[c] The bonding character follows from the nodal pattern; that is, does the MO possess a node which bisects the bond axis?

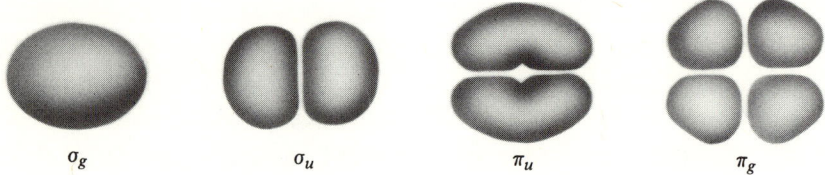

FIG. 5.5.4 Shapes of homonuclear diatomic molecular orbitals.

The shapes of the homonuclear diatomic molecular orbitals of several symmetry types are given in Fig. 5.5.4. This figure is analogous to Figure 4.1.3 of the atomic orbitals. From just the symmetry designations of the LCAO in Table 5.5.2 it is possible to write general LCAO–MO of each symmetry type: $\sigma_g = c_1(1s_a + 1s_b)$, or more generally, $\sigma_g = c_1(1s_a + 1s_b) + c_2(2s_a + 2s_b)$, or even more generally, $\sigma_g = c_1(1s_a + 1s_b) + c_2(2s_a + 2s_b) + c_3(2p_{za} + 2p_{zb})$, or etc. Clearly there is no limit to the number of AO which could be mixed to give an MO of a certain symmetry. Next we could add some $3d_{z^2}$ or $3s$ or $3p_z$. In fact the *expansion theorem* (Chapter 2) says that *a mathematical function (the MO) may be expressed as an infinite sum over a complete set (the AO basis set) of functions*. Fortunately, in molecular problems this expansion *converges rapidly*, so that it is not necessary to go outside the valence shell of the atoms for more AO, except for Hartree–Fock accuracy (see further). Furthermore, AO of very different orbital energies (i.e., $1s$ and $2s$) usually do not mix strongly. The situation for the first-row homonuclear diatomics is summed up in Figs. 5.5.5 and 5.5.6.

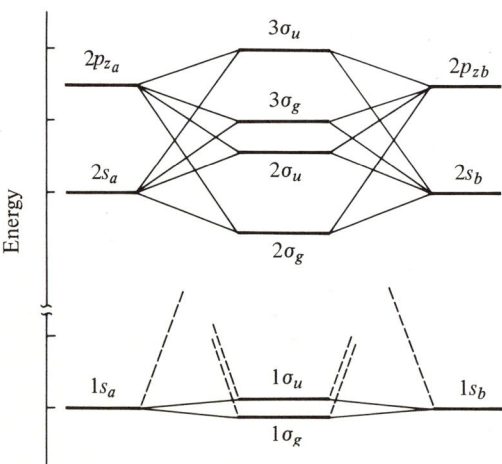

FIG. 5.5.5 σ MO as LCAO of valence shell AO in the first-row homonuclear diatomics (schematic). Dotted lines indicate weak mixing due to the large energy separation.

5.6 The Secular Equation

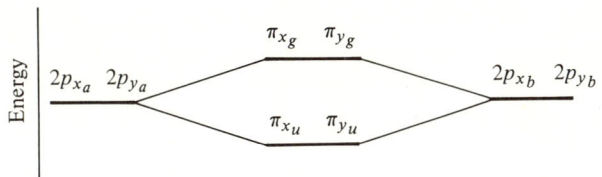

FIG. 5.5.6 π MO as LCAO of valence shell AO in the first-row homonuclear diatomics. Note the degeneracy of π_x and π_y MO.

5.6 THE SECULAR EQUATION

The occurrence of no less than six σ MO in Fig. 5.5.5 bears some explanation. Write a general σ_g MO as a linear combination of the LCAO of σ_g symmetry in Table 5.5.2 drawn from the $n = 2$ valence shell.

$$\sigma_g = c_1(1s_a + 1s_b) + c_2(2s_a + 2s_b) + c_3(2p_{z_a} + 2p_{z_b}) \quad (5.6.1)$$

σ_g has three unknown coefficients c_1, c_2, and c_3. The coefficients are determined from the variational principle by applying the minimization conditions, $\partial W/\partial c_1 = 0$, $\partial W/\partial c_2 = 0$, and $\partial W/\partial c_3 = 0$. These three conditions lead to three simultaneous equations, that is, a matrix equation which is 3×3, $(\mathbf{H} - S\mathbf{W})\mathbf{c} = 0$. For nontrivial solutions the secular determinant of the matrix equation must vanish.[14]

$$\det |\mathbf{H} - S\mathbf{W}| = 0 \quad \text{THE SECULAR EQUATION}$$

or
$$\begin{vmatrix} H_{11} - S_{11}W & H_{12} - S_{12}W & H_{13} - S_{13}W \\ H_{21} - S_{21}W & H_{22} - S_{22}W & H_{23} - S_{23}W \\ H_{31} - S_{31}W & H_{32} - S_{32}W & H_{33} - S_{33}W \end{vmatrix} = 0 \quad (5.6.2)$$

where $H_{12} = \langle(1s_a + 1s_b), H(2s_a + 2s_b)\rangle$, etc., are known integrals (matrix elements).

Multiplying out a 3×3 secular determinant gives a polynomial of degree 3 (the secular equation), with three roots. The lowest of the three roots is $W(1\sigma_g)$. It represents the energy of the first MO, $1\sigma_g$. By substitution of $W(1\sigma_g)$ back into the secular determinant we obtain the ratios of the coefficients[15] c_2/c_1, c_3/c_1, which define the MO[16] $1\sigma_g$ (normalization of the MO to unity completely determines the coefficients). The next higher root of (5.6.2), $W(2\sigma_g)$, when substituted back into equation (5.6.2) determines the ratio of the coefficients which define $2\sigma_g$, and so on for $3\sigma_g$. Again, the secular equation for the general (parent) MO of σ_u symmetry yields the

---

[14] See Appendix A.   [15] Appendix A, equation (A.2.16).
[16] Because of the large atomic energy differences between $1s$ and the other atomic orbitals H_{11} is much more negative than the other H_{ij}; such differences among the matrix elements in the secular determinant have the result that $1\sigma_g$ is largely $(1s_a+1s_b)$, that is, c_1 is much greater than c_2 or c_3 in $1\sigma_g$.

MO, $1\sigma_u$, $2\sigma_u$, and $3\sigma_u$ in the same way. This procedure is at the heart of all LCAO–MO calculations. Full details in matrix notation are given in Appendix A. For the present we summarize the steps in setting up and solving the secular equation.

1. The LCAO–MO is written as a sum of AO containing m unknown coefficients.
2. The variational principle applied to the Schördinger equation of the system yields m simultaneous equations, that is an $m \times m$ matrix equation.
3. The nontrivial solution of these equations requires that the secular determinant vanish. The determinant is $m \times m$; multiplying out the determinant yields a polynomial of degree m in W.
4. The polynomial has m roots, $W_1, W_2, \ldots, W_m$. The lowest root, W_1, is an upper bound to the energy of the lowest energy MO, $W_1 \geqslant E_1$, of that symmetry.
5. The other roots, taken in order, are upper bounds to the energies of the higher energy MO of that symmetry,[17] $W_2 \geqslant E_2, \ldots, W_m \geqslant E_m$.
6. The roots are substituted back into the secular determinant to get the ratio of the coefficients which define the MO having that energy. Normalization to unity completes the determination of the MO.

The result of the variational determination of the σ_g MO in H_2 from a trial LCAO–MO like equation (5.6.1) is presented in the next section.

5.7 THE HARTREE-FOCK-ROOTHAAN PROCEDURE

The linear variation method outlined above is universally applied for the approximate calculation of molecular wavefunctions because it uses the well-known techniques for solving secular equations. The accuracy of the wavefunction is improved by the judicious choice of larger and larger numbers of AO. In the *Hartree-Fock-Roothaan* procedure the molecular *Hartree-Fock equation*[18] for the molecular orbital ψ_i

$$\hat{H}_{\text{eff}}\psi_i(i) = E_i\psi_i(i) \qquad (5.7.1)$$

is solved by writing ψ_i as a general LCAO and solving the appropriate secular equation. Equation (5.7.1) results from minimization of the total energy with respect to variation of the molecular orbitals, subject to the condition that the molecular orbitals remain orthonormal. $\hat{H}_{\text{eff}}$ contains the *average potential* $\hat{V}_{\text{aver}}$, which accounts for the average (integrated or smeared out) interelectronic repulsion. If the variation is complete, and

---

[17] J. K. L. MacDonald, *Phys. Rev.* **43**, 830 (1933).
[18] Section 4.10 and Appendix C.

5.7 The Hartree-Fock-Roothaan Procedure

the equations are iterated until self-consistency is obtained, one then has the *best* set of occupied MO for the molecule, the Hartree-Fock MO. The analogy with the atomic Hartree-Fock and Hartree-Fock-Roothaan procedures discussed in Section 4.10 is complete; except, here we are to determine molecular orbitals rather than atomic orbitals. Equation (5.7.1) is analogous to (4.10.2), and analogously to (4.10.3) and (4.10.4) we have for the N-electron molecule with N occupied molecular spin-orbitals[19]

$$\hat{H}_{\text{eff}}(i) = \hat{h}(i) + \hat{V}_{\text{aver}}(i) \tag{5.7.2}$$

$$\hat{V}_{\text{aver}}(i) = \sum_{j=1}^{N} \left\langle \psi_j(j), \frac{(1-\hat{P}_{ij})}{r_{ij}} \psi_j(j) \right\rangle_j \tag{5.7.3}$$

however, in the molecule

$$\hat{h}(i) = -\tfrac{1}{2}\nabla_i^2 - \sum_{a=1}^{K} Z_a / r_{ia} \tag{5.7.4}$$

where there are K atoms in the molecule with nuclear charges Z_a. The total electronic energy is E_{HF}

$$E_{\text{HF}} = \sum_{i=1}^{N} E_i - \sum_{i>j}^{N} (J_{ij} - K_{ij}) \tag{5.7.5}$$

where the E_i are the orbital energies of equation (5.7.1), and the J_{ij} and K_{ij} are the Coulomb and exchange integrals of the molecular orbitals

$$J_{ij} = \langle \psi_i(1)\psi_j(2), (1/r_{12})\psi_i(1)\psi_j(2) \rangle$$
$$K_{ij} = \langle \psi_i(1)\psi_j(2), (1/r_{12})\psi_i(2)\psi_j(1) \rangle$$

Most often, Slater-type orbitals (STO) are used to build the trial LCAO–MO. For complete variation, leading to Hartree-Fock or near Hartree-Fock molecular orbitals, the effective nuclear charges in the STO [Z_{eff} in equation (4.9.2)] must be varied to minimize the energy, a procedure known as *optimization*.

The result of the Hartree-Fock-Roothaan procedure is molecular wavefunctions of high accuracy, molecular energies to within 1%, or better,[20] and ρ to about the same accuracy. The Hartree-Fock-Roothaan wavefunctions (or close approximations thereto) are available for a number of molecules; these accurate wavefunctions will form the basis for our discussion of the chemical bond.

---

[19] As in equation (4.10.4) we can include the term $j=i$ in $\hat{V}_{\text{aver}}(i)$ because it has zero contribution, $\hat{P}_{ii}=1$. The subscript j on the integral in equation (5.7.3) means integrate over the coordinates of the jth electron.

[20] The residual problem of electron correlation, which is represented by this 1% of E, makes Hartree-Fock energy *differences* of much lower accuracy (see Sections 4.11 and 5.15).

EXAMPLE H_2, $(1\sigma_g)^2$; $\Psi(H_2) = \hat{A} 1\sigma_g \alpha(1) 1\sigma_g \beta(2)$. For the trial $1\sigma_g$ MO we take an LCAO-MO of σ_g symmetry.

$$\widetilde{1\sigma_g}(1) = c_1[1s_a(1) + 1s_b(1)]$$
$$+ c_2[2s_a(1) + 2s_b(1)] + c_3[2p_{z_a}(1) + 2p_{z_b}(1)] \qquad (5.7.6)$$

or more generally,

$$\widetilde{1\sigma_g}(1) = \sum_{i=1}^{m} c_i \phi_i(1) \qquad \phi_i = nl_a + nl_b \qquad (5.7.7)$$

The AO $1s_a$, $2s_a$, $1s_b$, $2s_b$, ..., nl_a, nl_b are STO with $\lambda = 0$. Application of the variational principle to the total energy of the system yields the Hartree-Fock equation [(5.7.1), also Appendix C].

$$\hat{H}_{\text{eff}}(1) 1\sigma_g(1) = E(1\sigma_g) 1\sigma_g(1) \qquad (5.7.8)$$

where

$$\hat{H}_{\text{eff}}(1) = \hat{h}(1) + \hat{V}_{\text{aver}}(1)$$
$$\hat{h}(1) = -\tfrac{1}{2} \nabla_1^2 - 1/r_{1a} - 1/r_{1b}$$
$$\hat{V}_{\text{aver}}(1) = \langle 1\sigma_g(2), (1/r_{12}) \, 1\sigma_g(2) \rangle_2$$

Again, $\hat{H}_{\text{eff}}$ is the one-electron Hamiltonian for an electron moving in an H_2^+ bare nucleus potential $\hat{h}$ and the average potential due to the other occupied spin-orbital, $\hat{V}_{\text{aver}}$. Equation (5.7.8) is solved iteratively until self-consistency is obtained; thus the final MO is often called an SCF–LCAO–MO (see Section 4.10).

The Hartree-Fock $1\sigma_g$ MO will satisfy (5.7.8). The variational principle assures us that the trial LCAO–MO $\widetilde{1\sigma_g}$ can be made to approximate the Hartree-Fock MO. If the trial MO were a linear combination of a very large number of AO, there is evidence that a negligible difference would exist between the SCF–LCAO–MO and the true Hartree-Fock MO. However, most calculations are done with a restricted basis set of AO(STO).

Inserting $\widetilde{1\sigma_g}$ (5.7.7) into (5.7.8), multiplying from the left by ϕ_j, and integrating, we obtain

$$\sum_{i=1}^{m} c_i H_{ji} = E(1\sigma_g) \sum_{i=1}^{m} S_{ji}$$

or

$$\sum_{i=1}^{m} c_i (H_{ji} - S_{ji} E) = 0 \qquad (5.7.9)$$

where $H_{ji} = \langle \phi_j, \hat{H}_{\text{eff}} \phi_i \rangle$ and $S_{ji} = \langle \phi_j, \phi_i \rangle$. There is one such equation as (5.7.9) for each of the m ϕ_j; in matrix form the m simultaneous equations are

$$(\mathbf{H} - S\mathbf{E})\mathbf{c} = 0 \qquad (5.7.10)$$

$\mathbf{S}$ and $\mathbf{H}$ are the $m \times m$ matrices having elements S_{ji} and H_{ji}, respectively, and $\mathbf{c}$ is the column of c_i. The solutions to (5.7.10) follow from the secular equation, $\det |\mathbf{H} - S\mathbf{E}| = 0$. The secular equation, being a polynomial of degree m, has m roots. The lowest root E_1 is the orbital energy of $1\sigma_g$. Substitution of E_1 into the secular equation yields the ratio of the coefficients $c_2/c_1, c_3/c_1, \ldots, c_r/c_1$ [to give an example of equation (A.2.16)]. Then the normalization condition $\langle 1\sigma_g, 1\sigma_g \rangle = 1$ is $\sum_{i,j}^{m} c_i c_j S_{ij} = 1$, and therefore $1/c_1^2 = \sum_{i,j} (c_i/c_1)(c_j/c_1) S_{ij}$, which determines c_1 and all the coefficients.

5.7 The Hartree-Fock-Roothaan Procedure

Because **H** contains the c_i (see $\hat{V}_{\text{aver}}$) it is necessary to make an initial guess at the "field," that is, the coefficients in $\hat{V}_{\text{aver}}$, in order to find the H_{ji} and solve the secular equation. The new coefficients are used to define a new secular equation, and the process is repeated (iteration) until self-consistency is obtained. The result of this process is the SCF–LCAO–MO: For complete variation with a basis set of limited size, the Z_{eff} of the STO should also be optimized. Such a limited basis set, optimized SCF–LCAO–MO representation of the $1\sigma_g$ MO of H_2 at R_e is given in equations (5.7.11) and (5.7.12).[21]

$$1\sigma_g = c_1(1s_a + 1s_b) + c_2(2s_a + 2s_b) + c_3(2p_{za} + 2p_{zb}) \quad (5.7.11)$$

$$\begin{array}{ll} c_1 = 0.43262 & Z_{\text{eff}}(1s) = 1.378 \\ c_2 = 0.12384 & Z_{\text{eff}}(2s) = 1.176 \\ c_3 = 0.02827 & Z_{\text{eff}}(2p) = 1.820 \end{array} \quad (5.7.12)$$

Table 5.7.1 shows the convergence towards the Hartree-Fock $1\sigma_g$ MO

TABLE 5.7.1 Calculated Total Molecular Energy of H_2[a]

Configuration	Method	Basis	R (bohrs)	Total molecular energy (hartrees)
$(1\sigma_g)^2$	LCAO–MO	1s STO, $Z_{\text{eff}} = 1.0$	1.4	−1.09092
$(1\sigma_g)^2$	LCAO–MO	1s STO, $Z_{\text{eff}} = 1.2$	1.4	−1.12804
$(1\sigma_g)^2$	LCAO–MO–SCF	1s, 2s, $2p_z$ STO, $Z_{\text{eff}} = 1.2313$	1.4	−1.13211
$(1\sigma_g)^2$	LCAO–MO–SCF	eq.(5.7.12)	1.402	−1.13349
$(1\sigma_g)^2$	Hartree-Fock[b] SCF	9–term analytical basis	1.400	−1.13363
12 configurations (Sections 4.11 and 5.15)	CI	eq.(5.7.12)	1.402	−1.15919[c]

[a] S. Fraga and B. J. Ransil, *J. Chem. Phys.* **35**, 1967 (1961).
[b] W. Kolos and C. C. J. Roothaan, *Rev. Mod. Phys.* **32**, 219 (1960).
[c] $E(\text{exact}) = -1.174$ hartrees.

[21] S. Fraga and B. J. Ransil, *J. Chem. Phys.* **35**, 1967 (1961). Data used by permission.

of H_2 at $R = 1.4$ bohrs as the variation of the trial molecular orbital becomes more and more extensive. The last line of Table 5.7.1 shows the result of a configuration interaction study of H_2; the substantial improvement over the Hartree-Fock energy (also see Table 5.4.1) shows the severe limitation of even the best wavefunction of the single configuration form.

5.8 ANGULAR MOMENTUM

The spectroscopic states of diatomic molecules are classified (as for atoms) according to the orbital and spin angular momenta. However, because of the nonspherical symmetry of the diatomic molecule, only the component of orbital angular momentum along the internuclear axis λ is conserved; λ is analogous to the quantum number m of the atomic orbital because $m\hbar$ is the component of atomic orbital angular momentum in the z direction. The magnitude of λ is given in Table 5.5.2 for the various MO; the MO are labeled σ, π, δ, etc., according to $|\lambda| = 0, 1, 2, \ldots$. The λ values of the individual electrons are summed to give the *total orbital angular momentum* along the internuclear axis, Λ. The states are then given symbols

$$\Sigma, \Pi, \Delta, \Phi \ldots \quad \text{according to} \quad |\Lambda| = 0, 1, 2, 3, \ldots$$

The multiplicity is written as a left-hand superscript (as for atomic states). Homonuclear diatomics have the additional classification into g and u states which gives the overall symmetry of the state with respect to inversion ($g \times g = g$, $u \times u = g$, $g \times u = u$). Examples of the state symbols were given in Table 5.4.2 for H_2 and other small species.[22]

EXERCISE   Derive the state symbols used in Table 5.4.2

EXAMPLE He_2^+, configuration $(1\sigma_g)^2(1\sigma_u)$. $\lambda_1 = \lambda_2 = \lambda_3 = 0$, so $\Lambda = 0$, a Σ state. ($m_{s_1} = \pm\frac{1}{2}$, $m_{s_2} = \mp\frac{1}{2}$) and $m_{s_3} = \pm\frac{1}{2}$, therefore a doublet state. Finally, $g \times g \times u = u$, so the state symbol is $^2\Sigma_u$.

As in the case of atomic spectra a degenerate configuration of MO may be split into spectroscopic states of different energy by $\hat{V}(r_{ij})$. An important characteristic of the MO method is that it can correctly predict these spectroscopic states by addition of the angular momenta of individual electrons (see B_2, C_2, and O_2 in Section 5.10).

[22] Σ states are classified Σ^+ or Σ^- depending upon whether the wavefunction is symmetric or antisymmetric to reflection in the plane of the nuclei.

5.9 LINEAR TRANSFORMATIONS OF THE WAVEFUNCTION

We state and illustrate an important property of determinantal wavefunctions. Suppose the wavefunction of the molecule is the antisymmetrized product of molecular spin-orbitals $\psi_1, \psi_2, \ldots$

$$\Psi = \hat{A}\psi_1(1)\psi_2(2)\ldots = 1/\sqrt{N!}\,\det|\psi_1(1)\psi_2(2)\ldots| \qquad (5.9.1)$$

The MO have symmetry and are designated $1\sigma_g$, $1\sigma_u$, $1\pi_g$, etc.; the MO are either approximate MO or Hartree-Fock MO. A mathematical theorem says we can *take linear combinations of the columns of a determinant without changing the determinant except by a multiplicative constant* (which can be taken care of by normalization). This "linear combination" of columns (or rows) consists of adding together, or subtracting, the MO, with appropriate factors. The *linear transformation*[23] of the MO results in a new set of orbitals $\chi_1, \chi_2, \ldots$, that is, the new columns of the determinantal wavefunction.

$$X = \hat{A}\chi_1(1)\chi_2(2)\ldots = 1/\sqrt{N!}\,\det|\chi_1(1)\chi_2(2)\ldots| \qquad (5.9.2)$$

The two wavefunctions Ψ and X are physically indistinguishable because they predict the *same observables*. The total energy, the charge density ρ, and other observables are unchanged by the transformation. We actually applied such a linear transformation earlier in going from free-electron MO, ψ_i, to localized orbitals, χ_i (Section 3.8). In future applications this transformation property of the wavefunction will be used to obtain localized and chemically intuitive orbitals.[24] For an instructive example of the process consider the nonexistent molecule He_2.

EXAMPLE He_2, configuration $(1\sigma_g)^2(1\sigma_u)^2$. For simplicity, let $\langle 1s_a, 1s_b\rangle = 0$ so the normalized MO are $1\sigma_g = 1/\sqrt{2}(1s_a + 1s_b)$ and $1\sigma_u = 1/\sqrt{2}(1s_a - 1s_b)$. This is valid at large internuclear distances.

$$\Psi(He_2) = \hat{A}\,1\sigma_g\alpha(1)1\sigma_g\beta(2)1\sigma_u\alpha(3)1\sigma_u\beta(4)$$

Take linear combinations of the columns of the determinant as follows:

$$1/\sqrt{2}(1\sigma_g\alpha + 1\sigma_u\alpha) = 1s_a\alpha \qquad 1/\sqrt{2}(1\sigma_g\alpha - 1\sigma_u\alpha) = 1s_b\alpha$$

and similar linear combinations for the β spin-orbitals. The transformed wavefunction is

$$X(He_2) = \hat{A}\,1s_a\alpha(1)1s_a\beta(2)1s_b\alpha(3)1s_b\beta(4)$$

and the transformed configuration is clearly $(1s_a)^2(1s_b)^2$.

---

[23] C. A. Coulson, *J. Chim. Phys.* **46**, 198 (1949); J. Lennard-Jones, *Proc. Roy. Soc. (London)* **A198**, 1 (1949).

[24] Note, however, that the χ_i are not eigenfunctions of the Hartree-Fock Hamiltonian and thus cannot be used in discussing the ionization potentials or electronic transitions of the molecules.

EXERCISE   Show that the charge density in the above example is the same before and after transformation, that is show $\rho = 2(1\sigma_g)^2 + 2(1\sigma_u)^2 = 2(1s_a)^2 + 2(1s_b)^2$. Hint: Equation (4.4.1).

The charge density in He_2, $\rho = 2(1s_a)^2 + 2(1s_b)^2$, is obviously the charge density of two noninteracting helium atoms (nonbonded), a fact which is not as evident in the MO representation. Transformations from MO representations to representations of more intuitive chemical appeal can often give significant insight into a chemical situation. However, in general, wavefunctions are most easily obtained as MO through the use of symmetry and the variational principle, and transformations are a second step (see Section 5.14).

EXERCISE   Work out the above transformation for He_2 at small internuclear distances where $\langle 1s_a, 1s_b \rangle \neq 0$.

The Diatomic Molecule

5.10 BUILDING UP THE HOMONUCLEAR DIATOMIC MOLECULE

From the molecular orbitals of Figs. 5.5.5 and 5.5.6 we can write down the electronic configurations of all the homonuclear diatomic molecules of the first long row. As usual the configurations are written so that the most tightly held electrons are to the left (largest negative orbital energy).

Li_2 (6 electrons), Li $(1s)^2(2s)$

$$Li_2 \ (1\sigma_g)^2 (1\sigma_u)^2 (2\sigma_g)^2 \ ^1\Sigma_g$$

Bond order = $(2 - 2 + 2)/2 = 1$. $R_e(Li_2) = 2.67$ A [compare to $R_e(H_2) = 0.74$ A].

Bond energy[25] = $D_0(Li_2) = 24$ kcal/mole [compare to $D_0(H_2) = 103$ kcal/mole].

The large bond length of Li_2 is due to shielding of the nuclei by the electrons in the $(1\sigma_g)^2 (1\sigma_u)^2$ core. This shielding reduces the effective nuclear charge seen by the bonding $2\sigma_g$ electrons [optimized[26] Z_{eff} for $2\sigma_g$ are $2s$ (1.26) and $2p$ (1.52)]. As a result the bonding MO, $2\sigma_g$ is a big

---

[25] The "bond energy" can be determined from thermochemical measurements. It differs from D_e by a small quantity, the zero-point energy, of order 0.1 eV for diatomics. The bond energy is denoted D_0 and expressed in kilocalories per mole the conventional chemical energy unit, 1 eV = 23.06 kcal/mole. Zero-point energy is discussed after eq. (6.1.14).

[26] B. J. Ransil, *Rev. Mod. Phys.* **32**, 245 (1960).

loose MO giving a long, weak bond. Figure 5.10.1 compares the ρ contour maps of H_2 and Li_2, illustrating the difference between the two molecules.

Be_2 (8 electrons), Be $(1s)^2(2s)^2$

$$Be_2 \ (1\sigma_g)^2(1\sigma_u)^2(2\sigma_g)^2(2\sigma_u)^2 \ {}^1\Sigma_g.$$

Bond order = 0. R_e, D_0 not observed.

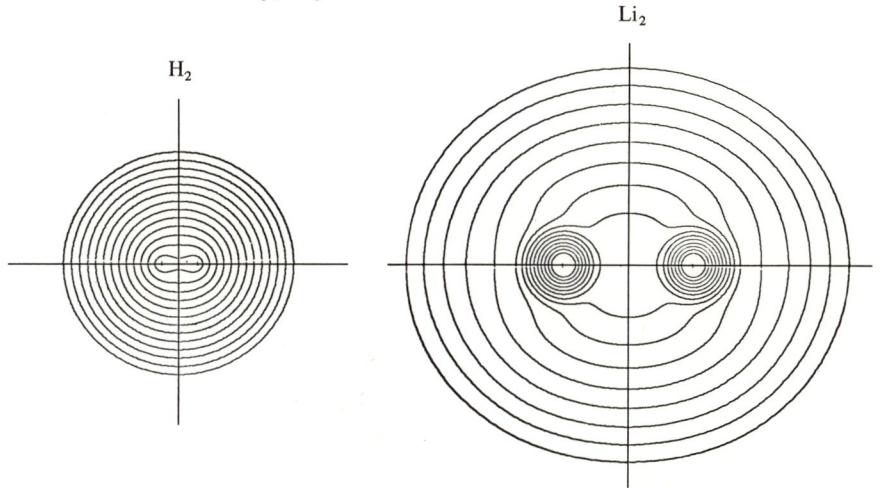

FIG. 5.10.1 Electronic density contour maps of H_2 and Li_2. Outermost contour is 6.1×10^{-5} in both cases. Innermost contour is 1.0 in the case of Li_2 and 0.25 in the case of H_2. After A. C. Wahl, *Science* **151**, 961 (1966). Copyright 1966 by the American Association for the Advancement of Science. Used by permission.

Even the most accurate calculations[27] show the ground state of Be_2 is repulsive and therefore a stable Be_2 molecule is predicted not to exist.

B_2 (10 electrons), B $(1s)^2(2s)^2(2p)$

$$B_2 \ (1\sigma_g)^2(1\sigma_u)^2(2\sigma_g)^2(2\sigma_u)^2(\pi_{xu})(\pi_{yu}) \ {}^3\Sigma_g, \ {}^1\Delta_g, \ {}^1\Sigma_g$$

Bond order = 1. $R_e(B_2) = 1.59$ A. $D_0(B_2) = 69$ kcal/mole.

B_2 has three spectroscopic states with this configuration. $\hat{V}(r_{ij})$ splits the energies of the states, and ${}^3\Sigma_g$ is the ground state. Remembering that we can ignore closed shells in determining the state symbol, we only consider the degenerate pair $(\pi_{xu})(\pi_{yu})$. Since π_x and π_y are made up of $2p_x$ and $2p_y$, which themselves can be reduced to $2p_{+1}$ and $2p_{-1}$ (see Chapter 4),

[27] C. F. Bender and E. R. Davidson, *J. Chem. Phys.* **47**, 4972 (1967).

the following combinations of quantum numbers will be allowed by the Pauli principle:

λ_1	λ_2	m_{s1}	m_{s2}	Λ	Spin component	State symbol
1	1	$\pm\frac{1}{2}$	$\mp\frac{1}{2}$	2	0	$^1\Delta$
-1	-1	$\pm\frac{1}{2}$	$\mp\frac{1}{2}$	-2	0	$^1\Delta$
1	-1	$\pm\frac{1}{2}$	$\mp\frac{1}{2}$	0	0	$^1\Sigma$
1	-1	$\pm\frac{1}{2}$	$\pm\frac{1}{2}$	0	± 1	$^3\Sigma$

That is, when both $2p$ electrons have the same λ value they must differ in $m_s(^1\Delta)$, but when the λ values differ they may, or may not, differ in $m_s(^1\Sigma, ^3\Sigma)$. Note that the $^1\Delta$ state is degenerate because $\Lambda = \pm 2$; this double degeneracy arises for all states having $\Lambda \neq 0$ and is called Λ doubling. For a physical picture of the degeneracy we might say the energy is the same whether the electrons move clockwise or counterclockwise about the internuclear axis. Hund's rule (Section 4.7) for the states of equivalent electrons says $^3\Sigma_g$ is the lowest state. This is explained by noting that the two electrons can minimize their mutual repulsion by going into different π MO, which gives a Σ state. Furthermore, by having the same spin (triplet state), the Pauli antisymmetrization will correlate their motion so they stay out of the same regions of space (see Chapter 4). A few years after Mulliken[28] predicted these states for B_2 on the basis of MO theory, they were confirmed by experiment.[29] The B_2 molecule is paramagnetic.

At this point we note that $(1\sigma_g)^2(1\sigma_u)^2 = (1s_a)^2(1s_b)^2$, because there is slight mixing of $n = 1$ and $n = 2$ shell AO to begin with, and it decreases as Z increases. Therefore we explicitly represent these four electrons in the configuration as two helium cores around their respective nuclei, that is, KK. These *atomic orbitals in molecules* are considered neither bonding nor antibonding, they are nonbonding.

EXERCISE   Topic for discussion. Can nonbonding orbitals be binding? Antibinding?

C_2 (12 electrons), C $(1s)^2(2s)^2(2p)^2$

$$C_2 \; KK(2\sigma_g)^2(2\sigma_u)^2(\pi_{x,yu})^4 \; ^1\Sigma_g$$

Bond order = 2. $R_e(C_2) = 1.24$ A. $D_0 = 140$ kcal/mole.

[28] R. S. Mulliken, *Rev. Mod. Phys.* **4**, 1 (1932).
[29] A. E. Douglas and G. Herzberg, *Can. J. Res.* **A18**, 179 (1940).

5.10 Building Up the Homonuclear Diatomic Molecule

The C_2 molecule, although unfamiliar to the chemist, is observed spectroscopically in outer space.[30] The bond energy of C_2 (two π bonds) is about twice that of B_2, which has one π bond. The degenerate pair of MO π_{xu} and π_{yu} (actually four molecular spin-orbitals) form a filled π shell with four electrons.

N_2 (14 electrons), N $(1s)^2(2s)^2(2p)^3$

$$N_2 \; KK(2\sigma_g)^2(2\sigma_u)^2(\pi_{x,yu})^4(3\sigma_g)^2 \; {}^1\Sigma_g$$

Bond order = 3. $R_e(N_2) = 1.1$ A. $D_0(N_2) = 225$ kcal/mole.

N_2 has two π bonds and one σ bond for a total bond order of 3. The large bond energy and the short R_e are indications of a strong bond. As an example of an LCAO–MO wavefunction which has been iterated to self-consistency in the Hartree-Fock-Roothaan method, consider the N_2 wavefunction in Table 5.10.1. The energy ordering of the MO in the

TABLE 5.10.1 LCAO–MO Wavefunction[a] of N_2

MO	c_1	c_2	c_3	Orbital energy (hartrees)
$1\sigma_g$	0.70480	0.00743	0.00203	−15.64705
$1\sigma_u$	0.70494	0.01502	0.00567	−15.64423
$2\sigma_g$	−0.16328	0.49029	0.24824	−1.42106
$2\sigma_u$	−0.15963	0.80527	−0.23244	−0.71370
$3\sigma_g$	−0.06640	0.38257	−0.58527	−0.55548

MO	c_π			
$\pi_{x,yu}$	−0.62139	—	—	−0.54540

[a] From B. J. Ransil, *Rev. Mod. Phys.* **32**, 245 (1960). Used by permission.

configuration of N_2 is based on the spectroscopic evidence that the ground state of N_2^+ molecular ion is $KK(2\sigma_g)^2(2\sigma_u)^2(\pi_{x,yu})^4(3\sigma_g)$. The σ orbitals are $\sigma_{g \text{ or } u} = c_1(1s_a \pm 1s_b) + c_2(2s_a \pm 2s_b) + c_3(2p_{za} \pm 2p_{zb})$, and, of course, $\pi_{(x \text{ or } y)u} = c_\pi(2p_{x,ya} + 2p_{x,yb})$. The Z_{eff} are 1s (6.67), 2s (3.83), $2p_z$ (4.50), and $2p_{x,y}$ (3.82), compared to $Z = 7$. The wavefunction in Table 5.10.1 is the "limited basis set" wavefunction (i.e., limited to valence shell AO). The wavefunction in the Hartree-Fock limit is obtained by judiciously expanding the basis set to contain AO of $n > 2$ and then further optimizing

[30] E. A. Ballik and D. A. Ramsey [*J. Chem. Phys.* **31**, 1128 (1959)] identified the ${}^1\Sigma_g$ state of C_2 as the ground state; it is about 0.1 eV below the ${}^3\Pi_u$ state of $KK(2\sigma_g)^2(2\sigma_u)^2(\pi_{x,yu})^3(3\sigma_g)$.

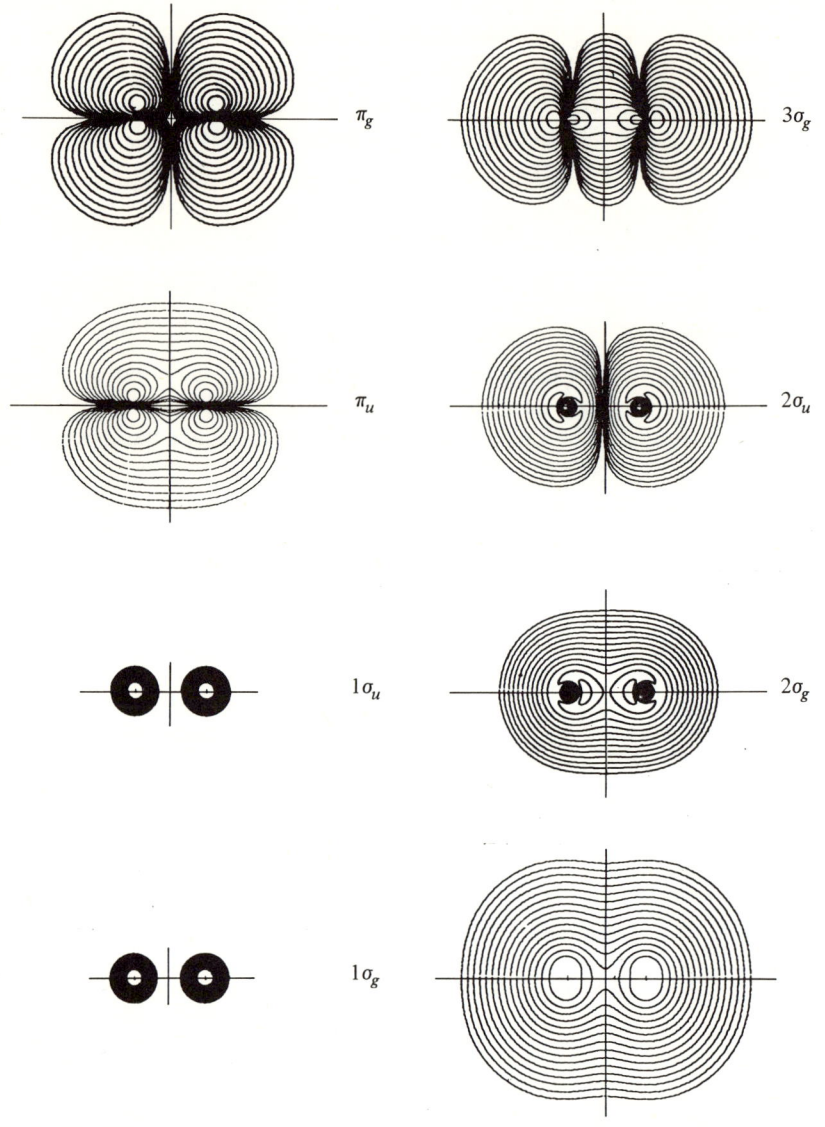

Fluorine Molecule Total Electron Density Contours

FIG. 5.10.2 ρ contour maps of the molecular orbitals of F_2. Total ρ can be obtained from the superposition of these MO charge densities. After A. C. Wahl, *Science* **151,** 961 (1966). Copyright 1966 by the American Association for the Advancement of Science. Used by permission.

the Z_{eff}.[31] The molecular energy given by the wavefunction of Table 5.10.1 is -108.633 hartrees which is over 99% of the experimental energy, -109.586 hartrees.

EXERCISE   Predict the ionization potential of N_2 from the data of Table 5.10.1. Give the results in electron volts and compare to the experimental I.P. of 15.77 eV.

O_2 (16 electrons), O $(1s)^2(2s)^2(2p)^4$

$$O_2 \ KK(2\sigma_g)^2(2\sigma_u)^2(\pi_{x,yu})^4(3\sigma_g)^2(\pi_{x,yg})^2 \ {}^3\Sigma_g, {}^1\Delta_g, {}^1\Sigma_g$$

Bond order = 2. $R_e(O_2) = 1.2$ A. $D_0(O_2) = 118$ kcal/mole.

The degeneracy of the configuration is removed by $\hat{V}(r_{ij})$, and ${}^3\Sigma_g$ is the resulting ground state (see B_2 discussion). O_2 is paramagnetic. The bond order of O_2 is reduced from that of N_2 by placing two electrons in the next available MO, $(\pi_{x,yg})$, which are antibonding as well as degenerate.

EXERCISE   Explain why the ionization of O_2 to O_2^+ yields an ion which has a larger D_0 and a shorter R_e than O_2 itself.

F_2 (18 electrons), F $(1s)^2(2s)^2(2p)^5$

$$F_2 \ KK(2\sigma_g)^2(2\sigma_u)^2(\pi_{x,yu})^4(3\sigma_g)^2(\pi_{x,yg})^4 \ {}^1\Sigma_g$$

Bond order = 1. $R_e(F_2) = 1.4$ A. $D_0(F_2) = 37$ kcal/mole.

Figure 5.10.2 gives the ρ contour maps of the individual MO of F_2 (Hartree-Fock wavefunction) as well as the total electronic density map of F_2. The MO of other first-row homonuclear diatomics have much the same nodal patterns and shapes as these MO of F_2, but differ in size and ρ value.[32]

Ne_2 (20 electrons), Ne $(1s)^2(2s)^2(2p)^6$

$$Ne_2 \ KK(2\sigma_g)^2(2\sigma_u)^2(\pi_{x,yu})^4(3\sigma_g)^2(\pi_{x,y})^4(3\sigma_u)^2 = KKLL$$

Bond order = 0. R_e, D_0 not observed.

5.11 SECOND-ROW AND OTHER HOMONUCLEAR DIATOMIC MOLECULES

Having reached Ne_2 of bond order zero, the entire set of first-row diatomics is complete. To proceed into the second row we actually start with Ne_2, but use it as the rare gas core (nonbonding). The MO are linear

---

[31] P. E. Cade, K. D. Sales, and A. C. Wahl [*J. Chem. Phys.* **44**, 1973 (1966)] give a calculated molecular energy, -108.9956 hartrees.

[32] A. C. Wahl, *Science*, **151**, 961 (1966). Also A. C. Wahl, P. J. Bertoncini, G. Das, and T. L. Gilbert, *Int. J. Quantum Chem.* **IS**, 123 (1967).

combinations of 3s, 3p, and 3d of appropriate symmetry (see Table 5.5.2). The configurations of the higher row molecules follow by analogy with the first row.

Na_2 (22 electrons), Na KL (3s)

$$Na_2 \; KKLL(1\sigma_g)^2 \quad \text{or} \quad (Ne)(Ne)(1\sigma_g)^2 \; {}^1\Sigma_g$$

Bond order = 1. $R_e(Na_2) = 3.07$ A. $D_0(Na_2) = 17.3$ kcal/mole.

K_2 (38 electrons), K (Ar) (4s)

$$K_2 \; (Ar)(Ar)(1\sigma_g)^2 \; {}^1\Sigma_g$$

Bond order = 1. $R_e(K_2) = 3.9$ A. $D_0(K_2) = 11.8$ kcal/mole.

Both K_2 and Na_2 use an orbital $(1\sigma_g)$ for bonding [Li_2 could be written $KK(1\sigma_g)^2$] but all these $1\sigma_g$ MO are solutions to quite different secular equations and are linear combinations of different AO. The number k in the MO $(k\sigma_g)$ or $(k\sigma_u)$ is just used to *order the roots* of the secular equation and has nothing to do with the principal quantum numbers of the AO in the LCAO. Now because of the large numbers of electrons involved [i.e., in $\hat{V}(r_{ij})$], these secular equations have not been solved to nearly the same accuracy as Li_2. However, just as the Li_2 bonding MO is largely $(2s_a + 2s_b)$, the bonding MO of Na_2 and K_2 will be mainly $(3s_a + 3s_b)$ and $(4s_a + 4s_b)$, respectively. The trends in the R_e and D_0 values (see Table 5.11.1) show that the bonding MO, and the bonds, get bigger and looser, as do the shielded AO, 3s, 4s, etc. (Chapter 4).

TABLE 5.11.1 Alkali Metal Molecules

Molecule	Re (A)	D_0 kcal/mole
Li_2	2.7	24.
Na_2	3.1	17.
K_2	3.9	12.
Rb_2	—	11.
Cs_2	—	10.

S_2 (32 electrons), S KL $(3s)^2(3p)^4$

$$S_2 \; KKLL(1\sigma_g)^2(1\sigma_u)^2(\pi_{x,yu})^4(2\sigma_g)^2(\pi_{x,yg})^2 \; {}^3\Sigma_g$$

Bond order = 2. $R_e(S_2) = 1.89$ A. $D_0(S_2) = 83$ kcal/mole.

The analogy to O_2 is evident and is substantially correct. However, in the absence of accurate calculations[33] the argument by analogy does not really give the detailed ordering of the MO energies. The bonding and antibonding MO of S_2 are mainly linear combinations of $n = 3$ AO.

EXERCISE Discuss the electronic configuration, MO, spectroscopic ground state, and bonding properties of P_2. $R_e(P_2) = 1.9$ Å and $D_0(P_2) = 116$ kcal/mole.

The halogen molecules form an interesting series of homonuclear diatomics. The electronic configuration of the halogen molecule is $(1\sigma_g)^2(1\sigma_u)^2(\pi_{x,yu})^4(2\sigma_g)^2(\pi_{x,yg})^4$, where the rare gas core is omitted. The bond lengths increase continuously from F_2 to I_2, but there is a very intriguing maximum in D_0 at Cl_2 (Table 5.11.2). It is easy to believe that

TABLE 5.11.2 Halogen Molecules

Molecule	$R_e(Å)$	D_0 (kcal/mole)
F_2	1.42	37.6
Cl_2	1.99	57.
Br_2	2.28	45.5
I_2	2.67	35.6

the increased stability of the heavier halogen molecules is due to the energetically available nd orbitals, but this is pure speculation in the absence of *highly accurate* calculations.[34]

5.12 HOMONUCLEAR DIATOMIC CHARGE DISTRIBUTIONS

Figure 5.12.1 presents the contour maps of the total electronic density in the homonuclear diatomic molecules. These ρ maps were plotted from Hartree-Fock-Roothaan wavefunctions.[35] The electronic densities, which

---

[33] There is no obstacle to obtaining LCAO–MO–SCF wavefunctions for molecules such as S_2 with its many electrons; in fact, completely automatic computer programs for performing such calculations on linear molecules (even polyatomic molecules) have been in use since the early 1960s.

[34] D_0 is sensitive to small effects in the wavefunction because it is a *small* difference between *large* energies. By "small effects in the wavefunction" we mean *electron correlation* (see Section 5.15 of this chapter).

[35] R. F. W. Bader, W. H. Henneker, and P. E. Cade, *J. Chem. Phys.* **46**, 3341 (1967); see also B. J. Ransil and J. J. Sinai, *J. Chem. Phys.* **46**, 4050 (1967); and A. C. Wahl, *Science* **151**, 961 (1966).

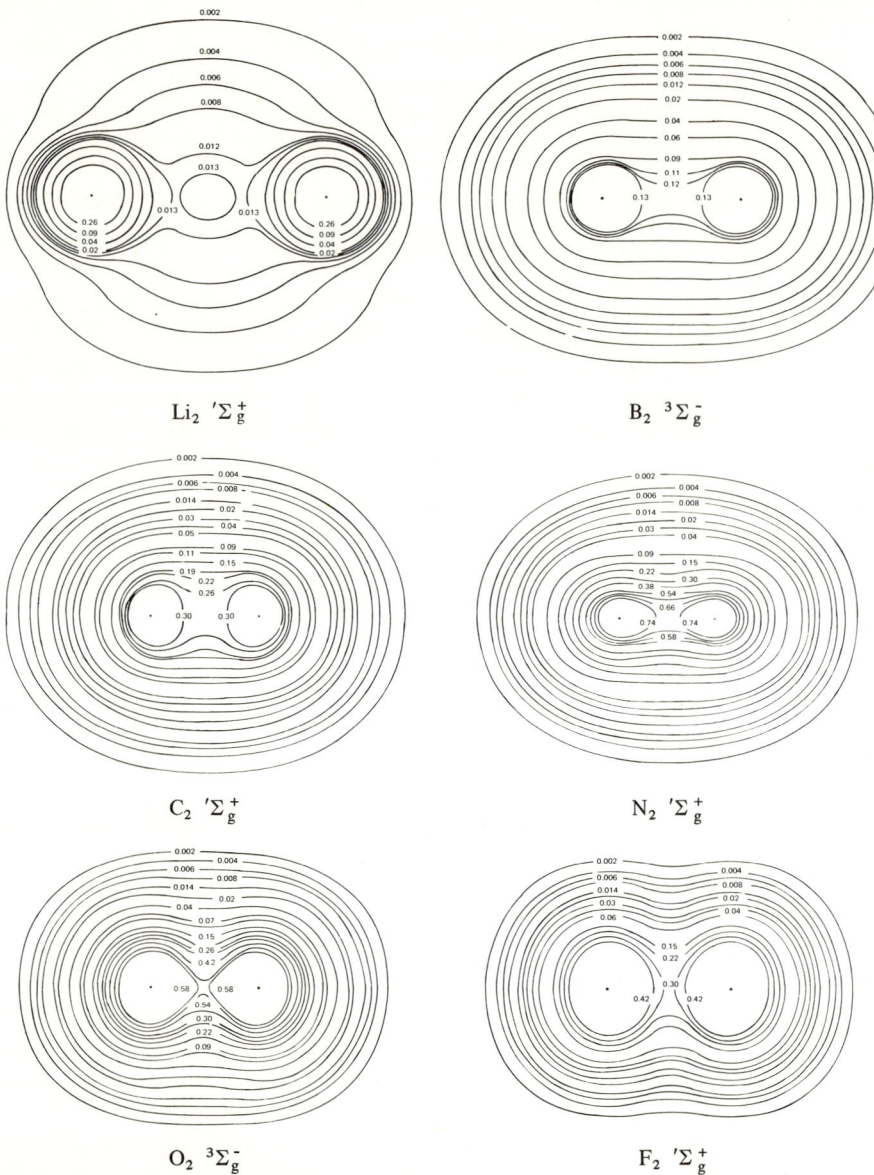

FIG. 5.12.1 Total molecular charge density contour maps of the first-row homonuclear diatomic molecules. The innermost circular contours centered on the nuclei have been omitted for the sake of clarity. The scale of length is the same for all molecules. From R. F. W. Bader, W. H. Henneker, and P. E. Cade, *J. Chem. Phys.* **46**, 3341 (1967). Reproduced by permission.

5.12 Homonuclear Diatomic Charge Distributions

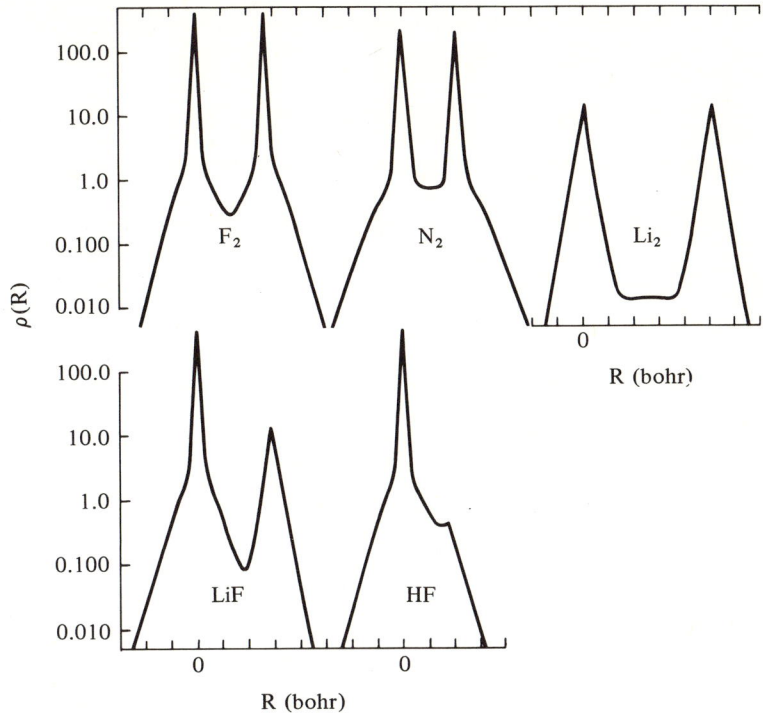

FIG. 5.12.2 Total charge densities (ρ profiles) as a function of internuclear distance. Units are electrons per atomic unit cubed versus internuclear distance in bohr. From B. J. Ransil and J. J. Sinai, *J. Chem. Phys.* **46,** 4050 (1967). Reproduced by permission.

are cylindrically symmetric about the bond, are mapped on a plane cutting through both nuclei. The electronic density piles up around the nuclei and in the binding region between the nuclei, then falls away at large distances. The ρ profiles of N_2, F_2, and Li_2, Fig. 5.12.2, are a clear demonstration of the peaks and valleys in density.

The contour maps give a good idea of molecular size and shape. For example, because the electronic density in fluorine is more tightly bound than in the other atoms, even though $R_e(F_2) > R_e(O_2)$, F_2 is the smaller molecule, a fact long known from experiment. However, the ρ contours do not always give a correct feeling for the tightness of binding (force) or the bond strength (energy). Li_2 is quite clearly a big loose molecule with diffuse ρ, and in the other extreme F_2 is a small tight molecule with a dense ρ, but the bond energies are comparable: $D_0(Li_2) = 24$ kcal/mole, $D_0(F_2) = 37$ kcal/mole.

Density difference contour maps are instructive in the study of chemical binding, that is, maps of

$$\Delta\rho = \rho_{\text{molecule}} - \rho_{\text{atoms}}$$

The "unperturbed" density[36] of the atoms at distance R_e is subtracted from the molecular density. The $\Delta\rho$ maps show the *redistribution of charge which occurs upon formation of a chemical bond.*[37]

In general, the amount of density placed into the *binding region* (see Section 5.3 of this chapter) by the unperturbed atoms at a distance R_e is insufficient to balance the repulsive force of the nuclei, $Z_a Z_b / R_e^2$. Binding results from a redistribution of electronic density so that the molecule has more density in the binding region than the unperturbed atoms at R_e. This binding charge density shows up as a positive $\Delta\rho$ in the binding region (or as a negative $\Delta\rho$ in the antibinding region)[38] (see Fig. 5.12.3). The amount of charge, its location, and its density will be important in bond formation.

EXERCISE   Prove that $\int \nabla\rho \, dV = 0$ for any molecule.

In Fig. 5.12.3 the dotted contours indicate where electronic charge density is removed upon bond formation relative to the atoms, the full contours indicate increased charge density. The polarization of the atomic charge densities is generally quadrupolar, that is, $- + -$ about each nucleus relative to the atom, with $+$ at the nucleus (decreased charge density at the nucleus). Similar quadrupolar polarizations of the atomic charge density should be observed upon placing the atom in an electric field directed along the internuclear axis. Quadrupolar polarization about the nucleus leads to increased electronic density in *both* the binding and the antibinding regions (separated by dotted lines in Fig. 5.12.3), but the binding charge density is concentrated on the internuclear axis where it is very effective in cancelling nuclear repulsion.

---

[36] The "unperturbed" atomic density must be realistically chosen. While Li and N have S ground states and are spherically symmetric, as are the densities, F, B, O, and C have P ground states and are not spherically symmetric. For the latter atoms a valence density is chosen. For O and C one electron is placed in p_z along the bond axis and the remaining electrons are averaged over p_x and p_y, while for B and C the p electrons are averaged over p_x and p_y. Of course, other choices of valence density are possible.

[37] Due to M. Roux et al., *J. Chim. Phys.* **54**, 218 (1956); *J. Chem. Phys.* **37**, 933 (1962)

[38] $\Delta\rho$ is not an experimental observable and like other nonobservables should be used cautiously as a *concept*, especially in regard to small $\Delta\rho$, or small $\Delta\rho$ differences between molecules. That is, small inaccuracies in ρ could produce gross errors in small $\Delta\rho$. These questions have been discussed by Ransil and Sinai, and Bader et al., footnote 35.

5.12 Homonuclear Diatomic Charge Distributions

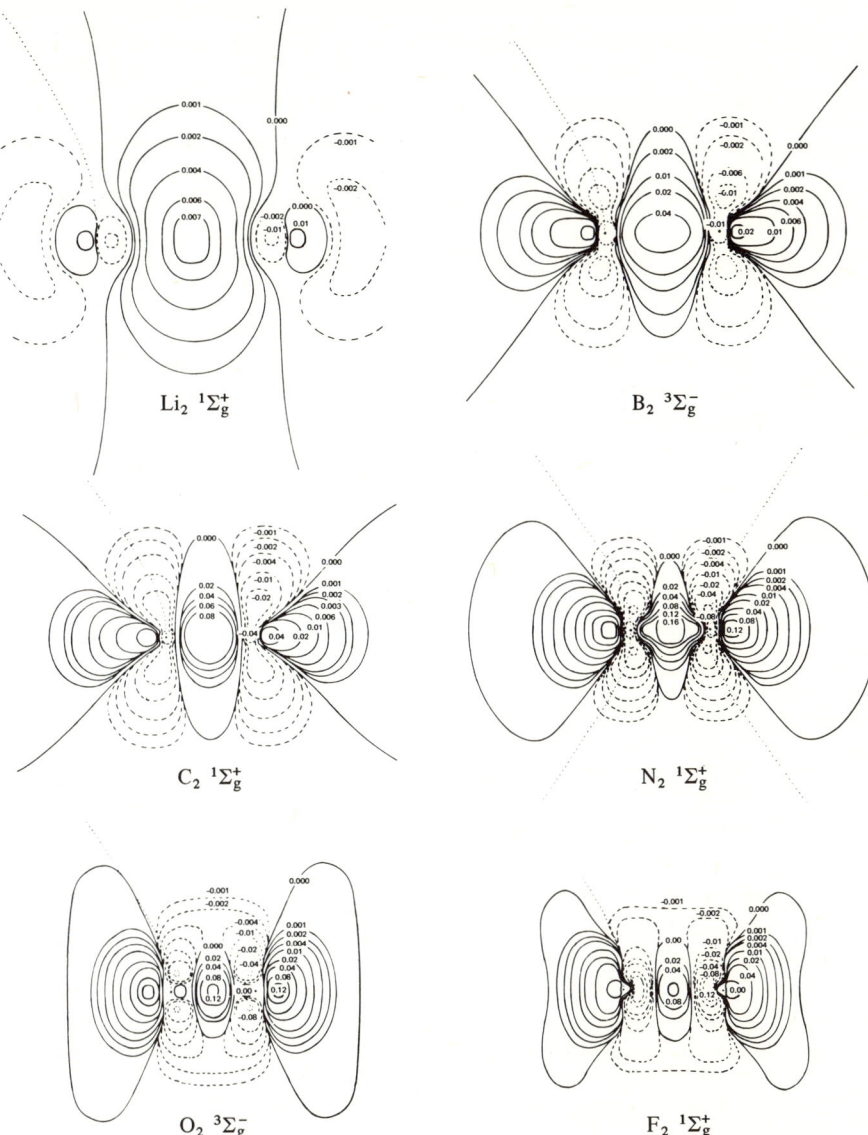

FIG. 5.12.3 Density difference maps for the first-row homonuclear diatomic molecules. The same scale of length applies to all the molecules. The dashed lines are negative $\Delta\rho$ contours. From R. F. W. Bader, W. H. Henneker, and P. E. Cade, *J. Chem. Phys.* **46**, 3341 (1967). Reproduced by permission.

All these molecules (except Li_2) have roughly similar shapes in their ρ maps, but the $\Delta\rho$ maps are sufficiently distinct to be molecular fingerprints. Again Li_2 is the most different, showing a large diffuse binding density due to its nondirectional bonding MO, which is largely $(2s_a + 2s_b)$. As for binding density in the other $\Delta\rho$ maps, they are clearly dominated by $\Delta\rho$ of p_z symmetry, leading to an interesting lack of positive $\Delta\rho$ of π symmetry even in C_2 KK $(2\sigma_g)^2(2\sigma_u)^2(\pi_{x,y u})^4$. The decreasing amount of charge accumulated in the binding regions of O_2 and F_2 (Fig. 5.12.3) can be ascribed to the occupation of the antibonding π_g MO in these molecules.

The individual occupied MO are classified as binding, antibinding, or nonbinding depending on the force they exert on the nuclei,[39] just as they are classified as bonding, antibonding, or nonbonding on the basis of energy and nodal pattern. As expected, the two classifications are in qualitative agreement.

5.13 THE HETERONUCLEAR DIATOMIC MOLECULE

In the heteronuclear diatomic molecule the inversion operator $\hat{I}$ is no longer a symmetry operator. From Table 5.5.2 the absence of $\hat{I}$ removes the distinction between g and u MO. However, σ, π_x, π_y, etc., are still symmetry distinctions. A homonuclear MO written $\sigma = c_1 1s_a + c_2 1s_b$ had the symmetry constraint $c_1 = \pm c_2$, so that $\hat{I}\sigma = \pm\sigma$. The removal of the symmetry constraint on these coefficients means that c_1 may differ from c_2 in magnitude as well as sign, so *polarity* is introduced into the MO. Another way of looking at the asymmetry of the situation is through the orbital energies, schematically illustrated in Fig. 5.13.1.

It is difficult to generalize about the bonding in heteronuclear diatomic molecules, primarily because the large number of hetero molecules cover the gamut of bonding types from nearly *homopolar covalent* to *ionic* bonds. It therefore seems worthwhile to present a few well-studied molecules in some detail. The hydrides are the simplest of all hetero molecules. Table 5.13.1 summarizes the experimental situation. The same R_e and D_e trends run through both rows of hydrides, HF and HCl having the greatest bond strengths.

The wavefunctions used in the calculation of the dipole moments are highly accurate. Writing the hydrides as AH, the 1σ and 2σ MO are linear combinations of many AO of σ symmetry, but principally ns_a and np_{za}, where $n = 2, 3$ depending on the row. The pair of filled MO shells $(1\sigma)^2(2\sigma)^2$ have a net binding effect in the heteronuclear diatomics, as in BH and AlH. The $(\pi_{x,y})$ MO are largely np_a, and being atomic-like they are nonbonding.

---

[39] Bader, Henneker, and Cade, footnote 35; see also Section 5.3.

5.13 The Heteronuclear Diatomic Molecule

The $(\pi_{x,y})$ electrons are the most easily removed, and therefore are responsible for the first ionization potential of the molecule. The shells represented in the configuration as K and KL are the nonbonding rare gas cores.

TABLE 5.13.1 The Hydrides[a]

Molecule	Configurations A	AH	R_e* (bohrs)	D_e* (eV)	Dipole moment μ_{calc} (D)	μ_{exptl} (D)	State
LiH	$K(2s)$	$K(1\sigma)^2$	3.02	2.52	-6.002	-5.882	$^1\Sigma$
BeH	$K(2s)^2$	$K(1\sigma)^2(2\sigma)$	2.54	(2.6)	-0.282	—	$^2\Sigma$
BH	$K(2s)^2(2p)$	$K(1\sigma)^2(2\sigma)^2$	2.34	3.58	1.733	—	$^1\Sigma$
CH	$K(2s)^2(2p)^2$	$K(1\sigma)^2(2\sigma)^2(\pi_{x,y})$	2.12	3.65	1.570	1.46	$^2\Pi$
NH	$K(2s)^2(2p)^3$	$K(1\sigma)^2(2\sigma)^2(\pi_{x,y})^2$	1.96	(3.80)	1.627	—	$^3\Sigma$
OH	$K(2s)^2(2p)^4$	$K(1\sigma)^2(2\sigma)^2(\pi_{x,y})^3$	1.83	4.63	1.780	1.66	$^2\Pi$
HF	$K(2s)^2(2p)^5$	$K(1\sigma)^2(2\sigma)^2(\pi_{x,y})^4$	1.73	6.12	1.942	1.82	$^1\Sigma$
NaH	$KL(3s)$	$KL(1\sigma)^2$	3.57	(2.3)	-6.962	—	$^1\Sigma$
MgH	$KL(3s)^2$	$KL(1\sigma)^2(2\sigma)$	3.27	(2.3)	-1.516	—	$^2\Sigma$
AlH	$KL(3s)^2(3p)$	$KL(1\sigma)^2(2\sigma)^2$	3.11	3.01	0.170	—	$^1\Sigma$
SiH	$KL(3s)^2(3p)^2$	$KL(1\sigma)^2(2\sigma)^2(\pi_{x,y})$	2.87	3.32	0.302	—	$^2\Pi$
PH	$KL(3s)^2(3p)^3$	$KL(1\sigma)^2(2\sigma)^2(\pi_{x,y})^2$	2.71	(3.34)	0.538	—	$^3\Sigma$
SH	$KL(3s)^2(3p)^4$	$KL(1\sigma)^2(2\sigma)^2(\pi_{x,y})^3$	2.55	3.70	0.861	—	$^2\Pi$
HCl	$KL(3s)^2(3p)^5$	$KL(1\sigma)^2(2\sigma)^2(\pi_{x,y})^4$	2.41	4.62	1.197	—	$^1\Sigma$

[a] From P. E. Cade and W. M. Huo, *J. Chem. Phys.* 47, 614, 649 (1967). Used by permission.
* Rounded to 3 figures, uncertain values in parenthesis.
[†] Calculated from Hartree-Fock-Roothaan wavefunctions by P. E. Cade and W. M. Huo, *J. Chem. Phys.* 45, 1063 (1966). Used by permission. Negative values of the dipole moment indicate A^+H^- and positive values indicate A^-H^+, for example, Li^+H^- and H^+F^-.

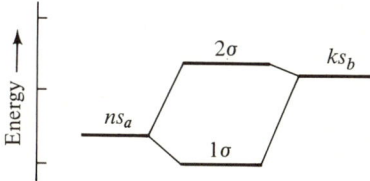

FIG. 5.13.1 The mixing of two s orbitals of different energy; k and n are the principal quantum numbers. The AO have combined into a bonding MO of lower energy and an antibonding MO of higher energy.

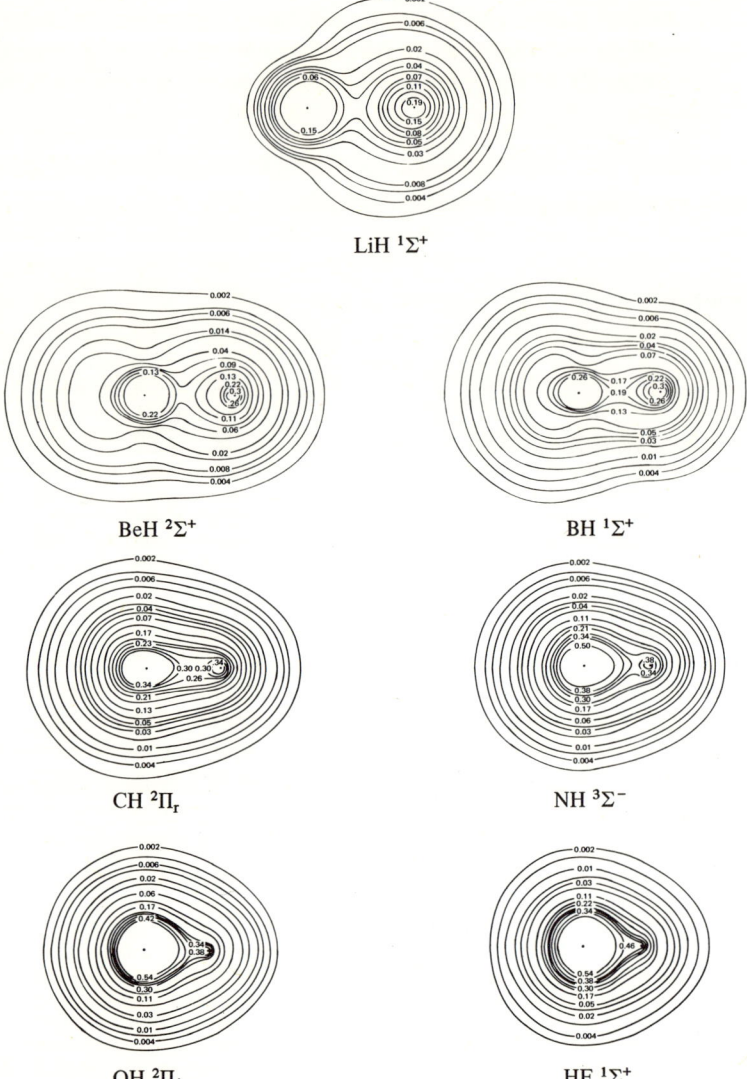

FIG. 5.13.2 Total molecular charge density contour maps of the first-row diatomic hydrides. All molecules are drawn to the same scale of length. The innermost circular contours encircling the A nucleus of the hydride HA have been omitted for clarity. The A nucleus is on the left. From R. F. W. Bader, I. Keaveny, and P. E. Cade, *J. Chem. Phys.* **47,** 3381 (1967). Reproduced by permission.

5.13 The Heteronuclear Diatomic Molecule

Figure 5.13.2 presents the electronic density contour maps of some hydrides.[40] The charge density is obviously unsymmetrically distributed between the nuclei, but further analysis is necessary. One of the important pieces of information in this analysis is the *dipole moment* μ. The dipole moment is an observable molecular property for which experimental values are sometimes available (Table 5.13.1)). The amount of *charge separation* in the molecule is measured by μ.

$$\mu = \mu_{electronic} + \mu_{nuclear}$$

$$\mu_{nuclear} = Z_a eR/2 - Z_b eR/2 \tag{5.13.1}$$

$$\mu_{electronic} = \left\langle \Psi, \sum_{i=1}^{N} -e\mathbf{r}_i \Psi \right\rangle = -\int \rho e \mathbf{r}_1 \, dV_1$$

Z_a, Z_b are the nuclear charges; R, the internuclear distance; e, the electronic charge; $\mathbf{r}_i$, the electronic position vector; and Ψ is the accurate electronic wavefunction. The electronic and nuclear distances are measured from the bond midpoint in equation (5.13.1). The units of μ are debyes (symbol D), 1 debye = 10^{-18} esu cm. For a homonuclear diatomic molecule the electronic density is symmetrically distributed between the nuclei a and b, so $\mu_{electronic} = 0$, and since $Z_a = Z_b$, $\mu_{nuclear} = 0$, for a net zero dipole moment. *Furthermore if one could bring two different neutral atoms* ($Z_a \neq Z_b$) *together to R_e without perturbing the spherical charge density around each atom one would still have $\mu = 0$.* That is, $\mu_{electronic} = -\mu_{nuclear}$ for this idealized case. In fact, however, the atoms perturb each other quite strongly in forming a bond, leading to a net shift of electronic charge density in one direction or the other, so $\mu \neq 0$ when $Z_a \neq Z_b$. In the extreme case a large amount of electronic density is shifted from a donor to an acceptor, yielding a *large dipole moment* and *ionic bonding*, A^+B^-. Examples are LiH ($\mu = -5.882$ D, sense Li$^+$H$^-$), LiF ($\mu = 6.3$ D, direction Li$^+$F$^-$), NaCl ($\mu = 9.00$ D), and KF ($\mu = 8.60$ D). In other cases the dipole moment of the heteronuclear diatomic is so small that the bonding is nearly *homopolar*, for example, CO ($\mu = 0.118$ D direction C$^-$O$^+$), NO ($\mu = 0.15$ D). Most heteronuclear diatomic molecules fall in the mean between these extremes, as illustrated in Table 5.13.1.

In our discussion we have implied essentially the following definitions of terms: *homopolar (covalent) bond* = equal sharing of the binding $\Delta\rho$ ($\mu = 0$), typically in homonuclear diatomic molecules; *heteropolar bond* = unequal sharing of $\Delta\rho$ ($\mu \neq 0$). The ionic bond is an extreme case of unequal sharing of shifting charge densities. Now that the charge density difference has been incorporated in the discussion, examine Fig. 5.13.3 for the hydrides. Comparison of Figs. 5.12.1 and 5.13.2 displays

---

[40] R. F. W. Bader, I. Keaveny, and P. E. Cade, *J. Chem. Phys.* **47**, 3381 (1967).

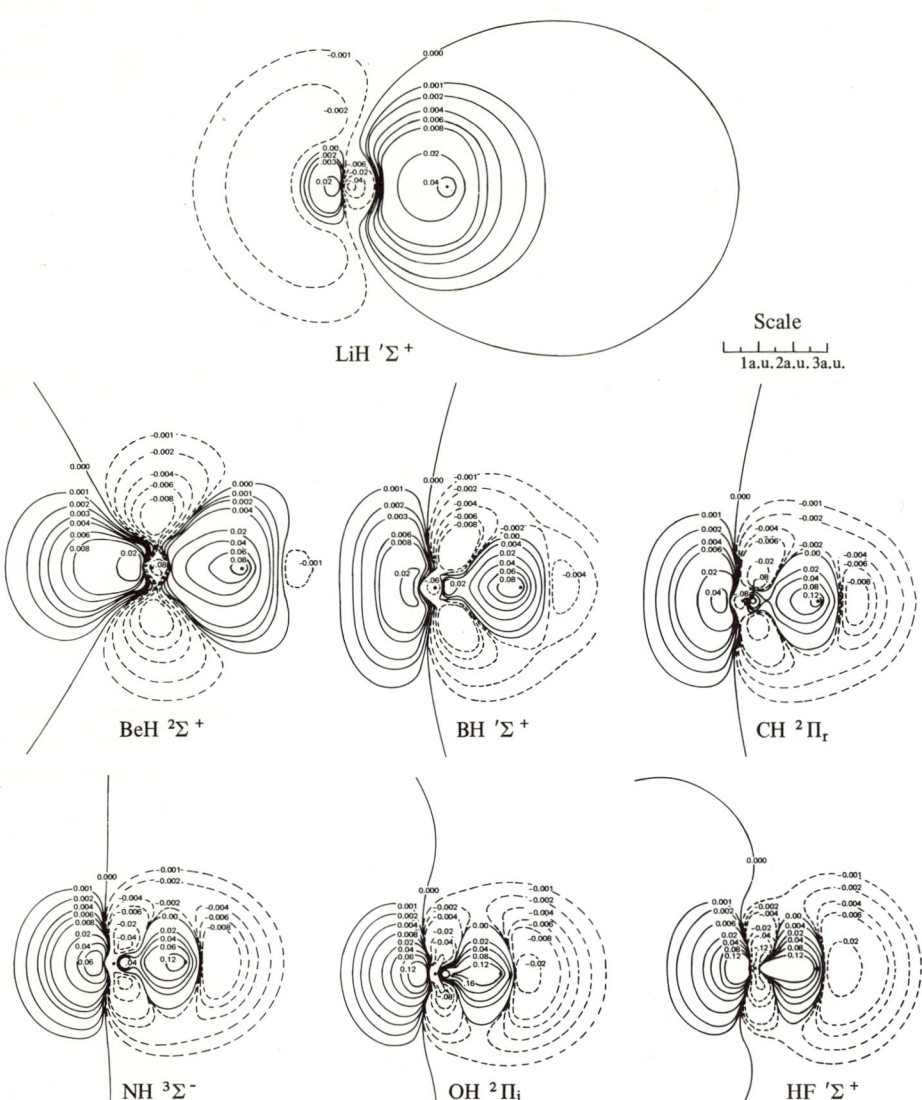

FIG. 5.13.3 Density difference maps of the first-row hydrides. See the caption of Fig. 5.12.3. From R. F. W. Bader, I. Keaveny, and P. E. Cade, *J. Chem. Phys.* **47**, 3381 (1967). Reproduced by permission.

5.13 The Heteronuclear Diatomic Molecule

the difference between the *homonuclear* and the *heteronuclear* molecule; comparison of Figs. 5.12.3 and 5.13.3 displays the difference between the *homopolar* and the *heteropolar* bond. The extreme heteropolarity of the ionic bond is illustrated by the $\Delta\rho$ of LiH, Fig. 5.13.3, where positive $\Delta\rho$ contours enclose the proton and resemble an s orbital polarized toward the Li^+ core.

The subtle distinction has been drawn between the asymmetry of the charge distribution in a heteronuclear molecule, which is due to $Z_a \neq Z_b$, and the unsymmetrical sharing of the charge transferred in bond formation. The latter determines the polarity of the bond.[41] The shifts in electronic density which result in binding of the first-row hydrides can be observed in the $\Delta\rho$ contour maps of Fig. 5.13.3. The maps illustrate the change in polarity from Li^+H^- to B^-H^+ and this polarity continues to the end of the period. In all cases there is accumulation of charge in the binding region between the nuclei.

OTHER HETERONUCLEAR DIATOMIC MOLECULES

LiF, BN (12 electrons), isoelectronic with C_2

$$\text{LiF } KK(1\sigma)^2(2\sigma)^2(\pi_{x,y})^4 \; ^1\Sigma$$

$R_e(\text{LiF}) = 1.5$ A. $D_0(\text{LiF}) = 137$ kcal/mole.

LiF is a prototype for the ionic bond. The Hartree-Fock-Roothaan wavefunction[42] shows that the K shells are very largely isolated about each nucleus, $1s_F$ and $1s_{Li}$, respectively. Furthermore 1σ and 2σ are mainly $2s_F$ and $2p_{z_F}$, and $\pi_{x,y}$ is mainly $2p_{x,y_F}$; Li^+F^-. But having simply said this we have not done justice to the full calculation, which shows appreciable $2s_{Li}$ and $2p_{zLi}$ in both 1σ and 2σ, and even appreciable amounts of $3d_{Li}$ and $3d_F$ in $\pi_{x,y}$! In any case, the object of interest is the electronic density. Figure 5.13.4 gives the $\Delta\rho$ contour maps and $\Delta\rho$ profiles of LiF and N_2 for comparison of ionic and covalent bonding. The transfer of charge from Li to F is evident in LiF, in contrast to the equal sharing of the transferred charge in N_2; $\mu(Li^+F^-) = 6.28$ debye.

$$\text{BN } KK(1\sigma)^2(2\sigma)^2(\pi_{x,y})^3(3\sigma) \quad ^3\Pi$$

$R_e(\text{BN}) = 1.28$ A. $D_0(\text{BN}) = 92$ kcal/mole.

In contrast to LiF, BN is electronically similar to C_2 and has a weakly polar bond. BN has a paramagnetic ground state, $^3\Pi$; for a time it was thought the ground state of C_2 had this configuration and was paramagnetic (see C_2 discussion).

---

[41] The "ability" of an atom to control this transfer of charge is given semiempirical quantitative expression in the various "electronegativity scales" (see Appendix F).

[42] A. D. McLean, *J. Chem. Phys.* **39**, 2653 (1963).

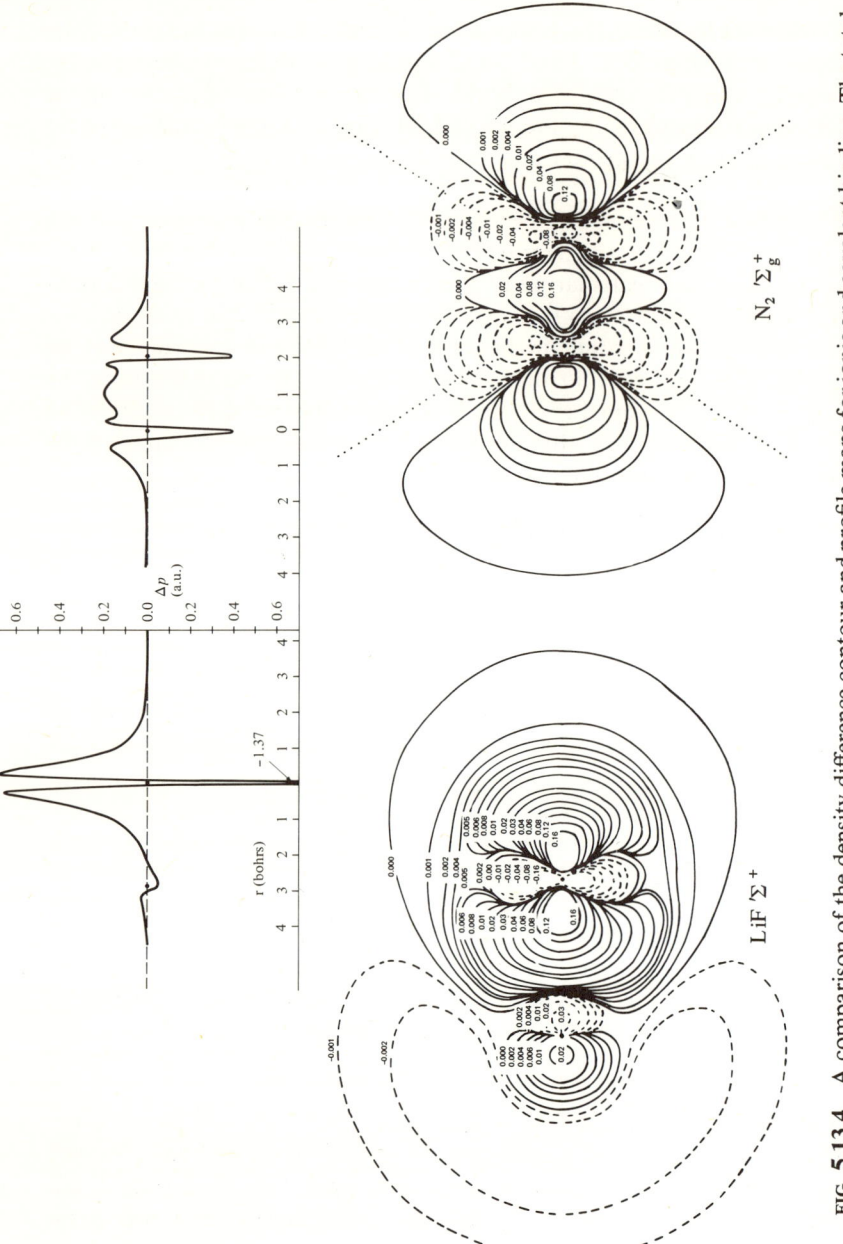

FIG. 5.13.4 A comparison of the density difference contour and profile maps for ionic and covalent binding. The total amount of charge within the zero contour encompassing the fluorine nucleus is 9.81 electrons. A total of 0.85 and 0.51 units of electronic charge migrate to the regions of charge increase in LiF and N_2, respectively. From R. F. W. Bader, W. H. Henneker, and P. E. Cade, *J. Chem. Phys.* **46**, 3341 (1967). Reproduced by permission.

BO, CN, CO$^+$ (13 electrons)

$$\text{Configuration } KK\,(1\sigma)^2(2\sigma)^2(\pi_{x,y})^4(3\sigma) \quad {}^2\Sigma$$

R_e about 1.2 A and D_0 about 185 kcal/mole.
Paramagnetic molecule.

BF, CO, NO$^+$, CN$^-$ (14 electrons), isoelectronic with N$_2$

$$\text{Configuration } KK(1\sigma)^2(2\sigma)^2(\pi_{x,y})^4(3\sigma)^2 \quad {}^1\Sigma$$

R_e about 1.1 A and D_0 about 200 kcal/mole.

The configuration and the large D_0 and small R_e indicate triple bonds exist in these molecules, as in N$_2$. BF with $R_e(\text{BF}) = 1.4$ A and $\mu = 1.04$ D differs most from the homopolar N$_2$, but N$_2$, CO, and BF form a continuous series with increasing polarity and roughly similar covalent bond order.[43,44]

NO (15 electrons)

$$\text{NO } KK\,(1\sigma)^2(2\sigma)^2(1\pi_{x,y})^4(3\sigma)^2(2\pi_{x,y}) \quad {}^2\Pi$$

$R_e(\text{NO}) = 1.2$ A. $D_0(\text{NO}) = 162$ kcal/mole.

The next available MO, $(2\pi_{x,y})$, is the antibonding companion to $(1\pi_{x,y})$, that is, the two roots of the secular equation for π MO. Occupation of $(2\pi_{x,y})$ has reduced the bond energy from 250 kcal/mole in NO$^+$ to 162 kcal/mole in NO.

EXERCISE   Discuss the MO configurations, states, and bonding in (1) BeO, (2) BeF, (3) PO, (4) ClF.

EXERCISE   Give the MO description of the bonding in HX molecules where X = F, Cl, Br, and I.

5.14 LOCALIZED ORBITALS

In Section 5.9 of this chapter it was shown that observables are invariant to certain linear transformations of the molecular orbitals, that is, ρ, the total energy, μ, etc., are the same before and after the MO are transformed. This very useful transformation property can be used to heighten the chemical visibility of the MO by transforming them to *localized orbitals* (LO).

---

[43] W. M. Huo, *J. Chem. Phys.* **43**, 624 (1965); C. Edmiston and K. Ruedenberg, *ibid.* **43**, S97 (1965); E. R. Davidson, *ibid.* **46**, 3320 (1967).

[44] The isoelectronic series, N$_2$, CO, and BF (14 electrons), and C$_2$, BeO, and LiF (12 electrons) are discussed from the point of view of their charge densities and $\Delta\rho$ maps by R. F. W. Bader and A. D. Bandrauk, *J. Chem. Phys.* **49**, 1653 (1968).

TABLE 5.14.1 Localized Orbitals of N_2. Table of Coefficients of the LCAO–LO[a]

LO	$1s_a$	$2s_a$	$2p_{z_a}$	$2p_{x_a}$	$2p_{y_a}$	$1s_b$	$2s_b$	$2p_{z_b}$	$2p_{x_b}$	$2p_{y_b}$
K_a	−0.994	0.114	−0.030	0	0	−0.004	−0.007	0.020	0	0
K_b	0.004	0.007	−0.020	0	0	0.994	−0.114	0.030	0	0
lp_a	0.127	0.892	−0.516	0	0	−0.029	−0.212	−0.191	0	0
lp_b	−0.029	−0.212	−0.191	0	0	0.127	0.892	−0.516	0	0
b_1	0.01	0.209	0.226	−0.239	0.448	0.01	0.209	0.226	−0.239	0.448
b_2	−0.01	−0.209	−0.226	−0.507	0.017	−0.01	−0.209	−0.226	−0.507	0.017
b_3	−0.01	−0.209	−0.226	0.268	0.431	−0.01	−0.209	−0.226	0.268	0.431

[a] After C. Edmiston and K. Ruedenberg, *J. Chem. Phys.* **43**, S97 (1965). Reprinted by permission. The coefficients are given here to three places.

5.14 Localized Orbitals

Unfortunately, while the MO are unique, the transformation to LO is not. Therefore there are many possible LO representations, and in order to choose a chemically significant set of LO an additional criterion must be imposed. A justifiable choice of LO is that set of orbitals which *minimizes* the total *interorbital* electronic interaction due to the $1/r_{ij}$ Coulombic repulsion. This choice of LO keeps the pair of electrons in a localized orbital as far away from other pairs as possible under the circumstance of being confined to the same molecule (see the case of the localized box orbitals of butadiene, Chapter 3).

Table 5.14.1 presents the localized orbitals of N_2. These coefficients were obtained by transforming the LCAO–MO of Table 5.10.1 with the minimization of interorbital repulsion.

MO Configuration of N_2 $(1\sigma_g)^2(1\sigma_u)^2(2\sigma_g)^2(2\sigma_u)^2(\pi_{x,yu})^4(3\sigma_g)^2$

LO Configuration of N_2 $(K_a)^2(K_b)^2(lp_a)^2(lp_b)^2(b_1)^2(b_2)^2(b_3)^2$

The equivalence of some of the LO of N_2 should be evident from the magnitudes of the coefficients in Table 5.14.1. The LO of N_2 come in three equivalent sets. $(K_a)^2(K_b)^2$ is certainly the $(1s_a)^2(1s_b)^2$ helium core (see Section 5.9) because the coefficients of $1s_a$ and $1s_b$ are $|0.994|$. The "*lone-pair*" orbitals of N_2 are lp_a and lp_b; they are nonbonding (and antibonding). The lone pairs of N_2 are mostly $2s_a$ or $2s_b$ but they have enough $2p_z$ to be localized on the back side of the N atom, away from the bond, in the antibinding region. Such localized linear combinations of s and p orbitals have directional properties (Fig. 5.14.1) and are called s–p hybrids.

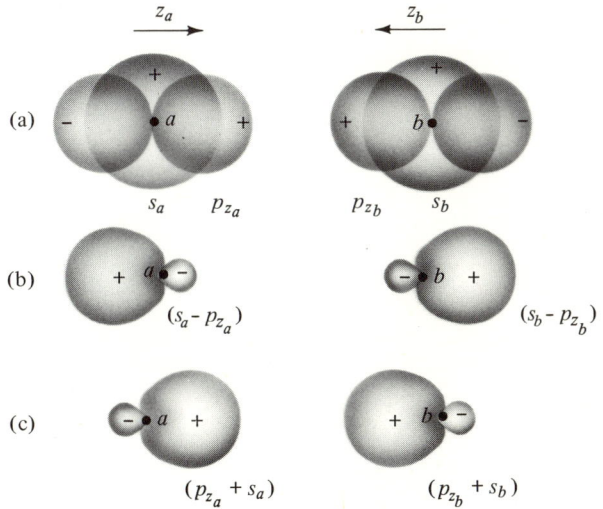

FIG. 5.14.1 *sp* hybridization: (a) the atomic orbitals on each center; (b) and (c) the resulting hybrids. The choice of sign between s and p AO depends on the directional localization desired and the original choice of phase in the AO.

EXERCISE Reverse the z_a and z_b axes (i.e., $z_a \to -z_a$); this has the effect of changing the phase of p_a and p_b. Now find the linear combinations of s and p AO which have the directional properties of Fig. 5.14.1b,c.

The equivalent lone-pair orbitals, lp_a and lp_b, resemble Fig. 5.14.1b. A localized "picture" of the bonding in molecules is often assembled by first forming the hybrids around each atom and then combining the overlapping hybrids into bonding LO, that is, the hybrids in Fig. 5.14.1c can be combined into a bonding orbital. However, it is often *not obvious* what hybridization to choose. The transformation of the unique MO quantitatively recovers this simple bonding picture, with the hybridization justified by the minimization of interorbital repulsion criterion. The bonding in N_2 is due to *three equivalent bonding orbitals*, with localizations schematically as shown in Fig. 5.14.2, which is how they got the name "*banana bonds.*" The three equivalent sets of LO are K_a, K_b, $1s$ cores; lp_a, lp_b, lone pairs; and b_1, b_2, b_3, banana bonds. Although it is not evident from this schematic, *remember that the total ρ of N_2 is the same in LO and MO representations* and *looks like Fig. 5.12.1*. In spite of the schematic illustration, the banana bonds fit together smoothly to give cylindrical charge density about the bond, and charge density between the nuclei on the internuclear axis. In fact, the ρ and $\Delta\rho$ contour maps are precisely the same for both the LO and MO configurations.

The banana bonds are neither π nor σ, that is, they are not symmetry orbitals at all, but go into each other under symmetry operations. For a convenient discussion of spectra and ionization potentials, the MO representation is irreplaceable. However, the LO representation has a satisfying resemblance to the Lewis structures.[45] For example, compare Fig. 5.14.2 to the Lewis structure for N_2, :N: : :N: .

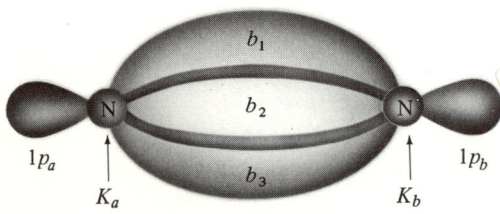

FIG. 5.14.2 Schematic illustration of the disposition of localized orbitals in the N_2 molecule. b_2 is behind the plane; b_1 and b_3 are in front of the plane.

[45] G. N. Lewis, *Valence and the Structure of Atoms and Molecules*, Dover, New York, 1966 (originally published in 1923).

5.14 Localized Orbitals

Lewis and Langmuir attempted to formulate a theory of valence based on first-row atoms taking up octets of electrons for maximum (rare gas) stability. To Lewis, the covalent bond consisted of a shared pair of electrons between two atoms (from the prototype, H_2). Sufficient pairs can be shared to make up an octet about each first-row atom. Lewis conjectured on the geometrical arrangements of the pairs, but quantum mechanics requires that we discuss the relative dispositions of charge densities rather than individual electrons. We found these charge densities can be obtained from LO (transformed MO). In sum, in the diatomic molecule the attraction of the nuclei, the action of the Pauli principle and the interelectronic repulsion, and the insensitivity of ρ to electron correlation, all make the closed-shell diatomic ground state representable to good accuracy by a wavefunction of spin-paired electrons in localized spatial orbitals which "resemble" Lewis structures.

F_2 is another interesting case. The LO representation has six equivalent lone-pair orbitals (*s*- and *p*-hybridized to point to the corners of a triangle; see Fig. 5.14.3), and one bond orbital, as well as the usual K_a, K_b.

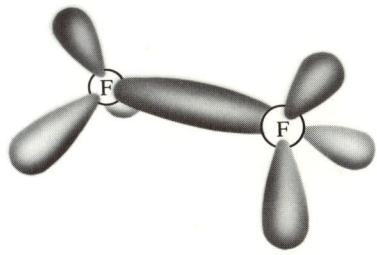

FIG. 5.14.3

EXERCISE   The LO configuration of F_2 is $(K_a)^2(K_b)^2$ (six lp)$^{12}(b)^2$. Compare the configuration and Fig. 5.14.3 to the Lewis structure, :F̈:F̈: .

The LO of homonuclear diatomics do not differ markedly from what one would expect from a knowledge of the MO and Lewis structures (i.e., compare the LO bond orders of F_2 and N_2 to the MO bond orders). LO representations of heteronuclear molecules, on the other hand, can offer interesting insights into the bonding, such as the multiple bonds in BF.[46]

EXERCISE   Give a schematic representation of the LO expected in CO and BF.

[46] C. Edmiston and K. Ruedenberg, *J. Chem. Phys.* **43**, S97 (1965).

5.15 BEYOND THE HARTREE-FOCK LIMIT

The Hartree-Fock wavefunction is the best wavefunction of the independent particle type.[47] Each electron moves nearly independently of the others in a potential due to nuclei and the averaged distribution of the other electrons. The wavefunction is an antisymmetrized product of MO, and this antisymmetry does tend to *correlate* the motion of electrons of like spin, that is, gives zero probability of finding them simultaneously close together. Unfortunately, we also need to correlate the motions of electrons of unlike spin, which the Hartree-Fock wavefunction does not. The trouble is compounded by placing electrons of unlike spin in the same spatial orbital[48] where they "see" each other much too often and "collide" frequently. Much of the correlation energy is due to these α-β pairs confined to the same spatial orbital, that is, $(1s)^2$ or $(1\sigma_g)^2$.

$$E_{\text{corr}} = E - E_{\text{HF}}$$

Since E_{corr} is a small quantity,[49] at first glance it might not be considered of chemical interest. Unfortunately, just the opposite is true. Usually, when two states of a system, or two systems, are involved in an observable, the independent particle model is not sufficiently accurate; examples are transition energies, transition intensities, and bond energies. These properties are sensitive to small corrections to the Hartree-Fock wavefunction due to the true correlated motion of the electrons. Under these circumstances it is often more fruitful to turn to *semiempirical* methods, that is, methods which contain theoretical parameters fixed by experiment. However, for small molecules it is possible to verify the power of the Schrödinger equation by solving it more exactly than just at the Hartree-Fock level; from these efforts will come more precise semiempirical parameters and a better understanding of empirical methods.

One very effective way of correlating the motion of electrons is by placing a factor of r_{12} (the interelectronic distance) directly into the trial wavefunction. Another way of allowing for electron correlation is to abandon the single determinant of MO which typifies the independent particle model of closed-shell molecules. Instead, use a linear combination of many determinants, that is, configuration interaction. For example, the H_2 ground state $^1\Sigma_g$ is improved by using, instead of just the single configuration $(1\sigma_g)^2$,

$$^1\Sigma_g = c_1(1\sigma_g)^2 + c_2(1\sigma_u)^2 \quad \text{Configuration Interaction (CI)} \quad (5.15.1)$$

---

[47] Much of the discussion of Section 4.11 is applicable to molecules.

[48] P. O. Löwdin, "Correlation Problem in Many-Electron Quantum Mechanics," *Advan. Chem. Phys.* **2**, 207 (1963).

[49] Strictly, E is the exact nonrelativistic energy.

5.15 Beyond the Hartree-Fock Limit

TABLE 5.15.1 Typical Calculated (LCAO–MO–SCF, Near Hartree-Fock) and Experimental Values of R_e, $\bar{\nu}_e$, and D_e

Molecule	R_e(bohrs) Calc.	Exptl.	$\bar{\nu}_e{}^*(cm^{-1})$ Calc.	Exptl.	$D_e(eV)$ Calc.	Exptl.	References
H_2	1.385	1.400	4561	4400	3.64	4.75	a
N_2	2.013	2.075	2730	2358	5.27	9.91	b
F_2	2.51	2.725	1257	892	−1.63	1.65	c
LiF	2.941	2.955	1033	964	4.08	5.95	d
CO	2.08	2.132	2431	2170	7.84	11.23	e
Cl_2	3.78	3.76	577	559	0.87	2.51	d
NaF	3.65	3.64	590	536	3.08	4.49	d
BF	2.355	2.391	1496	1402	6.18	8.58	e

[a] G. Das and A. C. Wahl, *J. Chem. Phys.* **44**, 87 (1966).
[b] P. E. Cade, K. D. Sales, and A. C. Wahl, *J. Chem. Phys.* **44**, 1973 (1966).
[c] A. C. Wahl, *J. Chem. Phys.* **41** 2600 (1964).
[d] Unpublished work reported by A. C. Wahl, P. J. Bertoncini, G. Das, and T. L. Gilbert, *Int. J. Quantum Chem.* IS, 123 (1967). Reprinted by permission.
[e] W. Huo, *J. Chem. Phys.* **43**, 624 (1965); **45**, 1554 (1966).
* $\bar{\nu}_e$ is the vibrational frequency in cm^{-1} (see Chapter 6).

Of course, $c_1 \gg c_2$, near R_e. Physically, CI allows the electrons a certain probability of avoiding each other (see Section 4.11).

The fact that the ordinary LCAO–MO wavefunction [based on $(1\sigma_g)^2$] has a space-part (unnormalized)

$$[1s_a(1) + 1s_b(1)][1s_a(2) + 1s_b(2)] = 1s_a(1)1s_a(2) + 1s_b(1)1s_b(2) \\ + 1s_a(1)1s_b(2) + 1s_b(1)1s_a(2) \quad (5.15.2)$$

should make us suspicious of this MO especially at large R, because the first two terms in (5.15.2) are $\Psi(H_a^- - H_b^+)$ and $\Psi(H_a^+ - H_b^-)$, that is, *ionic terms which enter the wavefunction with the same weight as the covalent terms*. This is forgivable in a trial wavefunction at R_e, but at *large R* it implies a certain probability for decomposition into $H^+ + H^-$. This is false; H_2 dissociates into two H atoms. Therefore, at large R the LCAO–MO wavefunction based on $(1\sigma_g)^2$ fails (i.e., the "correlation error" becomes huge,[50] 8 eV). The use of CI $(c_1(1\sigma_g)^2 + c_2(1\sigma_u)^2)$ at large R gives the correct result. Unhappily, the problem of dissociation limits is a

---

[50] Another way of facing the problem is to note that $1\sigma_g$ and $1\sigma_u$ become degenerate at very large R. These near-degeneracy problems have little, if anything, to do with real physical correlation of electronic motion. It has been suggested they be called "nondynamical" correlation effects [O. Sinanoğlu, *Proc. Nat. Acad. Sci.* **47**, 1217 (1961)].

common one in the LCAO–MO theory of molecules and the use of CI a vital remedy at large R.

The CI type of MO wavefunction is related to an early intuitive approximation to the H_2 ground-state wavefunction, known as the *valence bond* wavefunction.

$$\Psi_{VB} = 1s_a(1)1s_b(2) + 1s_a(2)1s_b(1) \qquad (5.15.3)$$

The VB wavefunction of (5.15.3) (which needs to be multiplied by an antisymmetric spin function for completeness) was used by Heitler and London,[51] one year after Schrödinger gave his equation to physics, to show that H_2 had a stable ground state. The Ψ_{VB} gives a better bond energy for H_2 than the MO wavefunction because Ψ_{VB} correlates the electrons to some extent. However, it is inferior to the CI wavefunction (5.15.1). Now if (5.15.3) is improved by the addition of ionic terms, $1s_a(1)1s_a(2)$ and $1s_b(1)1s_b(2)$, it can be shown to become identical to (5.15.1).

EXERCISE   Compare the generalized MO wavefunction (with CI) and the generalized VB wavefunction (with ionic terms) as follows:

$$\Psi_{MO} = c_1[1s_a(1) + 1s_b(1)][1s_a(2) + 1s_b(2)] + c_2[1s_a(1) - 1s_b(1)][1s_a(2) - 1s_b(2)]$$
$$\Psi_{VB} = k_1[1s_a(1)1s_b(2) + 1s_a(2)1s_b(1)] + k_2[1s_a(1)1s_a(2) + 1s_b(1)1s_b(2)]$$

Find the relation between c_1, c_2 and k_1, k_2 such that $\Psi_{MO} = \Psi_{VB}$.

It has proven difficult to generalize the VB type wavefunction to polyatomic molecules. The MO method, with its natural relation to the independent particle model, symmetry orbitals, electronic spectroscopy, and semiempirical methods, has come to be the most used and useful approach to polyatomic molecules.

Recently it has been shown possible to take another conceptual step beyond the Hartree-Fock wavefunction by including just correlation among *pairs* of electrons.[52] As for interactions involving more than two electrons, if three electrons come near each other at least two must have the same spin, but as these are correlated by the Pauli principle through antisymmetrization, the net result is to make three-electron and many-electron effects of doubtful importance.[53] The electron pairs can be correlated by r_{12} or by CI. The final result is the reduction of a very complex many-electron problem to a sum of two-electron problems each of which bears some resemblance to the H_2 correlation problem. The pair functions (or geminals) are expected to give the major portion of E_{corr}. In Section 4.11 it was claimed that the correlation energy is largely additive in electron pairs of unlike spin, with a major contribution arising from pairs occupy-

---

[51] W. Heitler and F. London, *Z. Phys.* **44**, 455 (1927).

[52] Oktay Sinanoğlu, *J. Chem. Phys.* **36**, 706, 3198 (1962); *Advan. Chem. Phys.* **6**, 315 (1964).

[53] Furthermore the "fluctuation potential" is of short range (see Section 4.11).

5.16 VALENCE BOND THEORY

ing the same spatial orbitals; at the moment this is a working hypothesis for molecules, and it is especially hypothetical for the delocalized electrons of conjugated molecules. The situation resembles the MO theory in the period 1938–1948; there is great hope for calculation and semiempirical generalization of correlation effects, but a paucity of results, and a lack of agreement on a specific method of attack.

Today the valence bond theory (VB) is not commonly used in actual calculations. It is difficult to apply to polyatomic molecules and does not easily explain molecular spectra. However, if the VB theory is not a calculational tool, it is historically the first method applied to the chemical bond[54]; furthermore, it is often used by chemists in a qualitative, or representational, sense in the absence of quantitative methods.

Like the LCAO–MO theory, the VB theory uses atomic orbitals, but in a different way. In the last section the VB wavefunction for the H_2 ground state was given; this wavefunction may be rationalized as follows. The ground state of the H atom is $1s$, specifically, for atom a and electron 1, $1s_a(1)$. If two hydrogen atoms are brought together without perturbing each other, the Hamiltonian is $[\hat{h}_a(1) + \hat{h}_b(2)]$, where $\hat{h}_a(1) = -\frac{1}{2}\nabla^2 - 1/r_{1a}$, and the eigenfunction is $1s_a(1)1s_b(2)$. It seems worthwhile to take $1s_a(1)1s_b(2)$ as a trial wavefunction at all R, except that it does not allow for the indistinguishability of the electrons. Now

$$(2 \pm 2S^2)^{-1/2}[1s_a(1)1s_b(2) \pm 1s_a(2)1s_b(1)]$$

treats the electrons indistinguishably. The factor $(2 \pm 2S^2)^{-1/2}$ is required for normalization, $S = \langle 1s_a, 1s_b \rangle$.

EXERCISE Assign the above spatial functions symmetries, Σ_u or Σ_g.

Finally, the wavefunction must be *antisymmetric* in the space-spin coordinates (Pauli principle). This is achieved by combining the symmetric space function $1s_a(1)1s_b(2) + 1s_a(2)1s_b(1)$ with an antisymmetric spin function, and by combining the antisymmetric space function $1s_a(1)1s_b(2) - 1s_a(2)1s_b(1)$ with a symmetric spin function. In Section 4.7 it was shown that the two-electron spin functions which are eigenfunctions to $\hat{S}^2$ are

$$\alpha(1)\beta(2) - \alpha(2)\beta(1) \quad M_S = 0 \quad S = 0, \text{ Singlet}$$

$$\left.\begin{array}{l} \alpha(1)\alpha(2) \quad M_S = 1 \\ \beta(1)\beta(2) \quad M_S = -1 \\ \alpha(1)\beta(2) + \alpha(2)\beta(1) \quad M_S = 0 \end{array}\right\} \quad S = 1, \text{ Triplet}$$

---
[54] Heitler and London, footnote 51.

Of these possible spin functions, only the first, the singlet, is antisymmetric. We now have

$$^1\Psi_{cov}(H_2) = (2 + 2S^2)^{-1/2}[1s_a(1)1s_b(2) + 1s_a(2)1s_b(1)]$$
$$\times (1/\sqrt{2})[\alpha(1)\beta(2) - \alpha(2)\beta(1)] \quad (5.16.1)$$

Equation (5.16.1) is the Heitler-London wavefunction for the ground state $^1\Sigma_g$ of H_2.

EXERCISE Write the wavefunction for the $^3\Sigma_u$ state of H_2. Set up the energy expressions for $^1\Sigma_g$ and $^3\Sigma_u$ in terms of the actual integrals involved. Show that

$$E(H_2) = 2E(H) + \frac{J \pm K}{1 \pm S^2}$$

where $J = \langle 1s_a(1)1s_b(2), \hat{H}'1s_a(1)1s_b(2)\rangle$ a $K = \langle 1s_a(1)1s_b(2), \hat{H}'1s_a(2)1s_b(1)\rangle$ $[\hat{H}' = \hat{H} - \hat{h}_a(1) - \hat{h}_b(2)]$. J and K are, *unfortunately*, called the VB Coulomb and exchange integrals, respectively.

EXERCISE How do the VB integrals J and K differ from the Coulomb and exchange integrals previously defined?

Equation (5.16.1) is a form of the wavefunction for the electron-pair bond which is generally applicable to diatomic molecules. Before proceeding with other molecules, remember (last section) that the covalent VB wavefunction of H_2 is improved by the addition of ionic terms (often called "ionic resonance structures").

$$^1\Psi_{ion}(H^+-H^-) = (2 + 2S^2)^{-1/2}[1s_a(1)1s_a(2) + 1s_b(1)1s_b(2)]$$
$$\times 1/\sqrt{2}\,[\alpha(1)\beta(2) - \alpha(2)\beta(1)] \quad (5.16.2)$$

Finally,

$$^1\Psi(H_2) = c_1\,^1\Psi_{cov}(H_2) + c_2\,^1\Psi_{ion}(H^+-H^-) \quad (5.16.3)$$

c_1 and c_2 are coefficients to be determined through the variational principle.

Equation (5.16.3) is of a form applicable to all the electron-pair bonds of homonuclear diatomic molecules. In general, the "atomic orbitals" to be paired are not simple AO, but *hybrids* (see Section 5.14) $\psi_a = N(2s_a + k2p_{z_a})$, $\psi_b = N(2s_b + k2p_{z_b})$. The hybrids ψ_a and ψ_b should overlap strongly to obtain a strong electron-pair bond.

$$^1\Psi_{cov}(A_2) = (2 + 2S^2)^{-1/2}[\psi_a(1)\psi_b(2) + \psi_a(2)\psi_b(1)]$$
$$\times 1/\sqrt{2}\,[\alpha(1)\beta(2) - \alpha(2)\beta(1)] \quad (5.16.4)$$

k is a parameter to be determined variationally.

5.16 Valence Bond Theory

If the molecule is multiply bonded, for example, N_2, there is a $^1\Psi_{cov}$ for each electron-pair bond; the bond wavefunctions are then multiplied together and antisymmetrized. Needless to say, the VB wavefunction looks complicated in comparison with the MO single-configuration wavefunction for the same multiply bonded molecule, but the VB wavefunction may be the more accurate. As for the "electron-pair bonds" of VB theory, transformation of the MO to localized orbitals (Section 5.14, especially Fig. 5.14.2 for N_2) offers an uncomplicated wavefunction with doubly occupied bonding orbitals, localized atomic cores, and nonbonded pairs.

It is not too difficult to show that MO and VB theories give identical results, even for approximate treatments of homonuclear systems. Consider the system He_2. The VB wavefunction must be constructed from $1s_a\alpha(1)$, $1s_a\beta(2)$, $1s_b\alpha(3)$, and $1s_b\beta(4)$. The antisymmetric VB wavefunction is therefore

$$\Psi_{VB}(He_2) = \hat{A} 1s_a\alpha(1) 1s_a\beta(2) 1s_b\alpha(3) 1s_b\beta(4) \tag{5.16.5}$$

but in Section 5.9 it was shown that this wavefunction is obtained from the MO wavefunction

$$\Psi_{MO}(He_2) = \hat{A} 1\sigma_g\alpha(1) 1\sigma_g\beta(2) 1\sigma_u\alpha(3) 1\sigma_u\beta(4) \tag{5.16.6}$$

by taking linear combinations of the columns of the determinant in (5.16.6). Therefore (5.16.5) and (5.16.6) are equivalent wavefunctions. It is generally true that filled atomic cores have equivalent approximate wavefunctions in MO and VB theories. In higher approximation it was shown in Section 5.15 (Exercise) that the VB and MO wavefunctions of H_2 become equivalent when the doubly excited configuration $(1\sigma_u)^2$ is added to the single-configuration MO wavefunction (CI), and ionic structures are added to the covalent VB function.

EXERCISE ψ_a and ψ_b are identical AO on nuclei a and b; $\phi_1 = N_1(\psi_a + k\psi_b)$ and $\phi_2 = N_2(\psi_b + k\psi_a)$, where N_1 and N_2 are normalization factors. $\Psi = N[\phi_1(1)\phi_2(2) + \phi_1(2)\phi(1)][\alpha(1)\beta(2) - \alpha(2)\beta(1)]$. Show that $\Psi = \Psi_{MO}$ (if $k = 1$) and $\Psi = \Psi_{VB}$ (if $k = 0$). Show that for intermediate values of k we have a wavefunction of the form MO with CI, or VB with ionic structures.[55]

The heteronuclear diatomic molecule offers an especially fertile field for VB speculation because the polarity of the bond finds expression in the addition of visible "ionic structures" to the covalent wavefunction, for example,

$$^1\Psi_{ion}(A^+B^-) = \psi_b(1)\psi_b(2)(1/\sqrt{2})[\alpha(1)\beta(2) - \alpha(2)\beta(1)]$$
$$^1\Psi_{ion}(A^-B^+) = \psi_a(1)\psi_a(2)(1/\sqrt{2})[\alpha(1)\beta(2) - \alpha(2)\beta(1)] \tag{5.16.7}$$
$$^1\Psi(AB) = c_1\,^1\Psi_{cov}(AB) + c_2\,^1\Psi_{ion}(A^+B^-) + c_3\,^1\Psi_{ion}(A^-B)^+$$

[55] I. Fischer, *Phil. Mag.* **40**, 386 (1949).

c_1, c_2, and c_3 are variational parameters, but one may suppose that the relative magnitude of these coefficients is related to "electronegativity" differences (see Appendix F) between the orbitals of A and B, or related to observables, such as the dipole moment. Generally, only one ionic structure is of importance in the VB wavefunction of a heteronuclear molecule, for example, for the HF molecule

$$^1\Psi(HF) = c_1\,^1\Psi_{cov}(HF) + c_2\,^1\Psi_{ion}(H^+F^-)$$

Although the VB ionic structures are explicit, visible indications of bond polarity, the VB structures have no reality of their own. Ionic structures offer another means of helping to represent the asymmetry of the charge distribution in the molecule. The MO wavefunction represents this asymmetry through the coefficients in the LCAO–MO.

REFERENCES

GENERAL REFERENCES

Herzberg, G., *Spectra of Diatomic Molecules*, Van Nostrand, New York, 1950.
Plllar, F. *Elementary Quantum Chemistry*, McGraw-Hill, New York, 1968.

SOURCES OF DATA[56]

Cotrell, T. L., *The Strengths of Chemical Bonds*, Butterworths, London, 1958.
Herzberg, G., *Spectra of Diatomic Molecules*, Van Nostrand, New York, 1950, Table 39.
Interatomic Distances, Special Publication No. 11 of the Chemical Society, London, 1958.
McClellan, A. L., *Tables of Experimental Dipole Moments*, Freeman, San Francisco, 1963.
Ransil, B. J., *Rev. Mod. Phys.* **32**, 239 (1960).

PROBLEMS

1. Sketch the MO charge density contour maps expected for all the occupied MO of O_2.
2. Sketch and compare the total charge density contour maps expected for Na_2, S_2, and Cl_2.
3. Describe the electronic structure of LiH^{2+}.
4. Assuming the MO of Be_2 are $1\sigma_g \approx 1s_a + 1s_b$, $1\sigma_u \approx 1s_a - 1s_b$, $2\sigma_g \approx 2s_a + 2s_b$, and $2\sigma_u \approx 2s_a - 2s_b$, give the linearly transformed orbitals and the configuration which help to explain the nonexistence of this molecule.
5. Discuss the bond properties of Cl_2 and Cl_2^+ using MO.
6. Discuss the fact that the bond length of NO is 1.15 A, which is longer than either CO or NO^+.

[56] Unless Author reference is given.

5.16 Valance Bond Theory

7. Discuss the bond properties of N_2, P_2, As_2, Sb_2, and Bi_2 in terms of their electronic structures.
8. Describe the localized orbital representations of all the molecules discussed in Sections 5.10 and 5.13.
9. Two atoms of different nuclear charge are brought together and form a stable bonded molecule. Given that there is no net electronic charge redistribution upon bond formation what is the magnitude of the dipole moment of this molecule?
10. Use the concept of electronegativity (Appendix) to explain the relative dipole moments of HCl (1.03 D), HBr (0.79 D), and HI (0.3 D).

6

MOLECULAR SPECTRA AND MOLECULAR SYMMETRY

The Great Architect seems to be
a mathematician.

SIR JAMES JEANS

MOLECULAR SPECTROSCOPY is the chief experimental means of obtaining information about molecular structure. Molecular symmetry is an essential element in the interpretation of molecular spectra because it makes possible the derivation of selection rules (Section 6.8) and the classification of molecular states *without* the calculation of energy eigenvalues.

Molecular Spectroscopy

The total energy of a molecule consists of the combined kinetic and potential energy of all its constituent particles. The arbitrary zero of energy is usually taken to be the electrons and nuclei separated to infinity and motionless. Relative to this zero the molecule possesses *translational*, *rotational*, *vibrational*, and *electronic* energy. In the present context we are interested in motion relative to the center of mass of the molecule, so we define the total *internal* energy of the molecule as

$$E_{\text{total}} = E_{\text{rot}} + E_{\text{vib}} + E_{\text{el}}$$

Strictly, this separation is inexact; vibration and rotation are not precisely separable (see further), nor are the electronic states precisely separable from the nuclear motion (see Appendix D). However, a good grasp of the interpretation of molecular spectra is first obtained by studying these energies separately, that is, by solving the appropriate Schrödinger equation for each motion. The coupling between vibration and rotation, for example, can be added at a later stage.

This chapter is not a comprehensive review of molecular spectroscopy, which would require several volumes.[1] Rather it is a selective summation and derivation of some of the important spectroscopic equations which lead to *structural* information about molecules. The chemist's natural desire to

---

[1] G. Herzberg, *Molecular Spectra and Molecular Structure*, vol. I, Spectra of Diatomic Molecules, 1950; vol. II, Infrared and Raman Spectra of Polyatomic Molecules, 1945; vol. III, Electronic Spectra and Electronic Structure of Polyatomic Molecules, 1966, D. van Nostrand, New York.

have structural knowledge of his molecules is the principal reason for the study of molecular spectra. These spectra can be classified according to the amount of energy $h\nu$ or the wavelength $c/\nu = \lambda$ involved in a typical transition, as well as the technique used in the observation.

To begin at the high-energy, or short-wavelength, end of the scale (see Table 2.2.1), the highest-energy transitions of chemical interest are between states of the nucleus. Extremely short wavelength radiation (γ *rays*) is emitted by the nucleus undergoing a transition, for example, ^{119}Sn (2.4×10^4 eV) or ^{57}Fe (1.4×10^4 eV). If the emitting atom is present in a crystalline lattice the nuclear γ radiation is extremely sharply peaked at a certain wavelength. Shifts in the wavelength of this sharp line due to the chemical environment, that is, ligands, constitutes a source of chemical information.[2] This is called *Mössbauer spectroscopy* because the sharpness of γ emission or absorption in a crystalline lattice is the *Mössbauer effect* (recoilless resonance emission and absorption).

An atom absorbing sufficient energy to eject an electron from an inner shell will emit a series of lines in the *x-ray region* (10^2–10^3 eV) of the spectrum. The lines result from emission by electrons cascading down from higher shells to fill the vacancy. For example, if a $1s$ (K shell) electron is removed, the K series of x-ray lines of the atom is emitted, K_α, K_β, etc., where K_α corresponds to a change in principal quantum number $n = 2 \rightarrow n = 1$, K_β corresponds to $n = 3 \rightarrow n = 1$, etc. Small differences in energy due to the quantum number l will be exhibited as a fine structure in the lines. Shifts in the lines serve to reveal information about the chemical environment of the atom.

The more important and highly developed molecular spectroscopic techniques use much lower energy than x-ray or γ-ray spectra. In the range of 1 to 10^2 eV there are molecular electronic transitions of the *visible*, *ultraviolet*, and *vacuum ultraviolet* regions. Instrumentation in these regions consists of optical spectrometers with prisms or gratings and photoelectric radiation detectors, or spectrographs with photographic plates as detectors. In the ultravoilet, *quartz* optics replace the ultraviolet-absorbing *glass* optics which are suitable for the visible. In the vacuum ultraviolet, where radiation-absorbing air must be excluded, optics are either dispensed with, or special crystals, for example, LiF, may be used as windows.

Infrared spectrometers generally operate in the wavelength range of 2 to 50 μ (1–10^{-2} eV) using glowing hot wires as a source of infrared radiation, NaCl optics, and a thermopile (heat-sensitive measuring device) as a detector. The molecular transitions involved are rotational, vibrational,

---

[2] R. H. Herber, *Ann. Rev. Phys. Chem.* **17**, 261 (1966); G. K. Wertheim, *Science* **144**, 253 (1964).

6.1 Vibration–Rotation Spectra

and simultaneous vibration–rotation transitions. Very low frequency modes of rotation occur at microwavelengths (10^{-1}–1 cm) and require nonoptical experimental techniques.

Magnetic resonance spectroscopy also requires nonoptical techniques. Magnetic resonance spectra result from electron or nuclear spin changes in an applied magnetic field. This is a very versatile and important modern spectroscopic technique which we deal with at length in Sections 6.9 to 6.11. The energy changes involved are exceedingly small and the wavelengths are correspondingly long, 1 to 10^2 cm and longer.

To proceed, we will first examine the rotational and vibrational motion of the diatomic molecule with the simple models of the rigid rotator and the harmonic oscillator (Section 6.1). The vibrational spectra of polyatomic molecules are treated in Section 6.7 after a discussion of the techniques of group theory (Sections 6.3 to 6.6). Electronic spectroscopy is very briefly treated in Section 6.2 and the basis for spectroscopic selection rules (electronic, infrared, etc.) is given in Section 6.8.

6.1 VIBRATION–ROTATION SPECTRA

The general features of diatomic spectra which we must explain are the sharp line spectra of heteronuclear diatomic molecules in the infrared and microwave regions (8000 A or 0.8 μ and longer wavelengths) and the band spectra of both homonuclear and heteronuclear molecules in the visible or ultraviolet regions (7000 A and shorter wavelengths). These spectra are remarkably different from atomic line spectra, which must be due to the fact that the molecule rotates as a whole about its center of mass, and the nuclei vibrate along the internuclear axis. In the present section we examine the Schrödinger equations for these motions in order to obtain the energy levels of E_{rot} and E_{vib} and the dependence on the *quantum numbers* for rotation and vibration.

RIGID ROTATOR

To understand the origin of the rotational energy levels in the diatomic molecule, it may be visualized as a *dumbbell* composed of two nuclei of mass M_1 and M_2, respectively, connected by a rigid weightless rod of length R_e. The classical mechanical expression for the energy of a rotating dumbbell is

$$E_{rot} = \tfrac{1}{2}I\omega_R^2 \qquad (6.1.1)$$

where $I = \mu R_e^2$ is the moment of inertia, $\mu = M_1 M_2/(M_1 + M_2)$ is the reduced mass, and ω_R is the angular velocity.

EXERCISE Derive (6.1.1) from the following: A particle moving on a circle of radius R, with velocity v, makes $v/2\pi R$ revolutions per second; ω_R is the angular velocity in radians per second. Motion is about the center of mass (CM) located at $R_1 = M_2 R_e/(M_1 + M_2)$ and $R_2 = M_1 R_e/(M_1 + M_2)$. See Fig. 6.1.1.

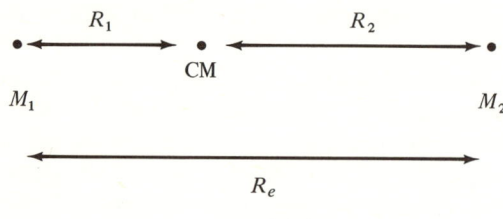

FIG. **6.1.1**

Equation (6.1.1) resembles the kinetic energy of a body of mass I, so we examine the angular part of the one-particle Schrödinger equation [eq. (4.1.4)] with $mr^2 = \mu R_e^2 = I$ and $V = 0$ (zero potential energy). The wavefunctions we seek are $\psi_{rot} = \psi(\theta, \phi) = \Theta(\theta)\Phi(\phi)$ with eigenvalues E_{rot}.

$$\frac{\sin\theta}{\Theta}\frac{\partial}{\partial\theta}\left(\sin\theta\frac{\partial\Theta}{\partial\theta}\right) + \frac{1}{\Phi}\frac{\partial^2\Phi}{\partial\phi^2} + \frac{2I\sin^2\theta}{\hbar^2}E_{rot} = 0 \qquad (6.1.2)$$

By the same reasoning that led to equations (4.1.5)–(4.1.7) we arrive at $\Phi(\phi) = \exp(im_J\phi)$, $m_J = 0, \pm 1, \pm 2, \ldots$, and (6.1.2) becomes

$$\frac{1}{\sin\theta}\frac{1}{\Theta}\frac{\partial}{\partial\theta}\left(\sin\theta\frac{\partial\Theta}{\partial\theta}\right) - \frac{m_J^2}{\sin^2\theta} + \frac{2I}{\hbar^2}E_{rot} = 0 \qquad (6.1.3)$$

Equation (6.1.3) is thrown into the form of the associated Legendre differential equation by the substitution $2IE_{rot}/\hbar^2 = J(J+1)$. The associated Legendre equation has acceptable solutions $\Theta_{Jm_J}(\theta)$ for $J = 0, 1, 2, 3, \ldots$ and $|m_J| \leq J$. Thus the rotational energy is quantized with *rotational quantum number J*. The functions $\Theta_{Jm_J}(\theta)$ are the normalized associated Legendre polynomials given in Chapter 4. The wavefunctions ψ_{rot} are the spherical harmonics $Y_{Jm_J}(\theta, \phi)$.

There are $2J + 1$ values of m_J; since E_{rot} depends on J, $2J + 1$ is the degeneracy of Θ_{Jm_J}.

$$E_{rot} = \frac{\hbar^2}{2I} J(J+1) \qquad J = 0, 1, 2, 3, \ldots \qquad (6.1.4)$$

6.1 Vibration–Rotation Spectra

$B = h/8\pi^2 cI$ is called the *rotational constant* of the molecule. In terms of B we can express E_{rot} in cm^{-1} as

$$E_{rot}/hc = BJ(J+1) \text{ cm}^{-1} \tag{6.1.5}$$

The pure rotational spectrum consists of lines having wavenumbers

$$\bar{\nu}_{rot} = B[J'(J'+1) - J''(J''+1)] \tag{6.1.6}$$

Single and double primes indicate upper and lower state, respectively. The selection rule for the rotational transition of the rigid rotator is

$$\Delta J = \pm 1 \quad \text{SELECTION RULE}$$

EXERCISE Use equation (6.1.6) and Fig. 6.1.2 to show that the selection rule $\Delta J = +1$ for absorption of infrared radiation leads to $\bar{\nu}_{rot}$ (cm^{-1}) $= 2BJ'$, and to equally spaced lines in the pure rotational spectrum with a spacing $2B$.

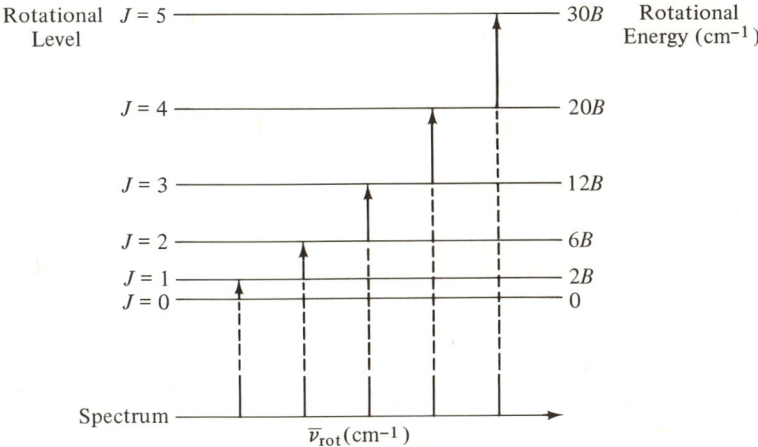

FIG. 6.1.2 Schematic representation of the pure rotational spectrum of a diatomic molecule in absorption. $\Delta J = +1$.

The rotational spectrum is an important source of structural information. The value of B obtained from the pure rotational spectrum is used to find the internuclear distance, as in the following exercise.

EXERCISE Given 20.7 cm^{-1} as the observed spacing in the pure rotational spectrum of H^{35}Cl, find R_e for this molecule. Also calculate the period of a single rotation, in seconds, in the $J = 1$ and $J = 2$ rotational states of HCl.

THE ROTATION OF POLYATOMIC MOLECULES

The rotational spectra of most polyatomic molecules are difficult to interpret. We can give only an incomplete outline of the simpler aspects of the problem. The polyatomic molecule has three moments of inertia I_1, I_2, and I_3, which may or may not be equal. It is convenient to classify the molecule according to the relative magnitudes of its moments of inertia. If $I_1 = I_2 = I_3$ the molecule is said to be a *spherical top*; example, CH_4. If $I_1 = I_2 \neq I_3$ the molecule is said to be a *symmetric top*; example, benzene. If the molecule has no two moments of inertia the same, $I_1 \neq I_2 \neq I_3$, it is said to be an *asymmetric top*; example, H_2O. The use of the word *top* should indicate that we are still drawing the analogy with the motion of rigid bodies, that is, the rigid rotator approximation. Correction terms (see further) may be added into the energy formula to allow for the nonrigidity of real molecules, but this is an exceedingly small correction.[3]

In the case of the symmetric top we can define two rotational constants $B_3 = h/(8\pi^2 c I_3)$ and $B_1 = B_2 = h/(8\pi^2 c I_1)$. Using the rigid rotator approximation the energy levels are

$$E_{rot}/hc = B_1 J(J+1) + (B_3 - B_1)K^2$$

J is the total angular momentum quantum number and K is the quantum number of the component of angular momentum in the z direction. $J = 0, 1, 2, 3, \ldots$ and $-J \leqslant K \leqslant J$.

EXERCISE   All rotational states of the symmetric top with $K \neq 0$ are doubly degenerate. Why?

Taking the molecular dipole moment to lie along the z direction, the selection rules for rotational transitions are $\Delta J = 0, \pm 1$ and $\Delta K = 0$. A symmetric top rotational spectrum thus contains the lines

$$\bar{v}_{rot} = B_1[J'(J'+1) - J''(J''+1)]$$

In the pure rotational spectrum $\Delta J = 0$ means no transition, so the spectrum is the same as in equation (6.1.6) for the diatomic molecule, that is, a set of equidistant lines having the separation $2B_1$. Thus no information is obtained about $B_3(I_3)$. Little structural knowledge results from polyatomic rotational spectra at this level of analysis. Furthermore, in the spherical top $B_1 = B_2 = B_3$ and E_{rot}/hc reduces immediately to the diatomic espression.[4] Finally, the asymmetric top wave equation can be

---

[3] G. Herzberg, footnote 1.
[4] The spherical top molecule, like the homonuclear diatomic molecule, has zero dipole moment; consequently both classes of molecule have *no pure rotational spectrum* (see further).

HARMONIC OSCILLATOR

solved only with difficulty and approximately, and the complex spectra are difficult to interpret. Unfortunately the majority of polyatomic molecules are asymmetric tops.

The nuclei of the diatomic molecule are not held rigidly apart at a distance R_e, but move along R in the potential supplied by their mutual repulsion $Z_a Z_b/R$, plus the changing electronic energy which is R-dependent. This total "nuclear potential energy" we call $E(R)$, $E(R) = Z_a Z_b/R + E_{\text{el}}(R)$. A plot of $E(R)$ versus R reveals a curve of the same general shape for all stable electronic states (solid line in Fig. 6.1.3). For many purposes it is sufficient to examine the $E(R)$ curve near R_e, because at ordinary temperatures the nuclei oscillate with internuclear distance reasonably close to R_e.

Near R_e $E(R)$ is closely fit by a parabola centered at R_e, $E(R) \approx \frac{1}{2}k(R - R_e)^2$. Replacing the true $E(R)$ with this parabola constitutes the *harmonic oscillator approximation*, that is, the displacement from equilibrium, $(R - R_e) = x$, results in a force, Force $= -dV/dR = -d(\frac{1}{2}kx^2)/dR = -kx$, which is a Hooke's law force, where k is the *force constant* for the oscillation.

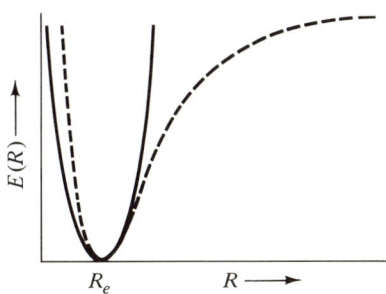

FIG. 6.1.3 A nuclear potential energy curve. The solid line is the parabola $\frac{1}{2}k(R - R_e)^2$.

EXERCISE Two masses are connected by a Hooke's law spring. What is the relation of k to the "stiffness" of the spring? Discuss the relation of k to the shape of the potential curve near R_e.

To obtain the quantum mechanical vibrational energy and wavefunction requires the solution of the equivalent one-body, one-dimensional Schrödinger equation for harmonic oscillation along R with displacement x from equilibrium, that is, quantum mechanical motion with potential energy operator $\frac{1}{2}kx^2$ and kinetic energy operator $(-\hbar^2/2\mu)d^2/dx^2$.

$$\left(-\frac{\hbar^2}{2\mu}\frac{d^2}{dx^2} + \tfrac{1}{2}kx^2\right)\psi(x) = E_{\text{vib}}\psi(x) \tag{6.1.7}$$

In order to simplify the many constants in this equation we make the change of variable $\xi = \sqrt{\beta}x$, (6.1.7) becomes

$$\frac{d^2\psi(\xi)}{d\xi^2} + \left(\frac{\alpha}{\beta} - \xi^2\right)\psi(\xi) = 0 \tag{6.1.8}$$

where $\beta = \sqrt{\mu k}/\hbar$ and $\alpha = E_{\text{vib}}(2\mu/\hbar^2)$. At large ξ (large displacements) (6.1.8) is of the form $d^2\psi/d\xi^2 = \xi^2\psi$ with *approximate* solutions $A\exp(\pm\xi^2/2)$. However, the solution $A\exp(+\xi^2/2)$ is rejected for poor behavior[5] at large ξ. This suggests that

$$\psi = u(\xi)\exp(-\xi^2/2) \tag{6.1.9}$$

is a desirable trial function; substituting (6.1.9) into (6.1.8) we obtain the differential equation which $u(\xi)$ must satisfy.

$$\frac{d^2 u}{d\xi^2} - 2\xi\frac{du}{d\xi} + \left(\frac{\alpha}{\beta} - 1\right)u = 0 \tag{6.1.10}$$

EXERCISE Derive equation (6.1.10).

Equation (6.1.10) is a known differential equation which carries the name Hermite. Well-behaved solutions of the Hermite differential equation exist for $[(\alpha/\beta) - 1] = 2v$, where $2v$ is any positive integer including zero. This restriction on v yields, from $\alpha/\beta = 2v + 1$,

$$E_{\text{vib}} = \hbar\sqrt{\frac{k}{\mu}}(v + \tfrac{1}{2}) \qquad v = 0, 1, 2, 3, \ldots \tag{6.1.11}$$

v is the *vibrational quantum number*. The $u(\xi)$ are the Hermite polynomials,[6] $H_v(\xi)$; the vibrational wavefunctions of the harmonic oscillator normalized to unity are

$$\psi_v(\xi) = \left(\frac{\sqrt{\beta/\pi}}{2^v v!}\right)^{1/2} H_v(\xi)\exp\left(-\frac{\xi^2}{2}\right) \tag{6.1.12}$$

---

[5] $\exp(+\xi^2/2)$ increases exponentially with increasing ξ^2, but we are looking for normalizable solutions on the interval $-\infty \leqslant \xi \leqslant \infty$.

[6] The properties of the Hermite polynomials and the Hermite differential equation are discussed by H. Margenau and G. M. Murphy, *The Mathematics of Physics and Chemistry*, D. van Nostrand, New York, 1956.

6.1 Vibration–Rotation Spectra

TABLE 6.1.1 The Hermite Polynomials

v	$H_v(\xi)$
0	1
1	2ξ
2	$4\xi^2 - 2$
3	$8\xi^3 - 12\xi$
4	$16\xi^4 - 48\xi^2 + 12$

A few of the Hermite polynomials are listed in Table 6.1.1; the harmonic oscillator wavefunctions and probability densities are plotted in Fig. 6.1.4.

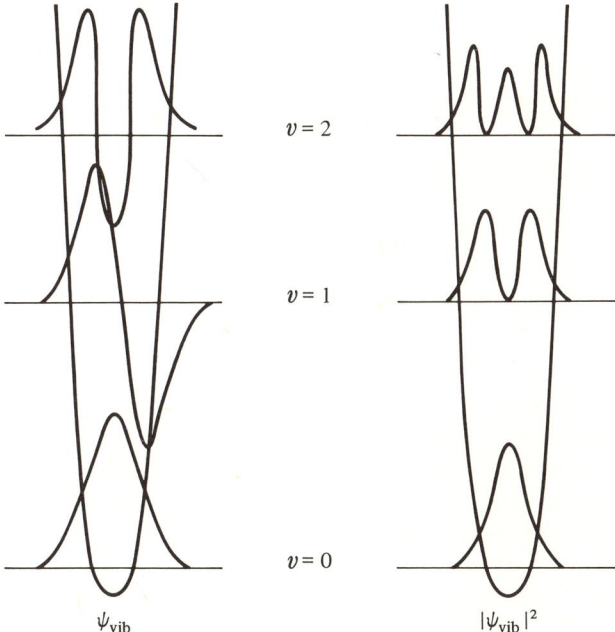

FIG. 6.1.4 Energy levels, wavefunctions, and probability density functions of the quantum mechanical harmonic oscillator superimposed on the harmonic oscillator potential.

EXERCISE Verify that equations (6.1.10) and (6.1.8) are of the form of Sturm-Liouville differential equations (Section 2.5). What are the properties expected of the harmonic oscillator eigenfunctions?

EXERCISE Show that the Hermite polynomials of Table 6.1.1 are solutions to equation (6.1.10).

EXERCISE Show that $\psi_v(\xi)$ of (6.1.12) satisfies (6.1.8), for example, for $v = 1$.

The *vibrational frequency* in $\sec^{-1}$ is

$$v = \frac{1}{2\pi}\sqrt{\frac{k}{\mu}} \qquad (6.1.13)$$

In terms of the vibrational frequency E_{vib} is

$$E_{\text{vib}} = hv(v + \tfrac{1}{2}) \qquad v = 0, 1, 2, 3, \ldots \qquad (6.1.14)$$

It is interesting to note that the lowest-energy vibrational state of the molecule ($v = 0$) does *not* have $E_{\text{vib}} = 0$, rather $E_{\text{vib}}(v = 0) = \tfrac{1}{2}hv$; this is the characteristic *zero-point energy* of the molecule. By comparison the rigid rotator has $E_{\text{rot}}(J = 0) = 0$. The zero-point energy is the difference between the two dissociation energies D_0 and D_e (Fig. 6.1.5).

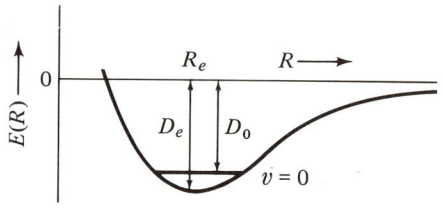

FIG. 6.1.5

This small residual energy of oscillation, $\tfrac{1}{2}hv$, is always present in the molecule and is a purely quantum mechanical phenomenon. It may be regarded as a consequence of the Heisenberg uncertainty principle, for if $E_{\text{vib}} = 0$, then $x = 0$ and $p_x = 0$, that is, both the position and momentum of the oscillator are precisely known. Again, in contrast with the rigid rotator which, when not rotating, still has unknown orientation.

The harmonic oscillator has the selection rule $\Delta v = \pm 1$ for optical transitions between vibrational energy levels. In wavenumbers the transition energy is ($\Delta v = +1$) for absorption

6.1 Vibration–Rotation Spectra

$$\bar{v}(\text{cm}^{-1}) = \frac{E_{\text{vib}}(v+1) - E_{\text{vib}}(v)}{hc} = \frac{v}{c}$$

where $\bar{v}$ is the *fundamental vibrational transition* (frequency) in cm^{-1}. Apparently, the selection rule leads to only the fundamental transition in the spectrum, $\bar{v}$, independent of v. But in fact, the deviation from the harmonic oscillator approximation in real molecules leads to deviations from the E_{vib} of (6.1.14) (see next exercise) and the selection rule becomes $\Delta v = \pm 1, \pm 2, \pm 3, \pm 4, \ldots$. As a result, the spectrum consists of a fundamental, $\bar{v}$, and a series of *overtone transitions* at $2\bar{v}$, $3\bar{v}$, etc., of decreasing intensity. Deviation from the harmonic oscillator approximation is called *anharmonicity*. Anharmonicity is just the lack of fit of the parabola to $E(R)$ in Fig. 6.1.3; the fit is poorest away from R_e, consequently anharmonicity becomes increasingly important with increasing v. It is evident, for instance, that real molecules fall apart if the energy exceeds D_e, while a pure harmonic oscillator can be in a state of arbitrarily high energy.

EXERCISE To account for anharmonicity, E_{vib} is expanded, $E_{\text{vib}}(\text{cm}^{-1}) = \bar{v}_e(v + \tfrac{1}{2}) - \bar{v}_e x_e (v + \tfrac{1}{2})^2 + \cdots$, where $\bar{v}_e$ is the vibrational frequency in the absence of anharmonicity and x_e is the quadratic correction constant due to anharmonicity. Given that the observed vibrational spectrum of H^{35}Cl consists of transitions at 2885.9, 5668.0, 8346.9, and 10923.1 cm^{-1}, calculate the vibrational constant $\bar{v}_e$ and anharmonicity constant x_e for HCl.

The force constant and fundamental frequency are also of interest with regard to chemical binding. Table 6.1.2 shows that N_2 has the highest

TABLE 6.1.2

Molecule	R_e (Å)	$\bar{v}_e$ (cm^{-1})
Li$_2$	2.67	351.
B$_2$	1.59	1051.
C$_2$	1.3	1641.
N$_2$	1.1	2358.
O$_2$	1.2	1580.
F$_2$	1.4	891.

vibrational frequency (greatest k) among the first-row homonuclear diatomics, and the frequency tends to fall off from N_2 in both directions.

EXERCISE   Compare the trends in $\bar{\nu}_e$, R_e, and D_0 for the molecules in Table 6.1.2; offer an explanation for any correlations observed.

EXERCISE   Correlate the data of Table 6.1.2 with the MO bond orders given in Chapter 4. Explain.

INTENSITY OF SPECTRA

In Section 6.8 it is shown that the transition probability (i.e., the intensity of the transition) and the selection rules are given by the *transition moment* $M_{ij} = \langle \psi_i, \boldsymbol{\mu}\psi_j \rangle$, where $\boldsymbol{\mu}$ is the dipole moment operator of the molecule and ψ_i and ψ_j are the wavefunctions of the initial and final states. To understand the infrared spectrum of the diatomic molecule we must examine the dependence of M_{ij} on the motion of the nuclei. Expand the dipole moment of the vibrating molecule in the displacement coordinate $(R - R_e) = x$, about equilibrium, $x = 0$.

$$\mu = \mu_0 + \left(\frac{d\mu}{dx}\right)_{x=0} x + \cdots$$

The first term in the expansion is the permanent dipole moment of the molecule. The second term (and higher terms) give the change in dipole moment upon displacement from equilibrium. The transition moment is then

$$M_{ij} = \langle \psi_i, \mu_0 \psi_j \rangle + \left\langle \psi_i, \left(\frac{d\mu}{dx}\right)_{x=0} x \psi_j \right\rangle + \cdots$$

The integral over μ_0 vanishes because μ_0 is a constant and ψ_i and ψ_j are orthogonal. Thus, neglecting higher terms, in order to obtain a vibrational spectrum $\langle \psi_i, (d\mu/dx)x\psi_j \rangle$ must not vanish. The rigid rotator ($x = 0$, $R = R_e$) has transition probabilities given by the transition moment

$$\langle Y_{J''m_{J''}}, \boldsymbol{\mu}_0 Y_{J'm_{J'}} \rangle.$$

However, it should be evident that for a homonuclear diatomic molecule both $\mu_0 = 0$ and $d\mu/dx = 0$. As a consequence the homonuclear diatomic molecule has *no* infrared spectrum. The vibrational levels of a homonuclear diatomic molecule may be obtained from its electronic spectrum.

We now see that the infrared selection rules $\Delta J = \pm 1$ and $\Delta v = \pm 1$ are applicable to heteronuclear diatomics. These selection rules are determined by the requirement that M_{ij} must not vanish if the transition is to be allowed. If $M_{ij} = 0$ the transition is said to be forbidden. There is a rather simple way of predicting when $M_{ij} = 0$; however, we can only go into this in Section 6.8 after discussing group theory in Sections 6.3–6.6.

6.1 Vibration–Rotation Spectra

Because of the appreciable energy separation between vibrational energy levels in diatomic molecules (see Table 6.1.2), at room temperature and thermal equilibrium the vast majority of molecules are in the $v = 0$ state.

EXERCISE Use the Boltzmann distribution law to determine the population ratio of $v = 0$ to $v = 1$ levels in Li_2, B_2, and N_2 at room temperature. Is there appreciable change as the temperature is raised to, say, 200°C?

VIBRATION–ROTATION

Optical transitions at room temperature occur with greatest intensity from $v = 0$ to $v = 1$. The molecule is also rotating, so the resultant spectrum is a *band* with various rotational transitions, $\Delta J = \pm 1$, superimposed on the vibrational fundamental. Figure 6.1.6 gives a schematic representation of a

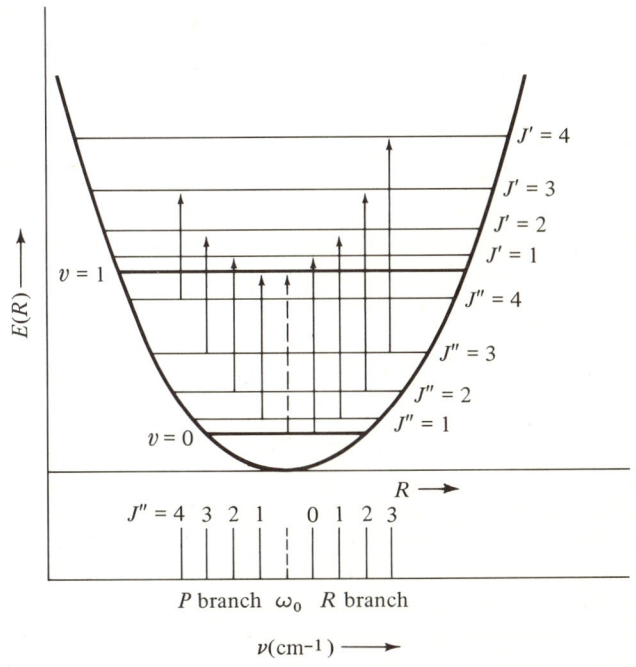

FIG. 6.1.6 Schematic representation of the vibration–rotation spectrum of a diatomic molecule. Part of the rotational fine structure of the vibrational transition is shown on a wavenumber scale. It consists of a P branch ($\Delta J = -1$) and an R branch ($\Delta J = +1$) with the band origin missing, $\omega_0 (\Delta J = 0)$.

vibrational band in the vibrational–rotational spectrum of a diatomic molecule. The transition between the rotationless states $J'' = 0$ to $J' = 0$ ($\Delta J = 0$) is forbidden; the missing line ω_0 is the *band origin*. The remaining lines are organized into two series of $2B$ spacing; the P branch ($\Delta J = -1$) and the R branch ($\Delta J = +1$) lie to either side of the band origin.

Just as anharmonicity becomes important with increasing vibrational quantum number, *centrifugal stretching* occurs with increasing J. Centrifugal stretching is easily visualized; as the molecule rotates faster and faster (higher J) the average distance between atoms increases, thus changing the moment of inertia. The more complete rotational energy is[7]

$$E_{\text{rot}}/hc = BJ(J+1) - DJ^2(J+1)^2$$

where D is the centrifugal stretching constant. The value of D depends on the force constant, because the larger the force constant the more difficult it is to stretch the bond. In any case, D is much smaller than B and constitutes a small correction.

The vibrational spectra of polyatomic molecules are more easily handled after we gain an appreciation of molecular symmetry and the methods of group theory. Therefore we postpone the treatment of polyatomic vibrations until Section 6.7.

6.2 ELECTRONIC SPECTRA

The electronic spectra of diatomic molecules are observed as absorption or emission bands in the visible and ultraviolet regions of the spectrum. The transitions comprising these bands take place between specific vibration–rotation levels of the electronic states of the molecule. If $\bar{\nu}_i$ is a frequency at which the molecule possesses a spectral transition, then

$$\bar{\nu}_i = \frac{1}{hc}(\Delta E_{\text{el}} + \Delta E_{\text{vib}} + \Delta E_{\text{rot}})$$

is the statement of the Planck condition for discrete molecular spectra.

The electronic states of molecules are classified according to orbital and spin angular momentum as explained in Section 5.8 and used in later sections. In Chapter 5 we discussed mainly the ground state. However, each molecule has *many* electronic states; a classification (or grouping) of the molecular electronic states is made according to their *atomic dissociation products*. As R increases the diatomic molecule eventually will dissociate into atoms, which are characterized by their spectroscopic

---

[7] G. Herzberg, footnote 1.

states. Thus, a complete description of the diatomic electronic state should also give the states of the atomic dissociation products. For example, Fig. 6.2.1 illustrates the $E(R)$ curves for the three lowest-energy states of O_2, $^3\Sigma_g$, $^1\Delta_g$, and $^1\Sigma_g$, which arise from the configuration $KK(2\sigma_g)^2(2\sigma_u)^2(\pi_{x,yu})^4(3\sigma_g)^2(\pi_{x,yg})^2$ and dissociate into two oxygen atoms, each in the ^3P ground state. Conversely, the collision of two oxygen atoms in their ground states can lead to the formation of O_2 in one of these states. All three states are stable bonding states with characteristic dissociation energies, vibrational and rotational constants ($\bar{v}_e$, $\bar{v}_e x_e$, B), and equilibrium internuclear distances (R_e). Calculated and experimental values of the spectroscopic constants are displayed in Table 6.2.1.

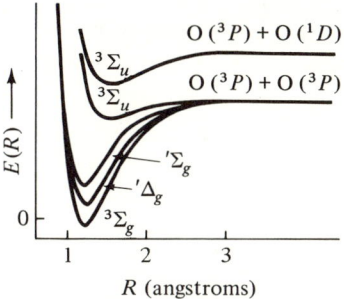

FIG. 6.2.1 Experimental nuclear potential energy curves of the low-lying states of O_2 including the ground state $^3\Sigma_g$.

TABLE 6.2.1 Spectroscopic Constants of Low-Lying States of O_2

State	D_e (eV)	D_0 (eV)	R_e (Å)	$\bar{v}_e$ (cm^{-1})	$\bar{v}_e x_e$ (cm^{-1})	B (cm^{-1})
$^3\Sigma_g$ Calc [a]	3.81	3.72	1.30	1582	14	1.25
Exptl [b]		5.08	1.21	1580	12	1.45
$^1\Delta_g$ Calc [a]	2.81	2.72	1.33	1406	16	1.19
Exptl [b]		4.10	1.22	1509	13	1.43
$^1\Sigma_g$ Calc [a]	2.44	2.36	1.34	1318	18	1.17
Exptl [b]		3.44	1.23	1433	14	1.40

[a] Calculated by obtaining the $E(R)$ curves at many values of R from CI wavefunctions (Section 5.15). The curves are then analyzed for the spectroscopic constants. From H. F. Schaefer and F. E. Harris, J. Chem. Phys. 48, 4946 (1968). Used by permission.
[b] G. Herzberg, Spectra of Diatomic Molecules, footnote 1.

Transitions among electronic states by absorption or emission of light are regulated by the optical selection rules

OPTICAL SELECTION RULES $\Delta \Lambda = 0, \pm 1$

$\Delta S = 0$

$g \leftrightarrow u$

Changes in the total oribtal angular momentum along the internuclear axis are restricted to units of 0 and ± 1; the multiplicity must be conserved, $\Delta S = 0$; and g states may only combine with u states and vice versa. Figure 6.2.2 shows the ground state and some lower excited states of Li$_2$ with their dissociation products.

EXERCISE   What are the transitions among which states of Figs. 6.2.1 and 6.2.2 seem to be allowed?

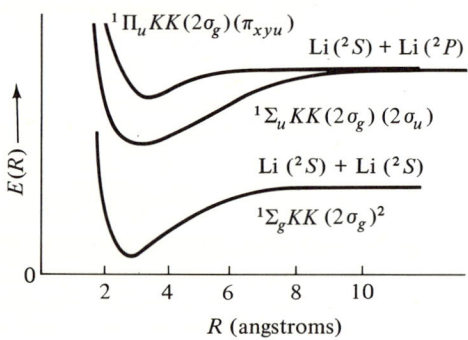

FIG. 6.2.2 Nuclear potential energy curves of Li$_2$ for several low-lying states.

THE FRANCK-CONDON PRINCIPLE

The $^1\Sigma_g - ^1\Sigma_u$ transition of Li$_2$ is an example of an allowed electronic transition. In the course of the transition the vibrational and rotational states of the molecule will also change. According to the Franck-Condon principle the intensity of the vibrational bands in an electronic transition is determined by the nature of the nuclear potential energy curves of the two electronic states. During an electronic transition the internuclear distance and nuclear velocities will change but very little. (Thus an arrow between the electronic states which indicates the transition should be drawn vertically, i.e., at constant R.)

Writing the vibrational–electronic (or vibronic) wavefunction for the ith electronic state as $\psi_{iv''}(\xi)\psi_i(y)$, where ξ and y are the nuclear and electronic

6.2 Electronic Spectra

coordinates, respectively, the transition probability is proportional to the vibronic transition moment,[8]

$$M_{ij}^{v''v'} = \langle \psi_{iv''}(\xi)\psi_i(y), \mu\psi_{jv'}(\xi)\psi_j(y) \rangle \tag{6.2.1}$$

If $\langle \psi_i(y), \mu\psi_j(y) \rangle$ is independent of R, the transition moment is

$$M_{ij}^{v''v'} = \langle \psi_i(y), \mu\psi_j(y) \rangle \langle \psi_{iv''}, \psi_{jv'} \rangle \tag{6.2.2}$$

Thus $M_{ij}^{v''v'}$ is directly proportional to the vibrational overlap, $\langle \psi_{iv''}, \psi_{jv'} \rangle$, between the electronic states, Figure 6.2.3 offers a schematic illustration of the variation of $M_{ij}^{0v'}$. with v'.

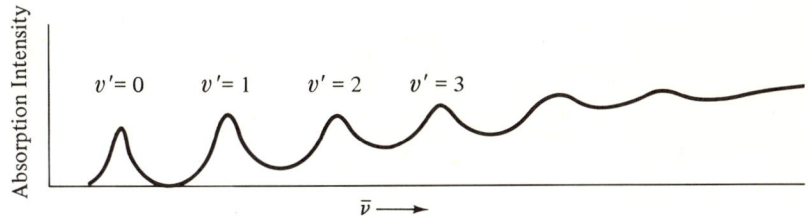

FIG. 6.2.3 Schematic of the vibrational structure in an electronic transition. The spectrum consists of a vibrational progression from the ground state ($v'' = 0$) to vibrational levels of the excited state ($v' = 0, 1, 2, ...$) and into the continuum.

There is a continuous probability of nuclear position along R, as can be seen from Fig. 6.1.4. The probability of finding the nuclei at R_e is $|\psi(\xi)|^2_{\xi=0}$, which is a sole maximum only in the $v = 0$ vibrational state.[9] Thus electronic transitions from the ground states of diatomic molecules are observed to occur over a range of R with the maximum probability of transition from R_e ($v = 0$). Specifically, the absorption spectrum of Li_2 ($^1\Sigma_g \to {}^1\Sigma_u$) consists of a series of bands corresponding to the transitions $^1\Sigma_g(v'' = 0) \to {}^1\Sigma_u(v' = 0, 1, 2, 3, ...)$. Each of these transitions is a *band* because it carries its own rotational structure; the rotational structure may, in principle, be observed by a high-resolution spectrograph.

In any molecule, at high values of v' the nuclear potential energy is far from that of a harmonic oscillator; many vibrational levels crowd in toward the dissociation limit. Transitions at or beyond the *dissociation*

---

[8] Section 6.8 and Appendix H.
[9] At large v the probability distribution of the nuclei along R becomes classical, that is, large probability only at the turning points of the oscillation.

limit (dotted line, Fig. 6.2.4), if allowed by the Franck-Condon principle, lead to continuous spectra as the molecule *photodissociates* into its component atoms. Photodissociation occurs quite readily from repulsive (unstable) excited states or with certain relative positioning of the nuclear potential energy curves. Figure 6.2.4 illustrates three photodissociation mechanisms which follow from the Franck-Condon principle and conservation of energy. In all three cases a certain amount of electronic excitation energy is converted into the relative kinetic energy of the nuclei as the molecule flies apart.

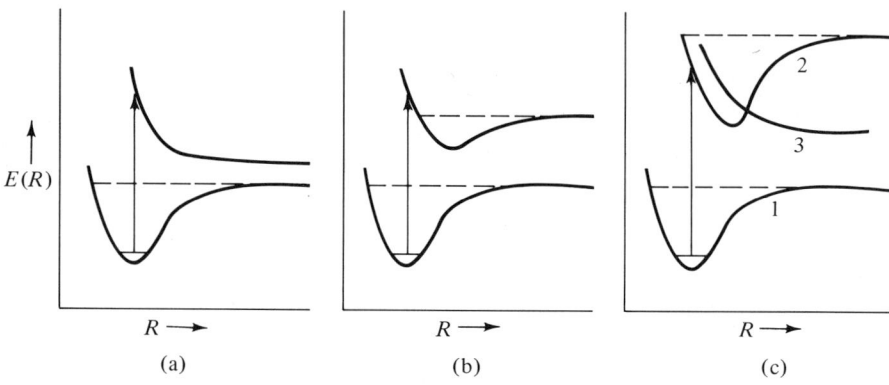

FIG. 6.2.4 Photodissociation mechanisms.
(a) Photoexcitation to a repulsive excited state. (b) Photoexcitation into the continuum of a stable excited state.
(c) Photoexcitation into a predissociative state, that is, radiationless transition into the repulsive state (3).

EXERCISE Redraw Fig. 6.2.4, indicating by vertical arrows the amount of electronic excitation energy converted into atomic kinetic energy (heat) by each mechanism.

The absorption spectra in (a) and (b) of Fig. 6.2.4 are continuous, that is, the nuclear motion in the excited state is unconfined by a harmonic oscillator or other attractive potential; this corresponds to the motion of free particles and has a continuous energy spectrum. In Fig. 6.2.4(c) a discrete spectrum is observed up to the energy of the crossing with the repulsive state, then the spectrum becomes fuzzy, that is, the lines broaden.[10]

---

[10] For a complete discussion of photodissociation and predissociation in diatomic molecules see G. Herzberg, vol. I, cited in footnote 1.

EXERCISE   Interpret the line broadening observed in predissociative spectra in terms of the uncertainty principle.

EXERCISE   Determine the dissociation energies D_e and D_0 of O_2 from the ground state $^3\Sigma_g$. Use the following spectroscopic observations. The Schumann-Runge bands of O_2 arise from $^3\Sigma_g \to {}^3\Sigma_u$, these bands terminate in a continuum at 1759 A. The $^3\Sigma_u$ state has dissociation products $O(^3P) + O(^1D)$. The energy difference between $O(^3P)$ and $O(^1D)$ is 15,868 cm^{-1}. Use Table 6.1.2, Fig. 6.2.1, and check your results against Table 6.2.1.

Molecular Symmetry

Molecules have symmetry. This is the beautiful deduction of structural chemistry, the branch of chemistry which came into prominence with the work of Kekulé, Le Bel, and van't Hoff in the late nineteenth century. Molecular symmetry follows naturally from the atomic hypothesis and Dalton's "chemical philosophy" (see Chapter 1). In the following sections the intuitive notion of symmetry is made more exact and useful through the mathematical discipline of *group theory*. Previously we used the symmetry elements of the diatomic molecule to *classify* the molecular orbitals according to their "symmetry." The usefulness of symmetry arose from the fact that *the total electronic density of the molecule must be invariant to the symmetry operations which send the molecule into itself*. For example, the inversion and reflections of Section 5.5 and Table 5.5.1—these operations served to classify the one-electron wavefunctions (molecular orbitals) as symmetric, $\hat{O}\psi = +\psi$, or antisymmetric, $\hat{O}\psi = -\psi$, to the operation $\hat{O}$. Of course, only the observable, total electronic density, need be invariant to the symmetry operations, so the occupied orbitals may be chosen as MO (symmetry orbitals) *or* LO (localized orbitals), the MO and LO being connected by a transformation which leaves the total electronic density invariant. This degree of freedom, MO or LO, is chemically very important, but does not lessen the utility of symmetry in the treatment of molecular spectra and structure. In order to extend such treatment to the structure and spectra of polyatomic molecules we need a more extensive symmetry notation and the precise methods of group theory. To begin with, what are the symmetry elements of the polyatomic molecule?

6.3 SYMMETRY ELEMENTS OF MOLECULES

The molecular symmetry elements of interest are associated with *point symmetry operations*, that is, the operations which send the molecule into itself. (The "molecule" is the three-dimensional array of point nuclei at their equilibrium positions.) All the point symmetry operations can be generated from just two types: rotations by a certain angle about an axis going through the origin, and reflections in planes containing the origin (all the point symmetry operations are relative to an invariant origin which remains at rest). The collection of point symmetry operations of the molecule define the *point group* (see Section 6.4) to which the molecule belongs, for example, H_2O is an example of a molecule belonging to the C_{2v} point group. The meaning of the point group symbols, C_{2v}, D_{6h}, etc., is explained in Appendix G and illustrated with molecular examples.

EXAMPLE H_2O. The symmetry elements of H_2O are illustrated in Fig. 6.3.1. The symmetry elements intersect at the origin, which is the oxygen nucleus in this molecule. The symmetry elements consist of

1. A *twofold axis* C_2, such that rotation by $2\pi/2$ sends the molecule into itself.
2. Two *vertical planes of reflection* (mirror planes) σ_v and σ_v' which contain the C_2 axis, such that reflection in these planes sends the molecule into itself.

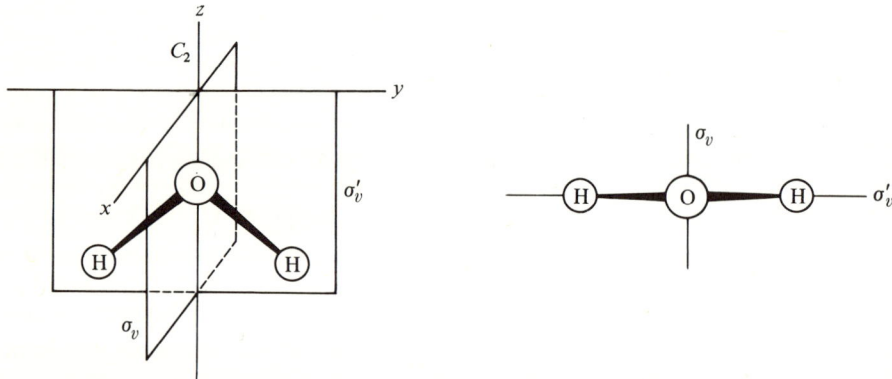

FIG. 6.3.1 H_2O molecule. A simple example of the C_{2v} point group. The origin of the coordinate system is at the oxygen nucleus. The mirror planes are $\sigma_v(\sigma_{xz})$ and $\sigma_v'(\sigma_{yz})$. There is also a twofold symmetry axis C_2.

EXAMPLE C_6H_6, Benzene. D_{6h} point group. Figure 6.3.2 illustrates *some* of the symmetry elements of benzene.

1. A sixfold axis C_6, such that rotation by $2\pi/6$ sends the molecule into itself. This is the principal symmetry axis.
2. Six twofold axes, $6C_2$, which are perpendicular to C_6.

3. Six vertical planes of reflection, $6\sigma_v$, which contain the principal axis.
4. A horizontal plane of reflection σ_h, which is perpendicular to the principal axis.

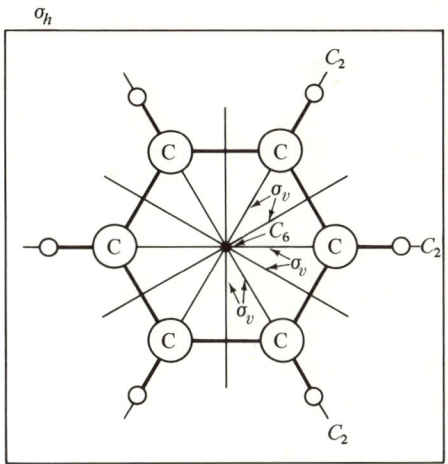

FIG. 6.3.2 Benzene. Point group $\mathbf{D}_{6h}$. The basic symmetry elements consist of a C_6 symmetry axis, 6 C_2 axes, 6 σ_v mirror planes, and a horizontal mirror plane σ_h.

In actuality, the possible symmetry elements of molecules are quite limited. The *types* of symmetry elements found in molecular point groups are listed in Table 6.3.1.

EXERCISE   Show that the existence of a center of symmetry is equivalent to the presence of a S_2 axis.

EXERCISE   Show that a linear molecule has a C_∞ rotation axis.

EXERCISE   Identify the symmetry elements of eclipsed ethane (see Fig. 6.3.3).

EXERCISE   If $\hat{C}_n \hat{C}_n = \hat{C}_n^2$ show that $\hat{C}_2^2 = \hat{C}_2^4 = \hat{E}$.

EXERCISE $\hat{\sigma}$ is the reflection operation. Show that $\hat{\sigma}^m = \hat{\sigma}$ if m is odd, and $\hat{\sigma}^m = \hat{E}$ if m is even.

EXERCISE   What are the symmetry elements which distinguish the homonuclear from the heteronuclear diatomic molecule? (See Figs. 5.5.2 and 6.3.4.)

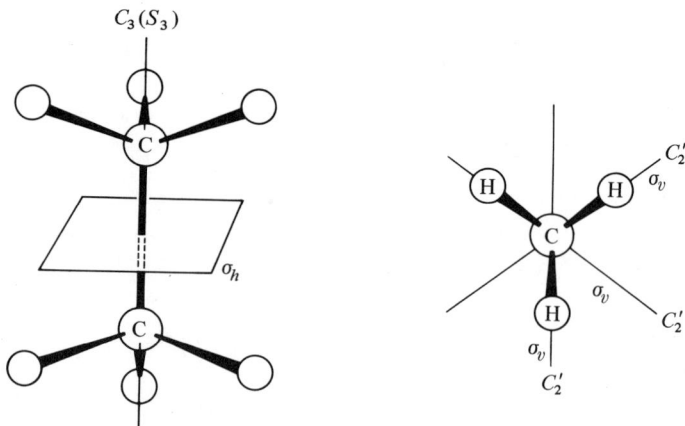

FIG. 6.3.3 Symmetry elements of eclipsed ethane. Point group D_{3h}.

TABLE 6.3.1 Point Symmetry Elements and Point Symmetry Operations[a]

Symbol	Symbol element	Symmetry operation
E	Identity[b]	No operation (or an n-fold operation by an element of order n)
C_n	Rotation axis[c] n-fold	A rotation about the C_n axis by an angle of $2\pi/n$
I	Center of symmetry	Inversion through the center of symmetry (origin), i.e., $x, y, z \rightarrow -x, -y, -z$
σ_h	Horizontal mirror plane, i.e., plane is perpendicular to the principal axis C_n	Reflection through the mirror plane
σ_v	Vertical mirror plane, i.e., plane contains the principal axis C_n	Reflection through the mirror plane
σ_d	Dihedral plane of reflection, contains the principal axis and bisects C_2 axes perpendicular to C_n	Reflection through the mirror plane
S_n	Improper rotation axis	A rotation of $2\pi/n$ followed by reflection in a plane perpendicular to the axis

[a] In addition to these point symmetry operations there are *translations* which are valid symmetry operations only for crystals.
[b] All molecules possess the identity symmetry element.
[c] Twofold axes perpendicular to the principal axis are primed, for example, C_2', C_2'', etc.

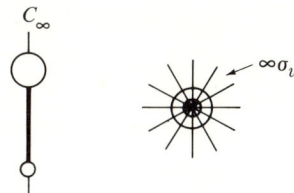

FIG. 6.3.4 Heteronuclear diatomic molecule. Point group $C_{\infty v}$.

6.4 GROUP THEORY

Molecules possess several symmetry elements; the symmetry operations form a very special *set* called a *group* (specifically, a *point group*).[11] *The point group of a molecule consists of the set of symmetry operations which send the molecule into itself.* The following statements define what is meant by a mathematical group.

1. The product of any two members of the set, and the square of each member, is another member of the set, that is, $AB = C$; A, B, and C are members of the set.
2. The set contains the identity E, that is $EA = AE = A$.
3. The associative law of algebra holds, that is, $(AB)C = A(BC)$; the product is independent of the order of multiplication.
4. Every member of the set has a reciprocal which is also a member of the set, that is, let B be the reciprocal of A, then $BA = E$ so $B = A^{-1}$ and B is a member of the set if A is.

If the set satisfies statements 1 to 4, the members of the set *form a group*. The following definitions and statements develop the terminology of abstract group theory.

1. The *order of the group*: The number of members of the group, including E.
2. *Commutation*: Two members of the group are said to commute if $AB = BA$ (only in special groups, called *Abelian*, do all the members of the group commute).
3. The *order of the member*: The lowest positive integer n such that $A^n = E$.
4. *Conjugate members*: If R and A are members of the group then $R^{-1}AR$ is also a member of the group, $R^{-1}AR = B$. A and B are said to be conjugate.

[11] See the general references on group theory at the end of this chapter.

5. The *class of a member*: The set generated by forming $R^{-1}AR$ for all R in the group is said to form the class of A. For example, if $R^{-1}AR = A$ or B for all R then A and B form a class by themselves.

EXERCISE   Prove that every member of the group must belong to its own class.

EXERCISE   Prove that if A belongs to the class of B then B belongs to the class of A.

6. Members of the same class have the same order, that is, if $B = R^{-1}AR$ then $B^n = (R^{-1}AR)^n = R^{-1}ARR^{-1}AR \cdots = R^{-1}A^nR$, so if $A^n = E$ then $B^n = E$.
7. The number in a class is an integral factor of the order of the group.[12]
8. The *group multiplication table*: A table built from all the products, AB, of all the members of the group, for example, see Table 6.4.1.

TABLE 6.4.1   Multiplication Table for Point Group C_{2v}

	$\hat{E}$	$\hat{C}_2$	$\hat{\sigma}_v$	$\hat{\sigma}_v'$
$\hat{E}$	$\hat{E}$	$\hat{C}_2$	$\hat{\sigma}_v$	$\hat{\sigma}_v'$
$\hat{C}_2$	$\hat{C}_2$	$\hat{E}$	$\hat{\sigma}_v'$	$\hat{\sigma}_v$
$\hat{\sigma}_v$	$\hat{\sigma}_v$	$\hat{\sigma}_v'$	$\hat{E}$	$\hat{C}_2$
$\hat{\sigma}_v'$	$\hat{\sigma}_v'$	$\hat{\sigma}_v$	$\hat{C}_2$	$\hat{E}$

THE POINT GROUPS

The point groups are given names (symbols) according to the symmetry elements present in the group. The notation used in this text (Schoenflies notation) is that used widely in molecular spectroscopy. The delineation of the symmetry elements present in the important point groups, together with molecular examples, are given in Appendix G.

EXAMPLE   Point group C_{2v} (see Fig. 6.3.1) has symmetry elements C_2, σ_v, σ_v', and E.

The multiplication table shows that the symmetry operations of a molecule (in this case H_2O) satisfy the four group postulates. The product of any two members of the set is another member of the set. The identity is included, and so is the inverse of each operator (in this group each

---

[12] Proofs of the theorems of this and the next section are often beyond the scope of this text. The interested student should consult one of the many specialized texts on group theory.

6.4 Group Theory

operation is its own inverse, that is, they are all of order 2). The student should verify that these symmetry operations form a group by mentally carrying out the operations corresponding to the symmetry elements in Fig. 6.3.1 and deriving the multiplication table for C_{2v}.

EXERCISE Prove that C_{2v} is an Abelian group and that each symmetry operation forms a class by itself.

EXAMPLE Point group C_{3v}. This is the point group of ammonia, NH_3, illustrated in Fig. 6.4.1.

The symmetry elements of NH_3 are a C_3 axis and three vertical planes of reflection, σ_v, σ_v', σ_v''. The symmetry operations of the group are reflections through the three mirror planes, and two rotations $\hat{C}_3$ (120°) and $\hat{C}_3^2$ (240°); of course, $\hat{C}_3^3 = \hat{E}$ is also in the group. The multiplication table is given in Table 6.4.2. The $\hat{\sigma}_v$ operations being of order 2 are their own inverse, but $\hat{C}_3^2$ is the reciprocal of $\hat{C}_3$.

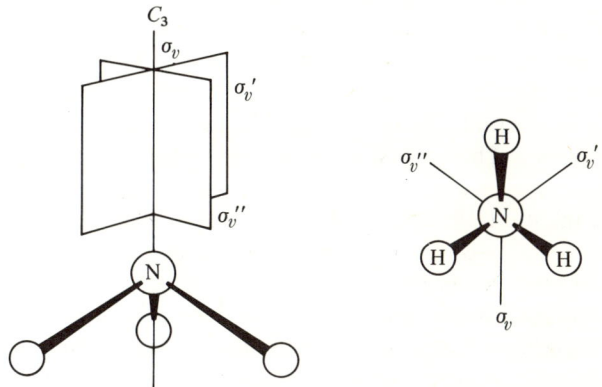

FIG. 6.4.1 Ammonia, NH_3. A simple example of point group C_{3v}. Symmetry elements consist of a C_3 symmetry axis and three vertical mirror planes.

TABLE 6.4.2 Multiplication Table for Point Group $\hat{C}_{3v}$

	$\hat{E}$	$\hat{C}_3$	$\hat{C}_3^2$	$\hat{\sigma}_v$	$\hat{\sigma}_v'$	$\hat{\sigma}_v''$
$\hat{E}$	$\hat{E}$	$\hat{C}_3$	$\hat{C}_3^2$	$\hat{\sigma}_v$	$\hat{\sigma}_v'$	$\hat{\sigma}_v''$
$\hat{C}_3$	$\hat{C}_3$	$\hat{C}_3^2$	$\hat{E}$	$\hat{\sigma}_v''$	$\hat{\sigma}_v$	$\hat{\sigma}_v'$
$\hat{C}_3^2$	$\hat{C}_3^2$	$\hat{E}$	$\hat{C}_3$	$\hat{\sigma}_v'$	$\hat{\sigma}_v''$	$\hat{\sigma}_v$
$\hat{\sigma}_v$	$\hat{\sigma}_v$	$\hat{\sigma}_v'$	$\hat{\sigma}_v''$	$\hat{E}$	$\hat{C}_3$	$\hat{C}_3^2$
$\hat{\sigma}_v'$	$\hat{\sigma}_v'$	$\hat{\sigma}_v''$	$\hat{\sigma}_v$	$\hat{C}_3^2$	$\hat{E}$	$\hat{C}_3$
$\hat{\sigma}_v''$	$\hat{\sigma}_v''$	$\hat{\sigma}_v$	$\hat{\sigma}_v'$	$\hat{C}_3$	$\hat{C}_3^2$	$\hat{E}$

EXERCISE   By examination of Table 6.4.2 show that $\hat{E}$, $\hat{C}_3$, and $\hat{C}_3^2$ form a group by themselves (subgroup of $\mathbf{C}_{3v}$).

EXERCISE   The classes of $\mathbf{C}_{3v}$. Show that $\hat{E}$ forms a class by itself. Show that $\hat{C}_3$ and $\hat{C}_3^2$ form a class by themselves. Finally show that $\hat{\sigma}_v$, $\hat{\sigma}'_v$, and $\hat{\sigma}''$ each forms a class by itself.

The designations of the point groups as $\mathbf{C}_{2v}$, $\mathbf{C}_{3v}$, etc., are explained in Appendix G. In addition to the point groups $\mathbf{C}_{nv}$, two examples of which have been given, other important point groups are $\mathbf{C}_{nh}$ (e.g., *trans*-dichloroethylene, $n = 2$), $\mathbf{D}_{nh}$ (e.g., benzene, $n = 6$), $\mathbf{D}_{nd}$ (e.g., cyclohexane, chair, $n = 3$) and the two highly symmetric cubic point groups $\mathbf{T}_d$ (e.g., CH_4) and $\mathbf{O}_h$ (e.g., SF_6).

6.5 GROUP REPRESENTATIONS

In Section 5.5 symmetry was used to classify the molecular orbitals of the homonuclear diatomic molecule. If $\hat{O}$ is a symmetry operator, then $\hat{O}\psi = \pm \psi$ is the eigenvalue relation which assures us that $|\psi|^2 = |\hat{O}\psi|^2$, that is, the electronic density due to the one-electron function ψ is the same before and after the symmetry operation. When dealing with finite point groups the group multiplication table severely limits the possible number of unique combinations of eigenvalues among the symmetry operations of the group. This is best explained by means of an example. Consider the finite point group $\mathbf{C}_{2v}$. $\mathbf{C}_{2v}$ has symmetry elements E, C_2, σ_v, and σ'_v, and the multiplication table, Table 6.4.1. If ψ is a MO of H_2O (Fig. 6.3.1) then $\hat{E}\psi = +\psi$, $\hat{C}_2\psi = +\psi$, $\hat{\sigma}_v\psi = +\psi$, and $\hat{\sigma}'_v\psi = +\psi$, is one possible set of eigenvalues ($+1, +1, +1, +1$) which satisfies the group multiplication table, for example, the table requires $\hat{C}_2\hat{\sigma}_v\psi = \hat{\sigma}'_v\psi$, which is satisfied by $\hat{C}_2\hat{\sigma}_v\psi = +\psi$, $\hat{\sigma}'_v\psi = +\psi$. This ψ is said to *form a basis for the totally symmetric representation of* $\mathbf{C}_{2v}$, the totally symmetric representation being $+1, +1, +1, +1$. There are only four sets of eigenvalues ± 1 for the point group $\mathbf{C}_{2v}$ which satisfy the group multiplication table. They are given in Table 6.5.1.

TABLE 6.5.1   Irreducible Representations of $\hat{C}_{2v}$ (Character Table)

	E	C_2	σ_v	σ'_v
A_1	1	1	1	1
A_2	1	1	-1	-1
B_1	1	-1	1	-1
B_2	1	-1	-1	1

6.5 Group Representations

EXERCISE Show that each set of eigenvalues in Table 6.5.1 satisfies the C_{2v} group multiplication table, Table 6.4.1.

A set of quantities which obey the same multiplication table as the elements of a group are said to form a representation of the group. The representation of the group need not be one-dimensional. It may be a set of matrices.

EXAMPLE

$$\hat{E} = \begin{pmatrix} 1 & 0 \\ 0 & 1 \end{pmatrix} \quad \hat{C}_2 = \begin{pmatrix} 1 & 0 \\ 0 & -1 \end{pmatrix}$$
$$\hat{\sigma}_v = \begin{pmatrix} 1 & 0 \\ 0 & 1 \end{pmatrix} \quad \hat{\sigma}'_v = \begin{pmatrix} 1 & 0 \\ 0 & -1 \end{pmatrix} \tag{6.5.1}$$

The claim is that (6.5.1) is a two-dimensional representation of C_{2v}. For example, according to Table 6.4.1, $\hat{\sigma}_v \hat{\sigma}'_v = \hat{C}_2$, which is satisfied by

$$\begin{pmatrix} 1 & 0 \\ 0 & 1 \end{pmatrix}\begin{pmatrix} 1 & 0 \\ 0 & -1 \end{pmatrix} = \begin{pmatrix} 1 & 0 \\ 0 & -1 \end{pmatrix}$$

EXERCISE Show that the matrix representation (6.5.1) satisfies the group multiplication table of C_{2v}. Construct other matrix representations from the one-dimensional representations in Table 6.5.1.

Representations of a group are generated by allowing the symmetry operators to work on a *basis*. For example, consider x, y, and z in Fig. 6.3.1. Note that $\hat{C}_2 x = -x$, $\hat{C}_2 y = -y$, $\hat{C}_2 z = z$, $\hat{\sigma}_v x = x$, etc. In matrix form these relations are easily written as follows:

$$\hat{E}\begin{pmatrix} x \\ y \\ z \end{pmatrix} = \begin{pmatrix} 1 & 0 & 0 \\ 0 & 1 & 0 \\ 0 & 0 & 1 \end{pmatrix}\begin{pmatrix} x \\ y \\ z \end{pmatrix} = \begin{pmatrix} x \\ y \\ z \end{pmatrix}$$

$$\hat{C}_2\begin{pmatrix} x \\ y \\ z \end{pmatrix} = \begin{pmatrix} -1 & 0 & 0 \\ 0 & -1 & 0 \\ 0 & 0 & 1 \end{pmatrix}\begin{pmatrix} x \\ y \\ z \end{pmatrix} = \begin{pmatrix} -x \\ -y \\ z \end{pmatrix}$$

$$\hat{\sigma}_v\begin{pmatrix} x \\ y \\ z \end{pmatrix} = \begin{pmatrix} 1 & 0 & 0 \\ 0 & -1 & 0 \\ 0 & 0 & 1 \end{pmatrix}\begin{pmatrix} x \\ y \\ z \end{pmatrix} = \begin{pmatrix} x \\ -y \\ z \end{pmatrix} \tag{6.5.2}$$

$$\hat{\sigma}'_v\begin{pmatrix} x \\ y \\ z \end{pmatrix} = \begin{pmatrix} -1 & 0 & 0 \\ 0 & 1 & 0 \\ 0 & 0 & 1 \end{pmatrix}\begin{pmatrix} x \\ y \\ z \end{pmatrix} = \begin{pmatrix} -x \\ y \\ z \end{pmatrix}$$

The three-dimensional matrices of equations (6.5.2) form a representation of C_{2v} in the basis $\begin{pmatrix} x \\ y \\ z \end{pmatrix}$, as you can easily prove to yourself by showing that the matrices satisfy the multiplication table of C_{2v}. There are an infinite number of representations of a point group because there are an infinite number of possible *basis sets*, that is, *any set of functions which goes into a linear combination of said functions under the symmetry operations of the group forms a basis for a representation of the group.*

> *Definition*: The *irreducible representations* of the point group form that set of representations in terms of which all other representations may be expressed. A representation which may be expressed in terms of irreducible representations is said to be *reducible*.

Of the infinite number of possible representations of a finite point group there are only a few *irreducible representations*. The irreducible representations of C_{2v} are given in Table 6.5.1. The student should be able to see that the representation of equations (6.5.1) is reducible to A_1 and B_1 of Table 6.5.1; and the representation of equations (6.5.2) is reducible to A_1, B_1, and B_2 of Table 6.5.1. (Compare the elements in the matrices to Table 6.5.1.) One says that x *transforms as* the irreducible representation B_1 in C_{2v}, that is, x forms a basis for the B_1 representation; y *transforms as* the representation B_2 in C_{2v}; etc.[13] Most of the irreducible representations of the finite molecular point groups are one dimensional, that is, ± 1. The following list gives some of the properties and theorems of group representations:[14]

1. The number of irreducible representations of a group is equal to the *number of classes* of symmetry elements in the group. For example, in C_{2v} each element forms a class by itself, four classes, four irreducible representations.
2. The *character* (or trace) of a representation is the sum of diagonal elements in the transforming matrix. For example, the character of reducible representation in equations (6.5.2) is $\chi_{red}(\hat{E}) = 3$, $\chi_{red}(\hat{C}_2) = -1$, $\chi_{red}(\hat{\sigma}_v) = 1$, and $\chi_{red}(\hat{\sigma}'_v) = 1$. Symmetry operations in the same class are represented by matrices having the same character. One-dimensional representations are their own characters, so Table 6.5.1 is the *character table* for C_{2v}. Complete sets of character tables for the point groups may be found in advanced texts.

---

[13] It is not always so obvious into which irreducible representations a given reducible representation can be decomposed, in which case equation (6.5.3) is useful.

[14] These are offered without proof.

6.5 Group Representations

3. If a reducible representation is too complicated for its irreducible components to be seen by inspection, we can use (6.5.3)

$$n_i = \frac{1}{h} \sum_{\hat{O}} \chi_{\text{red}}(\hat{O}) \chi_i(\hat{O}) \qquad (6.5.3)$$

where n_i is the number of times the ith irreducible representation is contained in the reducible representation; $\chi_{\text{red}}(\hat{O})$ and $\chi_i(\hat{O})$ are the characters of the reducible and ith irreducible representations; and h is the order of the group.

EXAMPLE Using the reducible representation of equations (6.5.2) for C_{2v}:

$$n_i = \tfrac{1}{4}[\chi_{\text{red}}(\hat{E})\chi_i(\hat{E}) + \chi_{\text{red}}(\hat{C}_2)\chi_i(\hat{C}_2) + \chi_{\text{red}}(\hat{\sigma}_v)\chi_i(\hat{\sigma}_v) \\ + \chi_{\text{red}}(\hat{\sigma}_v')\chi_i(\hat{\sigma}_v')]$$

$$n_{A_1} = \tfrac{1}{4}(3 - 1 + 1 + 1) = 1 \qquad n_{B_1} = \tfrac{1}{4}(3 + 1 + 1 - 1) = 1$$
$$n_{A_2} = \tfrac{1}{4}(3 - 1 - 1 - 1) = 0$$

EXERCISE Find n_{B_2}, the number of times the irreducible representation B_2 is contained in the reducible representation of equations (6.5.2).

4. The characters of the irreducible representations of a group are orthogonal in the following sense (one-dimensional irreducible representations):

$$\sum_{\hat{O}} \chi_i(\hat{O}) \chi_j(\hat{O}) = h \delta_{ij} \qquad (6.5.4)$$

EXERCISE Derive (6.5.3) from (6.5.4). Hint: $\chi_{\text{red}}(\hat{O}) = \sum_i n_i \chi_i(\hat{O})$.

5. We may now define a *projection operator* $\hat{P}^i$:

$$\hat{P}^i = l_i/h \sum_{\hat{O}} \chi_i(\hat{O}) \hat{O} \qquad (6.5.5)$$

The sum in (6.5.5) is over all symmetry operators $\hat{O}$ of the group; $\chi_i(\hat{O})$ is the character of $\hat{O}$ in the ith irreducible representation which has dimension l_i.

The importance of the projection operator follows from

$$\hat{P}^i \psi_i = \psi_i \quad \text{and} \quad \hat{P}^i \psi_j = 0 \qquad j \neq i \qquad (6.5.6)$$

where ψ_i and ψ_j form a basis for irreducible representations i and j, respectively. We can prove that (6.5.6) follows from (6.5.4) for one-dimensional irreducible representations as follows, from $\hat{O}\psi_j = \chi_j(\hat{O})\psi_j$:

$$\hat{P}^i \psi_j = \frac{1}{h} \sum_{\hat{O}} \chi_i(\hat{O}) \hat{O} \psi_j = \frac{1}{h} \sum_{\hat{O}} \chi_i(\hat{O}) \chi_j(\hat{O}) \psi_j$$

so from (6.5.4)

$$\hat{P}^i \psi_j = \frac{1}{h} \delta_{ij} h \psi_j = \delta_{ij} \psi_j$$

For some arbitrary basis function f, which may be expressed as a linear combination of the set $\psi_1, \psi_2, \psi_3, \ldots,$

$$f = \sum_i c_i \psi_i \qquad \hat{P}^i f = c_i \psi_i$$

c_i is a constant. *$\hat{P}^i$ projects out the component of f in the basis of the ith irreducible representation to within a constant factor.*

These properties of the group representations are used in the application of group theory to quantum mechanics, especially 3 and 5. Precisely how these seemingly abstract principles can be used to get practical information about a molecular system is the subject of our next section.

6.6 GROUP THEORY IN QUANTUM MECHANICS

In a previous chapter symmetry was used to classify the LCAO of the homonuclear diatomic molecule according to their eigenvalues under certain reflection and inversion operators (Table 5.5.1). It should now be evident that the resulting designations σ_g, σ_u, π_g, π_u also identify the irreducible representations of the point group $\mathbf{D}_{\infty h}$ of the diatomic molecule, and the LCAO so identified form bases for these irreducible representations. The σ_g and σ_u are one dimensional, and π_g and π_u are two dimensional (degenerate) irreducible representations of $\mathbf{D}_{\infty h}$. Because $\mathbf{D}_{\infty h}$ is an infinite group it has an infinity of irreducible representations; each representation corresponds to a value of the angular momentum $|\lambda| = 0, 1, 2, 3, \ldots$ ($\sigma, \pi, \delta, \ldots$). However, our interest centers on the finite point groups of polyatomic molecules.

The symmetry operators of a molecule "send the molecule into itself," which is to say the molecule is the same before and after the operation. Mathematically, $|\psi|^2 = |\hat{O}\psi|^2$, $\hat{O}\psi = \pm \psi$, where $\hat{O}$ is the symmetry operator.[15] This means that ψ forms a basis for an irreducible representation of the group; ψ is also an eigenfunction to the Hamiltonian, so it follows (see Section 6.6) that $\hat{O}$ and $\hat{H}$ must commute. The set of operators which commute with the Hamiltonian of a particular system are said to

---

[15] In general, if degeneracy exists, $\hat{O}\psi_k = \psi_l$, where ψ_k and ψ_l are members of a degenerate set, for example, form the basis for a two-dimensional irreducible representation of the group.

6.6 Group Theory in Quantum Mechanics

form the *group of the Hamiltonian*. Specifically, the symmetry operators of the point group of the molecule commute with the molecular Hamiltonian.

$$\hat{H}\psi = E\psi \qquad \hat{O}\psi = \pm\psi$$
$$\hat{H}\hat{O}\psi = \pm E\psi = \hat{O}\hat{H}\psi$$

or

$$\hat{O}\hat{H} = \hat{H}\hat{O}$$

The fact that the exact wavefunction forms the basis of an irreducible representation of the group is an important bit of information, because often it is almost all we know about the exact solution of the Schrödinger equation. In the LCAO–MO method the form of the MO can be determined (to within a normalization factor) by the use of symmetry (see next example).

From the expansion theorem (2.5.11) the wavefunction can be represented as a linear combination of functions which form a complete set. In general, the expansion set of functions forms the basis of a reducible representation of the group, that is, the basis set $\phi_1, \phi_2, \ldots$ transforms as

$$\begin{pmatrix} \phi_1 \\ \phi_2 \\ \vdots \end{pmatrix} = \begin{pmatrix} a_{11} & a_{12} & \cdots \\ a_{21} & a_{22} & \cdots \\ \vdots & \vdots & \end{pmatrix} \begin{pmatrix} \phi_1 \\ \phi_2 \\ \vdots \end{pmatrix} = \begin{pmatrix} \phi'_1 \\ \phi'_2 \\ \vdots \end{pmatrix}$$

where $\phi'_1 = a_{11}\phi_1 + a_{12}\phi_2 + a_{13}\phi_3 + \cdots$. The trace of the matrix representation $a_{11} + a_{22} + a_{33} = \cdots$ can identify it as a reducible representation of the group and help to identify its component irreducible representations. The basic problem is to break the basis set into sets which form the basis of irreducible representations, and to use these *symmetry-adapted* functions to represent eigenfunctions of the Hamiltonian belonging to those irreducible representations. This process of taking an arbitrary basis set and symmetry-adapting it to represent molecular orbitals is illustrated in the next example.

EXAMPLE   Find the symmetry-adapted combinations (LCAO–MO) of three 1s functions located at the corners of an equilateral triangle. The LCAO–MO transform as irreducible representations of the group. The symmetry elements of the triangle form the group $\mathbf{D}_{3h}$. The symmetry elements (see Fig. 6.6.1) are a C_3 axis ($\hat{C}_3$ and $\hat{C}_3^2$), three twofold axes ($3\hat{C}'_2$), a horizontal mirror plane ($\hat{\sigma}_h$), three vertical mirror planes ($3\hat{\sigma}_v$), and an improper rotation axis $S_3(\hat{C}_3\hat{\sigma}_h$ and $\hat{C}_3^2\hat{\sigma}_h)$.

Table 6.6.1 is the character table of $\mathbf{D}_{3h}$. The elements are grouped according to their class; there are six classes and therefore six irreducible representations of $\mathbf{D}_{3h}$. Two of the irreducible representations are two dimensional.

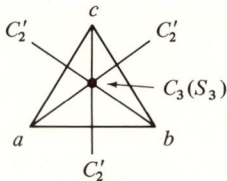

FIG. 6.6.1

TABLE 6.6.1 Character Table of D_{3h}

	E	$2C_3$	$3C_2'$	σ_h	$2S_3$	$3\sigma_v$
A_1'	1	1	1	1	1	1
A_1''	1	1	1	-1	-1	-1
A_2'	1	1	-1	1	1	-1
A_2''	1	1	-1	-1	-1	1
E'	2	-1	0	2	-1	0
E''	2	-1	0	-2	1	0

The set of atomic orbitals $1s_a$, $1s_b$, and $1s_c$ located at the corners of the triangle form a basis for a reducible representation of D_{3h}. This is easily seen by consideration of equations (6.6.1).

$$\hat{E}\begin{pmatrix}1s_a\\1s_b\\1s_c\end{pmatrix} = \begin{pmatrix}1 & 0 & 0\\0 & 1 & 0\\0 & 0 & 1\end{pmatrix}\begin{pmatrix}1s_a\\1s_b\\1s_c\end{pmatrix} = \begin{pmatrix}1s_a\\1s_b\\1s_c\end{pmatrix} \qquad \chi_{red}(\hat{E}) = 3$$

$$\hat{C}_3\begin{pmatrix}1s_a\\1s_a\\1s_c\end{pmatrix} = \begin{pmatrix}0 & 0 & 1\\1 & 0 & 0\\0 & 1 & 0\end{pmatrix}\begin{pmatrix}1s_a\\1s_b\\1s_c\end{pmatrix} = \begin{pmatrix}1s_c\\1s_a\\1s_b\end{pmatrix} \qquad \chi_{red}(\hat{C}_3) = 0$$

(6.6.1)

and so forth for the other operations. *Note that the character of the representation for a given operation is just the number of functions which are left undisturbed by the operation.* The characters for the other operations in this basis then follow immediately: $\chi_{red}(\hat{C}_2') = 1$, $\chi_{red}(\hat{\sigma}_h) = 3$, $\chi_{red}(\hat{S}_3) = 0$, and $\chi_{red}(\hat{\sigma}_v) = 1$. The student should verify these characters by writing out the transforming matrices.

We may now use (6.5.3) to determine the number of times each irreducible representation is contained in the representation with basis $1s_a$, $1s_b$, $1s_c$. The number of times A_1' is contained in this reducible representation is

$$n_{A_1'} = \tfrac{1}{12}\sum_{\hat{O}} \chi_{red}(\hat{O})\chi_{A_1'}(\hat{O}) \qquad (6.6.2)$$

Remember that the sum in (6.6.2) is over each member of the group even though the characters of operations in the same class are identical (see Table 6.6.1).

6.6 Group Theory in Quantum Mechanics

Equation (6.6.2) is then

$$n_{A_1'} = \tfrac{1}{12}(3 \times 1 + 2 \times 0 \times 1 + 3 \times 1 \times 1 + 3 \times 1 + 2 \times 0 \times 1 + 3 \times 1 \times 1)$$

$$= \tfrac{1}{12}(12) = 1$$

So A_1' is contained in the reducible representation only once. Proceeding in the same way the student will verify that $n_{A_1''} = n_{A_2'} = n_{A_2''} = n_{E''} = 0$, but $n_{E'} = 1$.

Finally, we identify (*project out*) the basis of the irreducible representation A_1' and the basis of E'. We can use the projection operators $\hat{P}^{A_1'}$ and $\hat{P}^{E'}$ of equation (6.5.5).

$$\hat{P}^{A_1'} = \tfrac{1}{12}\sum_{\hat{O}} \chi_{A_1'}(\hat{O})\hat{O}$$

$$\hat{P}^{A_1'} 1s_a = \tfrac{1}{12}(\hat{E} 1s_a + \hat{C}_3\, 1s_a + \hat{C}_3^2\, 1s_a + 3\hat{C}_2'\, 1s_a + \hat{\sigma}_h 1s_a + 2\hat{S}_3\, 1s_a + 3\hat{\sigma}_v 1s_a)$$

$$= \tfrac{1}{12}(1s_a + 1s_c + 1s_b + 1s_c + 1s_a + 1s_b + 1s_a + 1s_c + 1s_b$$
$$\quad + 1s_a + 1s_b + 1s_c)$$

$$= \tfrac{4}{12}(1s_a + 1s_b + 1s_c)$$

To within a constant factor (which should be determined by normalization to unity) the LCAO–MO which transforms as the *totally symmetric* irreducible representation is $\psi_{A_1'} = 1s_a + 1s_b + 1s_c$.

EXERCISE Show that the same function is obtained when $\hat{P}^{A_1'}$ works on $1s_b$ or $1s_c$.

The same procedure is used to project out the wavefunction which transforms as E'.

$$\hat{P}^{E'} 1s_a = \tfrac{2}{12}[2\, 1s_a - 1s_b - 1s_c + 2(1s_a) - 1s_b - 1s_c]$$

$$= \tfrac{4}{12}[2(1s_a) - 1s_b - 1s_c]$$

EXERCISE Show that $\hat{P}^{E'}\, 1s_b = \tfrac{4}{12}[2(1s_b)] - 1s_a - 1s_c]$ and $\hat{P}^{E'}\, 1s_c = \tfrac{4}{12}[2(1s_c) - 1s_a - 1s_b]$.

To within a normalization factor we have three (apparently different) functions which when taken two at a time form a basis for the irreducible representation E'. However, all three functions are *not* linearly independent,[16] that is, to within an arbitrary constant factor any one of the three functions can be expressed in terms of the other two, for example, $[2(1s_b) - 1s_a - 1s_c] + [2(1s_c) - 1s_a - 1s_b]$
$= -[2(1s_a) - 1s_b - 1s_c]$.

EXERCISE Show that

$$\begin{pmatrix} 2(1s_a) & -1s_b & -1s_c \\ 2(1s_b) & -1s_a & -1s_c \end{pmatrix}$$

forms a basis for the two-dimensional irreducible representation E' by actually transforming the vector with the operators of $\mathbf{D}_{3h}$ to obtain the characters of the transforming matrices.

---

[16] Appendix A, equation (A.1.13).

DIRECT PRODUCTS AND QUANTUM MECHANICAL INTEGRALS

The power of group theoretical analysis in quantum mechanics arises not only from the ability to reduce and classify functions into bases of irreducible representations, but in yielding information about integrals over products of these functions. Consider two sets of functions each of which forms a basis for a representation of the group $\phi_1, \phi_2, \phi_3, \ldots$ and $\phi'_1, \phi'_2, \phi'_3, \ldots$. The basis set $\phi_i \phi'_j$, which consists of products of these functions, also forms a basis for a representation of the group; it is called the *direct product representation*. In general, the direct product representation will be reducible even if the ϕ_i and the ϕ'_j each transformed as irreducible representations. The reduction of the direct product representation into its component irreducible representations is known if the characters of the direct product representation are known. The simple rule is that the character of the direct product representation is the product of the characters of the component functions.

PROOF

$$\hat{O}\phi_i = \sum_n c_{in} \phi_n \qquad \hat{O}\phi'_j = \sum_m d_{jm} \phi'_m$$

$$\hat{O}(\phi_i \phi'_j) = \sum_{n,m} c_{in} d_{jm} (\phi_n \phi'_m)$$

and $\hat{O}(\phi_k \phi'_l) = \sum_{n,m} c_{kn} d_{lm} (\phi_n \phi'_m)$ etc.

Therefore the trace (character) of the transforming matrix is

$$\chi(\hat{O})_{\text{direct product}} = \sum_{i,j} c_{ii} d_{jj} = \chi(\hat{O}) \chi'(\hat{O})$$

where $\chi(\hat{O})$ and $\chi'(\hat{O})$ are the characters of unprimed and primed representations.

EXAMPLE Use the character table of $\mathbf{D}_{3h}$ to find the component irreducible representations of the direct products in Table 6.6.2. In Table 6.6.2 the resultant irreducible representations follow from the application of (6.5.3) when the direct product is reducible.

TABLE 6.6.2 Direct Product of Irreducible Representations of $\mathbf{D}_{3h}$.

Product	E	$2C_3$	$3C'_2$	σ_h	$2S_3$	$3\sigma_v$	Resultant
$A'_1 \times A'_1$	1	1	1	1	1	1	A'_1
$A'_1 \times A''_1$	1	1	1	-1	-1	-1	A''_1
$A''_1 \times A''_1$	1	1	1	1	1	1	A'_1
$A'_1 \times E'$	2	-1	0	2	-1	0	E'
$A''_1 \times E'$	2	-1	0	-2	1	0	E''
$E' \times E'$	4	1	0	4	1	0	$E' + A'_1 + A'_2$

The direct product of a totally symmetric function with a nontotally symmetric function yields a nontotally symmetric function. On the other hand, the product of two functions belonging to the *same* irreducible representation yields a function which contains the totally symmetric representation, for example, $E' \times E'$ in Table 6.6.2.

The direct product rules have an immediate application in the evaluation of quantum mechanical integrals. The integral

$$\int_{-a}^{+a} \int_{-b}^{+b} \int_{-c}^{+c} f(x, y, z) \, dx \, dy \, dz$$

vanishes if there are cancelling positive and negative contributions from different regions of space. *Such is the case if $f(x, y, z)$ transforms as a nontotally symmetric irreducible representation.* This theorem permits us to examine the direct product of the functions in the integrand of the quantum mechanical integrals $\langle \psi_i, \psi_j \rangle$, $\langle \psi_i, \hat{H}\psi_j \rangle$, and in general $\langle \psi_i, \hat{M}\psi_j \rangle$, where $\hat{M}$ is any quantum mechanical operator, to see if the direct product contains a component transforming as the totally symmetric representation; if it does not, *the integral mush vanish*. This is a severe restriction on the integrand; for example in the overlap integral $\langle \psi_i, \psi_j \rangle$, both ψ_i and ψ_j must transform as bases of the *same* irreducible representation; in the energy integral $\langle \psi_i, \hat{H}\psi_j \rangle$, $\hat{H}$ is totally symmetric, so again the wavefunctions must transform in the same way. In Section 6.8 this theorem will determine the selection rules for spectral transitions between states of the molecule.

6.7 VIBRATIONAL SPECTRA OF POLYATOMIC MOLECULES

The polyatomic molecule consists of N nuclei which are undergoing small oscillations about their equilibrium positions. The positions of the nuclei are completely specified by the $3N$ coordinates $q_1, q_2, \ldots, q_{3N}$. It is convenient to choose the coordinates to represent the *displacement* from equilibrium; then $q_i = 0$ at equilibrium. In these coordinates the classical kinetic and potential energy (Section 2.1) of the nuclei are

$$T = \tfrac{1}{2} \sum_i m_i \left(\frac{dx_i}{dt} \right)^2 = \sum_i \sum_j \tfrac{1}{2} a_{ij} \frac{dq_i}{dt} \frac{dq_j}{dt} \tag{6.7.1}$$

$$V = V_0 + \sum_i \left(\frac{\partial V}{\partial q_i} \right)_0 q_i + \tfrac{1}{2} \sum_i \sum_j \left(\frac{\partial^2 V}{\partial q_i \, \partial q_j} \right)_0 q_i q_j + \cdots \tag{6.7.2}$$

In (6.7.2) the potential energy has been expanded in a Taylor series about the equilibrium position. The zero subscript means evaluated at equilibrium, but at equilibrium V is a minimum, so $(\partial V/\partial q_i)_0 = 0$ and the arbitrary constant V_0 may also be taken as zero. As a result

$$V = \sum_i \sum_j b_{ij} q_i q_j \qquad (6.7.3)$$

where $b_{ij} = \frac{1}{2}(\partial^2 V/\partial q_i \, \partial q_j)_0$ and terms higher than this quadratic term in V are ignored.

The classical equations of motion are now [equation (2.1.7)]

$$-\frac{dp_i}{dt} = \frac{\partial H}{\partial q_i} = \sum_j b_{ij} q_j \qquad (6.7.4)$$

There are $3N$ such Hamiltonian equations of motion ($i = 1, 2, 3, \ldots, 3N$) which are coupled together through the coefficients b_{ij}. Fortunately, it is possible to change the variables to a set $Q_1, Q_2, Q_3, \ldots$ such that each Q_i satisfies an independent equation.

$$-\frac{dP_i}{dt} = \lambda Q_i \qquad (6.7.5)$$

The proof of (6.7.5) is given elsewhere.[17]

The Q_i are the *normal coordinates* of the molecule and they have the following properties. Of the $3N$ degrees of freedom (independent modes of nuclear motion of N nuclei) six (five for a linear molecule) are non-periodic modes, that is, three translations and three rotations, the remaining $3N - 6$ ($3N - 5$ for a linear molecule[18]) are true vibrational modes. Since the potential energy expansion was cutoff at the quadratic term, the periodic motions described by the normal modes are simple *harmonic oscillations* (see diatomic vibrations, Section 6.1). In a normal mode of vibration all the nuclei carry out harmonic oscillations at the same frequency, and although their amplitudes of vibration may differ, they move, in general, in phase.[19]

Figure 6.7.1 displays the $3N - 5 = 4$ normal modes of CO_2, a linear triatomic molecule; v_1 is the *symmetric stretching* mode, v_3 is the *unsymmetric stretching* mode, and v_2 is the doubly degenerate *bending mode* of vibration. The arrows indicate the displacement from equilibrium of the nuclei on one-half of a complete cycle.

---

[17] E. B. Wilson, Jr., J. C. Decious, and P. C. Cross, *Molecular Vibrations*, McGraw-Hill, New York, 1955.

[18] The linear molecule has three translational and only two rotational degrees of freedom (zero moment of inertia about the symmetry axis).

[19] G. Herzberg, *Infrared and Raman Spectra*, Van Nostrand, New York, 1945, illustrates the normal modes of vibration of many simple polyatomic molecules.

6.7 Vibrational Spectra of Polyatomic Molecules

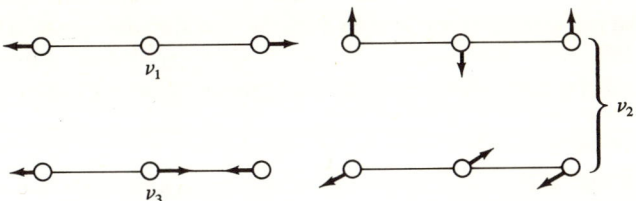

FIG. 6.7.1 Normal modes of vibration of a linear triatomic molecule such as CO_2. For CO_2, $v_1 = 1340$ cm^{-1}, $v_2 = 667$ cm^{-1}, and $v_3 = 2349$ cm^{-1}.

Figure 6.7.2 illustrates the normal modes of vibration of a molecule having the point group symmetry C_{2v}, such as H_2O: Again, symmetrical and unsymmettical stretching modes v_1 and v_3, and the bending mode v_2 ($3N - 6 = 3$) are present. An important property of the normal modes is that *each mode forms the basis of an irreducible representation of the molecular point group.*

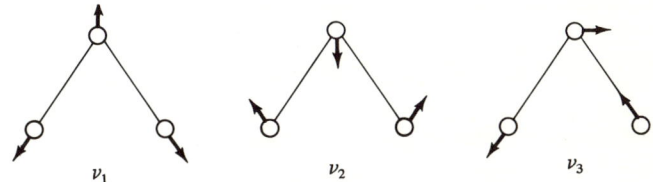

FIG. 6.7.2 Normal modes of vibration of a bent triatomic molecule, C_{2v}, such as H_2O. For H_2O, $v_1 = 3652$ cm^{-1}, $v_2 = 1595$ cm^{-1}, and $v_3 = 3755$ cm^{-1}. For SO_2, $v_1 = 1151$ cm^{-1}, $v_2 = 519$ cm^{-1}, and $v_3 = 1361$ cm^{-1}.

EXAMPLE Q_1, Q_2, and Q_3 are the normal mode displacements illustrated in Fig. 6.7.2. It is evident by inspection that Q_1 and Q_2 transform as the basis of an A_1 representation of C_{2v} (see Table 6.5.1), that is, $\hat{O}Q_1 = Q_1$ and $\hat{O}Q_2 = Q_2$ for each symmetry operator $\hat{O}$ of C_{2v}.

EXERCISE Show that Q_3, the unsymmetrical mode of Fig. 6.7.2, forms the basis of a B_1 representation of C_{2v}.

The normal modes and their assignments to irreducible representations of the point groups have been catalogued[20] for all the common small

[20] Wilson, Decious, and Cross, footnote 17; G. Herzberg, cited in footnote 19.

polyatomic molecules and many others of reasonable size. However, for large molecules of low symmetry the normal mode analysis is of great difficulty. One usually analyzes their infrared spectra in terms of characteristic frequencies of the bonds or groups of atoms present in the molecule.[21]

In principle, it is possible to determine the irreducible representations of the normal modes by decomposition of the reducible representation which arises from the transformation of the Cartesian coordinates of the nuclei. According to (6.5.3), all we need are the characters of the reducible matrix representation for which these Cartesian coordinates form a basis.

EXAMPLE H_2O. Use Cartesian coordinates on each nucleus with alignment as shown in Fig. 6.3.1. That is, vectors x_i, y_i, and z_i are attached to each of the nuclei. There is a contribution to the character of the transformation matrix only if the nucleus is unmoved by the symmetry operation. So, $\chi_{red}(\hat{E}) = 9$ because it leaves all the vectors unchanged. $\hat{C}_2$ leaves only oxygen unmoved, $z_0 \to z_0$, $x_0 \to -x_0$, $y_0 \to -y_0$, $\chi_{red}(\hat{C}_2) = -1$. $\hat{\sigma}_v$ also interchanges the hydrogen nuclei, leaving oxygen unmoved, $z_0 \to z_0$, $x_0 \to x_0$, $y_0 \to -y_0$, $\chi_{red}(\hat{\sigma}_v) = 1$. $\hat{\sigma}'_v$ leaves all the nuclei unmoved and sends $z \to z$, $y \to y$, and $x \to -x$ for each nucleus, $\chi_{red}(\hat{\sigma}'_v) = 3$. Using (6.5.3) to decompose the reducible representation we get

$$\chi_{red} = 3A_1 + A_2 + 3B_1 + 2B_2$$

From these nine irreducible representations we must subtract out the irreducible representations of the three translational and the three rotational motions.[22]

$$\chi_{trans} = B_2(x) + B_1(y) + A_1(z)$$
$$\chi_{rot} = B_1(R_x) + B_2(R_y) + A_2(R_z)$$

where R_x, R_y, R_z are rotations about these respective axes. By subtraction we obtain normal modes of C_{2v} triatomic molecule.

$$\chi_{vib} = A_1(Q_1) + A_1(Q_2) + B_1(Q_3)$$

Because each vibration in a polyatomic molecule is (in the harmonic oscillator approximation) an independent harmonic oscillation, the total vibration energy in the molecule is the sum of $3N-6$ harmonic oscillator energies.

$$E_{vib} = \sum_{i=1}^{3N-6} h\nu_i(v_i + \tfrac{1}{2}) \qquad v_i = 0, 1, 2, 3, \ldots \qquad (6.7.6)$$

The vibrational spectrum of the polyatomic molecule is complex because of the several allowed transitions between the ground state ($v_1 = v_2 = v_3 = \ldots 0$) and excited vibrational states with other possible values of the v_i. However, in contrast to the diatomic molecule, *not all* vibrational transitions with $\Delta v_i = \pm 1$ are allowed. This brings us to the discussion of selection rules.

---

[21] L. J. Bellamy, *Infra-red Spectra of Complex Molecules*, Wiley, New York, 1954.
[22] The x, y, and z translations transform as the x, y, and z coordinates [see eq. (6.5.2)]. The rotations R_x, R_y, and R_z transform as the coordinate products yz, xz, and xy, that is, as the direct products $B_1 \times A_1$, $B_2 \times A_1$, and $B_2 \times B_1$.

6.8 OPTICAL SPECTRA AND SELECTION RULES

We are concerned with the effect of electromagnetic radiation on an atom or molecule. The molecule has stationary states of energy E_i and E_j and the radiation is of frequency $v = (E_i - E_j)/h$. This satisfies the condition under which radiation is absorbed and emitted by the molecule. The *intensity* of absorption or emission depends on the *probability* of the radiative process for an isolated molecule and on the *population* of the *i*th and *j*th levels. Here we are specifically concerned with the *probability* for an isolated molecule.

The molecule has a probability of *spontaneous emission* and a probability of *induced emission*. (By induced emission we refer to the effect of the electromagnetic field in inducing emission; spontaneous emission occurs even in the absence of the field.) Induced emission is the basic mechanism of the LASER,[23] for if light enters a sample in which the population of the excited state exceeds that of the ground state, more light is emitted than absorbed by the sample, that is, amplification of the incident light to very high intensities occurs.

The probability that an isolated molecule will undergo an induced transition between levels i and j $[(E_i - E_j)/h = v]$ is proportional to the radiation density at frequency v, $\rho(v)$.

$$\text{Probability} = B_{ij}\rho(v) \tag{6.8.1}$$

B_{ij} is the constant of proportionality (the Einstein coefficient for absorption or induced emission). The relation between B_{ij} and the probability of spontaneous emission, A_{ij}, was derived by Einstein[24a] from an equilibrium argument given elsewhere,[24b] $A_{ij} = B_{ij} 8\pi h v^3/c^3$. Since our interest is mainly in the absorption of radiation we will now relate B_{ij} to experimental observables on the one hand, and to the wavefunctions and their group theoretical symmetries on the other.

THE BEER-LAMBERT LAW

From the observational point of view the absorption intensity is determined using the Beer-Lambert law. The Beer-Lambert law states that the intensity decrease after light of frequency v has traversed a length of solution containing absorbing molecules (solute concentration C moles/liter) is proportional to (1) the incident intensity, (2) the path length, and (3) the concentration of absorbing solute. For a differential path length dl, this implies

$$-dI = kCI\,dl$$

---

[23] *L*ight *A*mplification by *S*timulated (Induced) *E*mission of *R*adiation.
[24] (a) A. Einstein, *Phys. Z.* **18**, 121 (1917). (b) L. Pauling and E. B. Wilson, *Introduction to Quantum Mechanics*, McGraw-Hill, New York, 1935.

where I is the intensity, C the concentration, and k the constant of proportionality. Integrating

$$\ln(I/I_0) = -kCl \tag{6.8.2}$$

The constant of integration I_0 is the incident intensity on the sample ($I = I_0$ at $l = 0$); l is in centimeters. In log base 10 with $\varepsilon = k/2.303$ we have

$$I = I_0 10^{-\varepsilon C l} \qquad \log(I/I_0) = -\varepsilon C l \tag{6.8.3}$$

where ε is the *molar extinction coefficient*.

Electronic and other molecular transitions do not occur at a single frequency, but are spread over a range of frequencies which define a band. It follows that the absorption intensity varies with frequency over the band; ε is a function of v. The quantum mechanical transition probability is related to ε integrated over the frequencies which comprise the band, that is, the entire transition. It can be shown[25] that the relation is

$$B_{ij} = \frac{2302c}{hnN_0} \int \frac{\varepsilon}{v} dv \tag{6.8.4}$$

where n is the index of refraction of the medium and N_0 is Avogadro's number.

THE TRANSITION MOMENT

Light absorption occurs because the molecule is *perturbed* by the electromagnetic field. The energy of a system of charged particles in an electric field E_x parallel to the x axis is increased by an amount $E_x \sum_j e_j x_j$, where e_j is the charge and x_j is the x coordinate of the jth particle. This constitutes a perturbation of the molecule expressible in terms of $\sum_j e_j x_j$, the x component of the dipole moment, $\sum_j e_j x_j = \mu_x$. Application of time-dependent perturbation theory to a molecule arbitrarily oriented in the field yields[26]

$$B_{ij} = 8\pi^3/3h^2(|\langle \psi_i, \mu_x \psi_j \rangle|^2 + |\langle \psi_i, \mu_y \psi_j \rangle|^2 + |\langle \psi_i, \mu_z \psi_j \rangle|^2)$$
$$B_{ij} = 8\pi^3/3h^2 |\langle \psi_i, \boldsymbol{\mu} \psi_j \rangle|^2 \tag{6.8.5}$$

ψ_i and ψ_j are the wavefunctions of the ith and jth states of the molecule, and $\boldsymbol{\mu}$ is the dipole moment.

The quantity

$$\mathbf{M}_{ij} = \langle \psi_i, \boldsymbol{\mu} \psi_j \rangle \tag{6.8.6}$$

is the *transition moment*. Thus the transition moment controls the transition probability for an isolated molecule.

[25] S. J. Strickler and R. A. Berg, *J. Chem. Phys.* **37**, 814 (1962).
[26] Appendix H.

6.8 Optical Spectra and Selection Rules

For electronic transitions the transition probability is often expressed in terms of the *oscillator strength, f*.

$$f = \tfrac{2}{3} G \, \Delta E \, |\langle \psi_i, \boldsymbol{\mu} \psi_j \rangle|^2 \tag{6.8.7}$$

ΔE is the transition energy in hartrees, $\langle \psi_i, \boldsymbol{\mu} \psi_j \rangle$ is in atomic units, and G is the final-state degeneracy.

SELECTION RULES

The selection rules for optical transition are obtained by applying the group theoretical theorem to the integrand of the transition moment integral: *the direct product of ψ_i, $\boldsymbol{\mu}$ and ψ_j must contain the totally symmetric representation or else the integral is zero and the transition is forbidden.*[27]

Selection Rule: $\quad \psi_i \times \boldsymbol{\mu} \times \psi_j$ transforms as the totally symmetric representation

The dipole moment operator has three components μ_x, μ_y, and μ_z which transform as the x, y, and z coordinates in the molecular point group. In the electronic spectra of polyatomic molecules the selection rule requires that the direct product of irreducible representations for which ψ_i and ψ_j form the basis must contain the irreducible representation for which either x or y or z forms the basis.[28] The transition is then polarized in either the x, y, or z direction, that is, occurs under the influence of the component of the electric vector of light in the x, y, or z direction.

We complete the discussion of vibration spectra by presenting two examples of vibrational selection rules.

EXAMPLE H_2O. Possesses three normal modes of A_1 and B_1 symmetry (Section 6.7); x, y, and z form the basis of irreducible representations B_2, B_1, and A_1, respectively. Because the "vibrationless state"[29] ($v_1 = v_2 = v_3 = 0$) has symmetry A_1, the fundamental transitions are all allowed.

$$A_1(Q_0) \times A_1(z) \times A_1(Q_1) = A_1$$
$$A_1(Q_0) \times A_1(z) \times A_1(Q_2) = A_1$$
$$A_1(Q_0) \times B_1(y) \times B_1(Q_3) = A_1$$

Q_0 is the vibrationless ground state, and Q_1, Q_2, and Q_3 are excited vibrational states with one quantum of these normal modes.

---

[27] Perturbation theory (Appendix H) leading to (6.8.5) limits itself to the largest interaction between radiation and matter (electric dipole interaction). A transition may be forbidden as electric dipole radiation but may be allowed as electric quadrupole or magnetic dipole radiation. The transition may then appear, but very, very weakly. Such observations are rare.

[28] Selection rules for electronic transitions of atoms (Section 4.8) and diatomic molecules (Section 6.2) follow from the same theorem.

[29] Actually all three modes have their zero-point vibrational energy when $v = 0$.

EXAMPLE C_6H_6, benzene, $\mathbf{D}_{6h}$. Benzene has $3N - 6 = 30$ vibrational degrees of freedom and 20 normal modes of vibration (10 modes are doubly degenerate) of symmetry A_{1g}, A_{2g}, A_{2u}, B_{1u}, B_{2g}, B_{2u}, E_{2g}, E_{2u}, E_{1g}, and E_{1u}. Because (x, y) and z transform as E_{1u} and A_{2u} the only excited vibrational states which can combine with the vibrationless ground state A_{1g} must be of u symmetry ($g \times u \times u = g$). Furthermore only the normal modes which transform as x, y, or z will have allowed infrared transitions, which limits the infrared spectrum of benzene to A_{2u} and E_{1u} modes.

Magnetic Resonance Spectra

6.9 INTRODUCTION TO MAGNETIC RESONANCE SPECTROSCOPY

Spin angular momentum always plays an important role in atomic and molecular spectroscopy. The selection rule $\Delta S = 0$ (Sections 4.8 and 6.2) for atomic and molecular transitions follows from the vanishing transition moment $\langle \psi_s, \boldsymbol{\mu}\psi_t \rangle = 0$, where ψ_s and ψ_t belong to different eigenvalues of the operator $\hat{S}^2$ and as a result have *orthogonal* spin functions. However, there are nonoptical spectroscopic techniques which are more directly concerned with the spin properties of electrons and nuclei. These are the magnetic resonance techniques, NMR (*N*uclear *M*agnetic *R*esonance) and ESR (*E*lectron *S*pin *R*esonance).[30, 31]

The proton and the neutron occur as elementary particles in the nucleus and form the class of particles called *nucleons*. The nucleon has a spin $\frac{1}{2}$, and, as in the case of the electron, the nucleon spin is an intrinsic angular momentum detectable from the resultant magnetic moment of the nucleon. In nuclei the nucleon spins couple, yielding a resultant *nuclear spin I*. Nuclei may have integral, half-integral, or zero spin depending on the mass number A (the number of nucleons) and the atomic number Z (number of protons), as shown in Tables 6.9.1 and 6.9.2.

TABLE 6.9.1 Relation Between Atomic Number Z, Mass Number A, and the Nuclear Spin I^a

$A =$ Even	Odd	Even	Odd
$Z =$ Even	Odd	Odd	Even
$I =$ Zero	h.i.	i.	h.i.

[a] h.i., half-integral; i., integral.

[30] See A. Carrington and A. D. McLachlan, *Introduction to Magnetic Resonance With Applications to Chemistry and Chemical Physics*, Harper & Row, New York, 1967.
[31] Equivalently, EPR (*E*lectron *P*aramagnetic *R*esonance).

TABLE 6.9.2 Properties of Some Atomic Nuclei

Isotope	A	Z	I	g_N
^{1}H	1	1	$\frac{1}{2}$	5.58554
^{2}H	2	1	1	0.85738
^{6}Li	6	3	1	0.8221
^{10}B	10	5	3	0.6003
^{12}C	12	6	0	—
^{13}C	13	6	$\frac{1}{2}$	1.4044
^{14}N	14	7	1	0.40358
^{16}O	16	8	0	—
^{19}F	19	9	$\frac{1}{2}$	5.2546
^{23}Na	23	11	$\frac{3}{2}$	1.478
^{119}Sn	119	50	$\frac{1}{2}$	2.082
^{31}P	31	15	$\frac{1}{2}$	2.2610
^{207}Pb	207	82	$\frac{1}{2}$	1.1674

The elementary magnetic properties of electrons and nuclei are summarized in Table 6.9.3 in terms of two fundamental quantities, the magneton β and the g factor. Although the electronic magnetic moment has only two possible orientations in a magnetic field, corresponding to $m_s = \pm\frac{1}{2}$ and components of μ_e in the field direction $-m_s g_e \beta_e$, the nuclear magnetic moment has, in general, several orientations, with components of $\mu_N = m_I g_N \beta_N$. Analogous to m_s, m_I is the magnetic quantum number of the nucleus, $m_I = -I, -I+1, -I+2, \ldots, I$. From Table 6.9.2 the empirical g factor g_N has a *characteristic value* for each nucleus, for example, g_N(proton) = 5.585; however the nuclear magnetic resonance spectrum does depend on the molecular environment and thus provides molecular information. The electronic g factor g_e departs from its characteristic free-electron value (2.0023) in molecules; thus the ESR spectrum, like the NMR spectrum, provides information about the molecular electronic distribution and its magnetic properties.

How a magnetic resonance spectrum arises will be dealt with next. However, we should bear in mind that most molecules have diamagnetic ground states (zero magnetic moment) and therefore do *not* have an ESR spectrum. Species which can be studied with ESR are limited to organic and inorganic free radicals and other paramagnetic species such as the paramagnetic ions (e.g., Fe^{3+}). These paramagnetic species may be studied in solution or in the solid state. On the other hand, NMR is applicable to

TABLE 6.9.3 Table of Quantities Associated with Spin

Electrons	Quantity	Nuclei
g_e	The spectroscopic splitting factor (the g factor). An empirical quantity having the value 2.0023 for the free electron.	g_N
$\beta_e = e\hbar/2mc$ $\quad m$, electron mass $\beta_e = 9.2732 \times 10^{-21}$ erg/gauss	β_e, the Bohr magneton β_N, the nuclear magneton	$\beta_N = e\hbar/2Mc$ M, proton mass $\beta_N = 5.050 \times 10^{-24}$ erg/gauss
$\gamma_e = g_e \beta_e/\hbar$	Magnetogyric ratio	$\gamma_N = g_N \beta_N/\hbar$
$\boldsymbol{\mu}_e = -\hat{S}\beta_e g_e/\hbar$ $\quad = -\hat{S}\gamma_e$	Magnetic moment operator	$\boldsymbol{\mu}_N = \hat{I}\beta_N g_N/\hbar$ $\quad = \hat{I}\gamma_N$
$\mu_{ez} = -m_s \beta_e g_e$	Component of magnetic moment in the z direction. $m_s = \pm\frac{1}{2}$ $m_I = -I, -I+1, \ldots, I$	$\mu_{Nz} = m_I \beta_N g_N$
$E = -\boldsymbol{\mu}_e \cdot \mathbf{H}$ $E = m_s g_e \beta_e H$	Energy in a magnetic field $\mathbf{H}$ (direction of $\mathbf{H}$ defines the z axis).	$E = -\boldsymbol{\mu}_N \cdot \mathbf{H}$ $E = -m_I g_N \beta_N H$

a wider variety of chemical problems, as should be evident from the nuclei listed in Table 6.9.2 and the discussion of chemical shielding to follow in Section 6.11.

From Table 6.9.3 the excess energy of a particle of spin $\frac{1}{2}$ ($S = \frac{1}{2}$ or $I = \frac{1}{2}$) in a magnetic field is

$$E = \pm\tfrac{1}{2}g\beta H \tag{6.9.1}$$

For electrons, the upper and lower signs refer to parallel (α) and antiparallel (β) spin orientations to the field direction, and vice versa for nuclei.

The electron or nucleus may be induced to reorient itself in the magnetic field by application of an oscillating electromagnetic field. The arrow in Fig. 6.9.1 indicates the transition which occurs when an electromagnetic

6.9 Introduction to Magnetic Resonance Spectroscopy

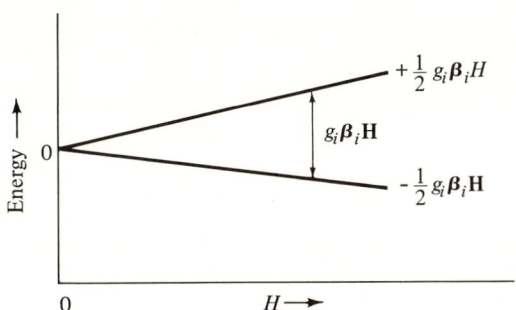

FIG. 6.9.1 Energy of a spin ½ particle in a magnetic field.

field of frequency $v = g\beta H/h$ is applied to the sample in the magnetic field. Electromagnetic radiation of this frequency satisfies the *resonance condition* for the transition, ergo *magnetic resonance*. The use of the word "resonance" in quantum mechanics is derived from certain semiclassical analogies with the theory of oscillation. In classical theory two coupled oscillators of similar natural frequency are said to be in resonance. If one oscillator is excited it soon transfers its excitation to the other oscillator. A simple example consists of two tuning forks of the same frequency coupled via a common base. If one fork is struck it soon ceases to vibrate but the other fork begins to oscillate. "Resonance" in quantum chemistry incorporates these ideas to help explain what is really a quantum mechanical phenomenon; the word was popularized by Linus Pauling.[32] It behooves us to explain (by means of a semiclassical model) what oscillators are in resonance in "magnetic resonance."

When an electron or nucleus is placed in a magnetic field the field exerts a torque or twisting force on the magnetic dipole moment of the particle. The net result is a precession of the dipole moment about the direction of the applied field. This is entirely analogous to the precession of a child's spinning top about the direction of the earth's gravitational field. Figure 6.9.2 illustrates the precession of a nuclear magnetic moment about the magnetic field direction. The energy of the precessing magnetic moment is $E = -\boldsymbol{\mu} \cdot \mathbf{H}$ and depends on the quantum number m_s or m_I (Zeeman effect). In any case the spin system is in a stationary state and is spectroscopically uninteresting unless it can be induced to undergo transitions between states of different m_s or m_I. Such transitions are induced by generating an electromagnetic field with oscillating magnetic vector

---

[32] L. Pauling and E. B. Wilson, Jr., *Introduction to Quantum Mechanics*, pp. 314–325, McGraw-Hill, New York, 1935.

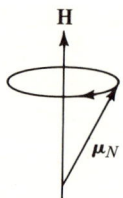

FIG. 6.9.2 Precessional motion of the nuclear magnetic moment about the magnetic field direction.

perpendicular to the direction of the applied field **H**. If the frequency of the radiation is varied until it precisely equals the frequency of precession, one has satisfied the *resonance condition* (i.e., magnetic resonance). In this resonance condition energy is absorbed from the electromagnetic field and the particle undergoes a transition, $m_s = -\frac{1}{2}$ to $m_s = \frac{1}{2}$ for the ESR experiment, and $\Delta m_I = -1$ for the NMR experiment.

That resonance occurs at radio or micro wavelengths follows from equation (6.9.1) (or the selection rule $\Delta m_I = \pm 1$), for the transition energy is $\Delta E = g\beta H$, which with the Planck condition, $\Delta E = h\nu$, yields the frequency for resonance at field strength H.

$$\nu = g\beta H/h \qquad (6.9.2)$$

EXAMPLE   Find the NMR frequency for the proton in a field of 10^4 gauss.

ν (proton) $= 5.585 \times 5.05 \times 10^{-24}$ erg gauss$^{-1} \times 10^4$ gauss/6.625×10^{-27} erg sec

$\nu = 42.5 \times 10^6$ sec$^{-1} = 42.5$ megahertz (MHz)

$\lambda = c/\nu = 3 \times 10^{10}$ cm sec$^{-1}/42.5 \times 10^6$ sec$^{-1} = 7.05$ meters

EXAMPLE   Find the NMR frequency of ^{23}Na in a field of 10^4 gauss.

ν (^{23}Na) $= 1.478 \times 5.05 \times 10^{-24}$ erg gauss$^{-1} \times 10^4$ gauss/6.625×10^{-27} erg sec

$\nu = 11.3$ MHz

$\lambda = 26.6$ meters

EXAMPLE   Find the ESR frequency of the free electron in a field of 10^4 gauss.

ν (free electron) $= 2.0023 \times 9.2732 \times 10^{-21}$ erg gauss$^{-1} \times 10^4$ gauss/6.625×10^{-27} erg sec

$\nu = 28 \times 10^9$ sec$^{-1} = 28$ kMHz

$\lambda = 3 \times 10^{10}/28 \times 10^9 \approx 1$ cm

Thus the NMR frequencies are of radio frequency rather than optical frequency and ESR occurs in the microwave region (centimeter wavelengths) in a field of 10,000 gauss. The instrumentation for NMR and ESR has no resemblance to ordinary optical spectroscopic equipment with its gratings and prisms which operate in the wavelength region 10^{-8}–10^{-4} cm. In fact, ESR and NMR spectrometers must differ because of the great difference in resonance frequency in fields of the same strength. An ESR X-band spectrometer operates at about 3-cm wavelength ($\bar{\nu}$ about 9.5 kilomegahertz); the field for resonance of a free electron is then about 3400 gauss. ESR Q-band spectrometers operate at shorter wavelength (35.0 kilomegahertz, 12,500 gauss) and are commonly employed for low-temperature, that is, solid-state, studies.

Figure 6.9.3 gives a schematic of a nuclear magnetic resonance spectrometer. The sample tube containing the material under investigation rotates rapidly between the poles of a large magnet to assure a homogeneous

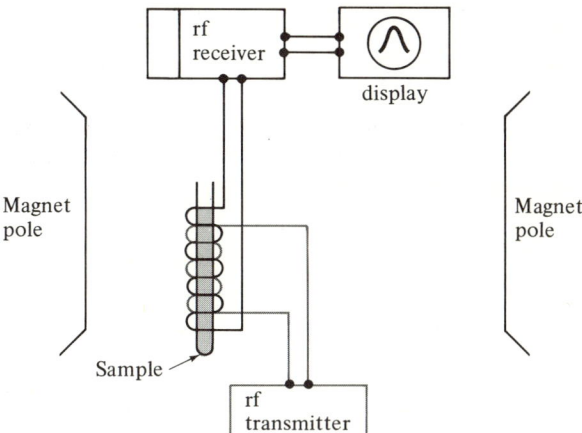

FIG. **6.9.3** Nuclear magnetic resonance spectrometer.

applied field. A radiofrequency field (rf) is applied to induce transitions; the induced signal is received, amplified, and displayed. In actual experimental practice the only practical procedure is to hold the radiofrequency constant and vary the applied field to produce resonance. Figure 6.9.4 presents the NMR signal received from the protons of ethyl alcohol as the magnitude of the magnetic field H is scanned at a constant radiofrequency of 40 mHz. Before explaining the NMR spectrum of C_2H_5OH we examine the energy levels of two simple species in a magnetic field, namely, hydrogen and helium.

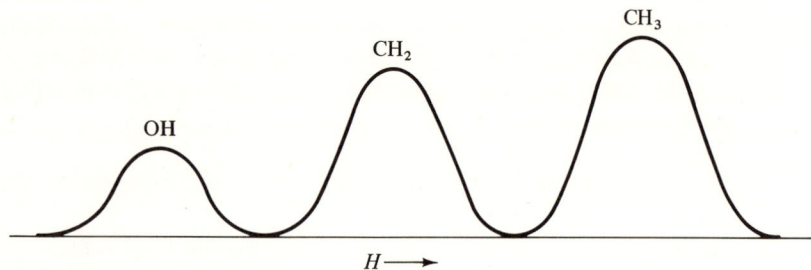

FIG. 6.9.4 Schematic NMR spectrum of C_2H_5OH at low resolution.

6.10 MAGNETIC RESONANCE IN HYDROGEN AND HELIUM

For a fuller appreciation of the NMR–ESR experiment we now give a discussion of the spectrum of these simple systems in terms of the usual quantum mechanical formalism. The hydrogen atom contains an electron and a proton both having spin. This in itself complicates the Zeeman spectrum because there is both a nuclear and an electron Zeeman effect in the spectrum of the H atom.

There are two possible orientations of the electron spin and two possible orientations of the proton spin, so four total-spin combinations arise. Let us denote the two spin functions for each particle as $\alpha(e)$, $\beta(e)$, $\alpha(N)$, and $\beta(N)$. From equation (4.6.9)

$$\hat{S}_z \alpha(e) = \tfrac{1}{2}\hbar \alpha(e) \qquad \hat{S}_z \beta(e) = -\tfrac{1}{2}\hbar \beta(e)$$

$$\hat{I}_z \alpha(N) = \tfrac{1}{2}\hbar \alpha(N) \qquad \hat{I}_z \beta(N) = -\tfrac{1}{2}\hbar \beta(N)$$

The four possible electron-nuclear spin functions (which are degenerate, to zero order, in the absence of a magnetic field) are $\alpha(e)\alpha(N)$, $\alpha(e)\beta(N)$, $\beta(e)\alpha(N)$, and $\beta(e)\beta(N)$. However, upon being placed in a magnetic field the energies of these states differ. From Table 6.9.3 it follows that if H is in the z direction the Hamiltonian for the Zeeman energy is

$$\hat{H}_0 = g_e \boldsymbol{\beta}_e H \hat{S}_z - g_N \boldsymbol{\beta}_N H \hat{I}_z \tag{6.10.1}$$

The energies of the electron-nuclear spin states follow from the eigenvalue relations,

$$\begin{aligned}
\hat{H}_0 \alpha(e)\alpha(N) &= \tfrac{1}{2}\hbar(g_e \boldsymbol{\beta}_e - g_N \boldsymbol{\beta}_N)H\alpha(e)\alpha(N) \\
\hat{H}_0 \alpha(e)\beta(N) &= \tfrac{1}{2}\hbar(g_e \boldsymbol{\beta}_e + g_N \boldsymbol{\beta}_N)H\alpha(e)\beta(N) \\
\hat{H}_0 \beta(e)\alpha(N) &= \tfrac{1}{2}\hbar(-g_e \boldsymbol{\beta}_e - g_N \boldsymbol{\beta}_N)H\beta(e)\alpha(N) \\
\hat{H}_0 \beta(e)\beta(N) &= \tfrac{1}{2}\hbar(-g_e \boldsymbol{\beta}_e + g_N \boldsymbol{\beta}_N)H\beta(e)\beta(N)
\end{aligned} \tag{6.10.2}$$

6.10 Magnetic Resonance in Hydrogen and Helium

Thus in a magnetic field the four spin states of the H atom would be characterized by the energies of (6.10.2) *but* for the existence of an important perturbing effect called the *Fermi contact interaction*. This *hyperfine coupling* effect represents the energy of the nuclear moment in the magnetic field produced at the nucleus by the electric currents associated with the *spinning* electron. The perturbing Hamiltonian has the form

$$\hat{V} = a\hat{I} \cdot \hat{S} \tag{6.10.3}$$

where

$$a = \frac{8\pi}{3} g_e \beta_e g_N \beta_N |\psi(0)|^2$$

is the hyperfine coupling constant, and $|\psi(0)|^2$ is the electronic density at the nucleus. It follows from equation (2.7.7) that the energy to first order is given by the expectation value of $\hat{V}$ over the zero-order wavefunction. For example, for the state $\alpha(e)\alpha(N)$ we have

$$\varepsilon^{(1)} = \langle \alpha(e)\alpha(N), a\hat{I} \cdot \hat{S}\alpha(e)\alpha(N) \rangle$$
$$= \langle \alpha(e)\alpha(N), a\hat{I}_z \hat{S}_z \alpha(e)\alpha N \rangle$$
$$= \tfrac{1}{4}a$$

The zero-order and first-order energy shifts in the energy of a hydrogen atom in a magnetic field are summarized in Fig. 6.10.1.

EXERCISE   Derive the remaining hyperfine coupling shifts shown in Fig. 6.10.1.

If the hydrogen atom in the applied field H is subjected to an electromagnetic field perpendicular to the direction of the applied field, and if the frequency v is correct for resonance, several kinds of spin transitions can be induced. In an ESR transition the frequency is correct for resonance inducing an electron spin change, that is, $\alpha(e)\alpha(N) \leftrightarrow \beta(e)\alpha(N)$; while in an NMR transition the frequency is correct for resonance inducing a nuclear spin change, that is, $\alpha(e)\alpha(N) \leftrightarrow \alpha(e)\beta(N)$. A third category of transition involves the simultaneous spin change of electron and nucleus, $\alpha(e)\alpha(N) \to \beta(e)\beta(N)$; it has a very low probability and is termed a "forbidden" transition. The allowed ESR transitions of hydrogen are shown in Fig. 6.10.1.

EXERCISE   Determine the transition energies in the ESR spectrum of hydrogen in ergs. What is the separation of the two ESR lines?

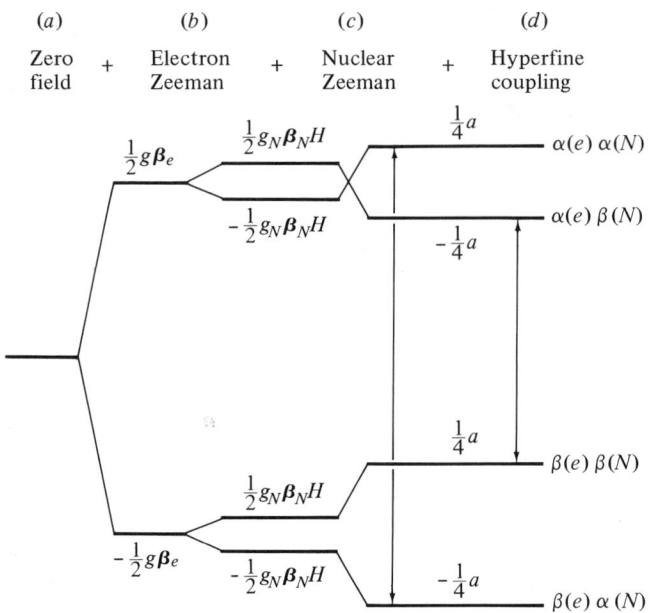

FIG. 6.10.1 First-order spin energies of the hydrogen atom and the allowed ESR transitions. From A. Carrington and A. D. McLachlan, *Introduction to Magnetic Resonance*, p. 17, Harper & Row, New York, 1967. Reproduced by permission.

HELIUM ATOM

Because of hyperfine coupling, the NMR spectrum of hydrogen and other paramagnetic species is unusual. The helium atom has a much more typical magnetic resonance spectrum. The electron spins in helium are coupled to give a singlet state with zero magnetic moment; thus only the nuclear magnetic moment need be considered. However, the prevalent ^{4}He nucleus has $I = 0$ (although the less abundant ^{3}He does have $I = \frac{1}{2}$). ^{3}He has two spin states in a magnetic field, $\alpha(N)$ and $\beta(N)$, so resonance for the bare ^{3}He nucleus occurs at $\nu = g_N \beta_N H/h$. However, the important fact is that the observed NMR frequency of ^{3}He (and other nuclei in atoms and molecules) differs from the theoretical bare nucleus value. The difference is due to the *chemical shift*. The chemical shift arises from the magnetic field of electronic currents *induced* by the external field. The induced field is proportional to the applied field H, so the net result is an effective magnetic field at the nucleus

$$H_{\text{eff}} = (1 - \sigma)H$$

where σ is called the *shielding coefficient* (a small number of order 10^{-6}). The nuclear Zeeman Hamiltonian must now be written

$$\hat{H}_0 = -g_N \beta_N (1-\sigma) H \hat{I}_z \qquad (6.10.4)$$

The chemical shift is discussed in the next section and eventually a numerical value for σ(helium) is derived.

A similar effect is operational in ESR spectra so that even in hydrogen the ESR lines do not occur precisely at the frequency predicted for the free electron ($g_e = 2.0023$). Now, while in NMR spectroscopy one takes g_N to be an inherent property of the nucleus and ascribes corrections of the Zeeman energy to shielding, in ESR spectra one ascribes changes in the Zeeman energy to changes in the magnetic moment of the electron, that is, changes in g_e. Thus, $\mu_{ez} = -m_s \beta_e g_e$ differs in different atomic and molecular environments so g_e is not a constant. The Zeeman energy is still written $E = m_s g_e \beta_e H$.

6.11 THE CHEMICAL SHIFT AND SPIN–SPIN COUPLING IN NMR

Although each nucleus with nonvanishing spin has a unique g factor and thus a unique resonance line, the usefulness of the NMR spectrometer would be very limited if this is all there was to an NMR spectrum. Figure 6.9.4 indicates there is much more to NMR spectroscopy. First of all, scanning the applied field near proton resonance has produced *three* resonance signals, one for each set of nonequivalent protons in the molecule. Furthermore, in an experiment with an instrument of greater resolution each maximum is further split into several components. Molecules having more than one magnetic nucleus often give NMR spectra showing the interaction of the magnetic moments of the different nuclei. The interaction is also characteristic of the molecule.

The different resonances for chemically nonequivalent protons arise because the magnetic field at a proton is due to both the externally applied field *and* the field due to electronic circulations in the molecule *induced* by the external field. Diamagnetic electronic distributions react to the application of the external field by a circulation which gives rise to a magnetic field *opposed* to the applied field. Since the electronic environment of chemically nonequivalent nuclei differs, the effective field strength differs. If H is the applied field, the field at the ith nucleus is $H(1-\sigma_i) = H_i$, and resonance occurs at

$$\nu = g_N \beta_N H_i / h = g_N \beta_N H(1-\sigma_i)/h \qquad \text{RESONANCE CONDITION} \qquad (6.11.1)$$

for the ith nucleus having a *shielding coefficient* σ_i.

From Fig. 6.9.4 we see that the protons of the methyl group are most highly shielded by the induced electron currents because the highest field must be applied for these protons to obtain resonance. This effect constitutes the *chemical shift* which makes NMR spectroscopy such a sensitive and interesting probe of the molecular structure. In general, the area under a resonance peak due to chemically equivalent protons is proportional to the number of equivalent protons (as in the C_2H_5OH spectrum).

The *chemical shift* δ_j of the jth nucleus in a given molecule is

$$\delta_j = \frac{H_j - H_r}{H_r} = \frac{H(1 - \sigma_j) - H(1 - \sigma_r)}{H(1 - \sigma_r)} \qquad (6.11.2)$$

or $\delta_j = \sigma_r - \sigma_j$ because $\sigma_r \ll 1$, H_j is the field strength at resonance for the jth nucleus [eq. (6.9.3)], and H_r is the field at resonance for the reference compound [usually tetramethylsilane, $(CH_4)_4Si$]. It is necessary to measure the chemical shift relative to a reference because resonance occurs at different field strengths at different v.

Modern NMR spectrometers are of such high resolution that additional splittings in the NMR spectra are revealed as a fine structure which replaces the broad maxima with groups of closely spaced lines. These splittings arise from the coupling of nuclear spins on adjacent atoms.

NMR SPIN–SPIN COUPLING

Consider the NMR spectrum of acetaldehyde, CH_3CHO. Figure 6.11.1 illustrates the proton resonance spectrum of this molecule. Two chemically distinct types of proton are evident in the two distinct chemically shifted groups of lines. But why the fine structure and intensity variation among the separated lines? In order to account for the structure in the spectrum the chemically shifted Zeeman effect, $\hat{H}_0 = -g_N \beta_N H(1 - \sigma)\hat{I}_z$, must be supplemented by a coupling Hamiltonian between each pair of proton spins in the molecule.

$$\hat{V}_{ij} = J\hat{I}_{z_i} \cdot \hat{I}_{z_j} \qquad \text{Coupling of } i\text{th and } j\text{th Proton Spins} \qquad (6.11.3)$$

J is called the coupling constant (we work with E/h in units of sec^{-1}, so J has units of sec^{-1}). To analyze the acetaldehyde spectrum divide the protons into two types according to their shielding coefficient, A type (CH_3) and B type (CHO). Thus the Zeeman Hamiltonian is ($v_0 = g_N\beta_N H/h$)

$$\hat{H}_0 = -v_0(1 - \sigma_A)(\hat{I}_{z_1} + \hat{I}_{z_2} + \hat{I}_{z_3}) - v_0(1 - \sigma_B)\hat{I}_{z_4} \qquad (6.11.4)$$

where $\hat{I}_{z_1}$, $\hat{I}_{z_2}$, and $\hat{I}_{z_3}$ are the components of the methyl proton spins. Ignoring coupling between methyl protons,[33] the aldehyde proton spin couples equally with the three methyl spins giving a Hamiltonian

$$\hat{V}_{AB} = J(\hat{I}_{z_1} + \hat{I}_{z_2} + \hat{I}_{z_3}) \cdot \hat{I}_{z_4} \qquad (6.11.5)$$

---

[33] Carrington and McLachlan, footnote 30, Section 4.4.4, p. 51.

6.11 *The Chemical Shift and Spin-Spin Coupling in NMR* 249

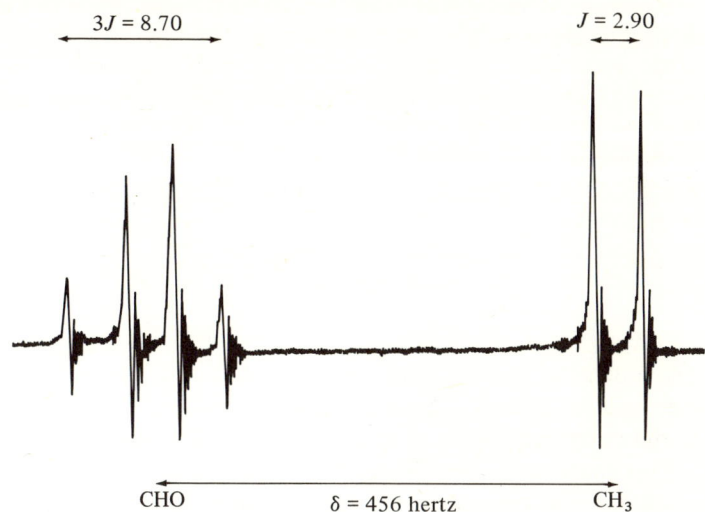

FIG. **6.11.1** The NMR spectrum of acetaldehyde. The magnetic field increases to the right. From A. Carrington and A. D. MacLachlan, *Introduction to Magnetic Resonance*, p. 41, Harper & Row, New York, 1967. Reproduced by permission.

Defining total spins for the A and B type

$$\hat{F}_{z_A} = \hat{I}_{z_1} + \hat{I}_{z_2} + \hat{I}_{z_3} \qquad \hat{F}_{z_B} = \hat{I}_{z_4}$$

and magnetic quantum numbers of the A and B type protons

$$M_A = m_{I_1} + m_{I_2} + m_{I_3} \qquad M_B = m_{I_4}$$

Thus $\hat{H}_0 + \hat{V}_{AB}$ is the operational Hamiltonian.

$$\hat{H}_0 + \hat{V}_{AB} = -v_0(1 - \sigma_A)\hat{F}_{z_A} - v_0(1 - \sigma_B)\hat{F}_{z_B} + J\hat{F}_{z_A} \cdot \hat{F}_{z_B} \quad (6.11.6)$$

To first order the proton spin energy levels are

$$E/h = -v_0[(1 - \sigma_A)M_A + (1 - \sigma_B)M_B] + JM_AM_B \quad (6.11.7)$$

Now examine the two kinds of allowed NMR transitions in type A and B protons. If one of the protons of group A spin-flips, M_A changes by ± 1 so there are two lines at

$$E/h = v = v_0(1 - \sigma_A) - JM_B = v_0(1 - \sigma_A) \pm \tfrac{1}{2}J$$

Each of these lines is equally intense, that is, both orientations $M_B = \pm\tfrac{1}{2}$ are equally probable (see Fig. 6.11.1).

However, if the aldehyde proton (type B) spin-flips there are *four* lines at frequencies

$$E/h = \nu = \nu_0(1 - \sigma_B) - JM_A$$

because there are *four* possible values of M_A corresponding to the four methyl spin arrangements.

$$\begin{array}{ll} \alpha\alpha\alpha & M_A = \tfrac{3}{2} \\ \beta\alpha\alpha, \alpha\beta\alpha, \alpha\alpha\beta & M_A = \tfrac{1}{2} \\ \alpha\beta\beta, \beta\alpha\beta, \beta\beta\alpha & M_A = -\tfrac{1}{2} \\ \beta\beta\beta & M_A = -\tfrac{3}{2} \end{array}$$

The aldehyde proton resonance consists of four lines of which $M_A = \pm\tfrac{1}{2}$ are the most intense. The intensity ratios are $1:3:3:1$, which correspond to the number of ways of arranging the methyl spins.

The analysis of the CH_3CHO NMR spectrum in Fig. 6.11.1 is now complete. The analysis was based on the fact that the coupling constant J is much smaller than the relative chemical shift $\delta = \nu_0(\sigma_A - \sigma_B)$. In more complex spectra J is comparable in magnitude to δ; the resulting couplings give spectra which must be unraveled by detailed calculations. Programs for high-speed digital computers exist to help with the analysis of complex NMR spectra.

THE INTERPRETATION OF SHIELDING COEFFICIENTS AND COUPLING CONSTANTS

Even more interesting than the analysis of complex NMR spectra is the interpretation of the observed chemical shielding coefficients and coupling constants. These are important observables related to the electronic wavefunction and its magnetic properties.

For simplicity consider a closed-shell atom, for example, helium, ^3He. When placed in the applied magnetic field the nuclear magnetic moment precesses about the direction of the applied field, but in addition, the entire spherical electron cloud also precesses about the field direction with angular velocity $\omega = eH/2mc$. An electron at a distance $\mathbf{r}$ from the nucleus has velocity $\mathbf{v} = \boldsymbol{\omega} \times \mathbf{r}$ and produces a magnetic field at the nucleus as follows:

$$H' = -\frac{e}{c}\frac{\mathbf{r} \times \mathbf{v}}{r^3} = -\frac{e^2}{2mc^2}\frac{\mathbf{r} \times (\mathbf{H} \times \mathbf{r})}{r^3} \tag{6.11.8}$$

Averaging this field over the electronic distribution $\rho(r)$, $\langle H' \rangle = \int H'\rho(r)\,dV$ and examining the component of $\langle H' \rangle$ in the z direction, which was previously defined as σH, we obtain an expression for σ.

$$\sigma = \frac{e^2}{3mc^2}\int \frac{\rho(r)}{r}\,dV \tag{6.11.9}$$

6.11 The Chemical Shift and Spin-Spin Coupling in NMR

This is the Lamb formula. The formula says that the shielding coefficient is given by the expectation value of $1/r$. Thus the shielding coefficient of the closed-shell atom is directly calculable from the electronic density. As an example, for helium, equation (6.11.9) yields

$$\sigma(\text{helium}) = 59.93 \times 10^{-6}$$

Unfortunately, shielding coefficients in molecules are considerably more complicated. First, because the electron density is no longer spherically symmetric about the nucleus, σ is no longer *isotropic* (the same in all directions). Second, because there is an additional term in the shielding coefficient which is often of opposite sign to the Lamb term.

$$\sigma = \sigma_d + \sigma_p$$

The *diamagnetic* (Lamb) term σ_d corresponds physically to the precession of the original stationary electron density about the field direction; σ_p is the *paramagnetic* term and it arises from *changes* in the original electron density *induced* by the applied field. The paramagnetic term vanishes for s orbitals but it can be quite large when there are p or d orbitals near the nucleus, as in fluorine.

Carrington and McLachlan[34] divide the shielding coefficient of the proton in a diamagnetic molecule into four contributions:

1. The diamagnetic shielding due to currents induced in $\rho(r)$ on the hydrogen atom containing this proton. An increase in $\rho(r)$ on the hydrogen atom increases the shielding and therefore the chemical shift can be related to the electron-withdrawing power of the substituent (see Fig. 6.11.2).
2. Paramagnetic contribution due to changes in $\rho(r)$ on the hydrogen atom, which should be quite small.
3. Combined contribution of the diamagnetic and paramagnetic terms on other atoms in the molecule.
4. Finally, a contribution due to electronic currents flowing around closed rings.[35] For example, the six π electrons of the benzene ring are sufficiently free to *circulate* when a magnetic field is applied perpendicular to the ring. Ring circulation gives rise to a magnetic field opposed to the applied field, which changes the proton shielding coefficients of benzene by about -1.75×10^{-6}. The ring current is believed to account for the proton shift of -1.95×10^{-6} in going from ethylene to benzene.

Spin–spin coupling arises from the magnetic interaction of a nucleus with the surrounding electronic density, thus perturbing said density and changing the magnetic field at all other nuclei in the molecule. Specifically

---

[34] Carrington and McLachlan, footnote 30, Section 5.3, p. 58.
[35] L. Salem, *The Molecular Orbital Theory of Conjugated Systems*, Chaps. 4 and 5. Benjamin, New York, 1966.

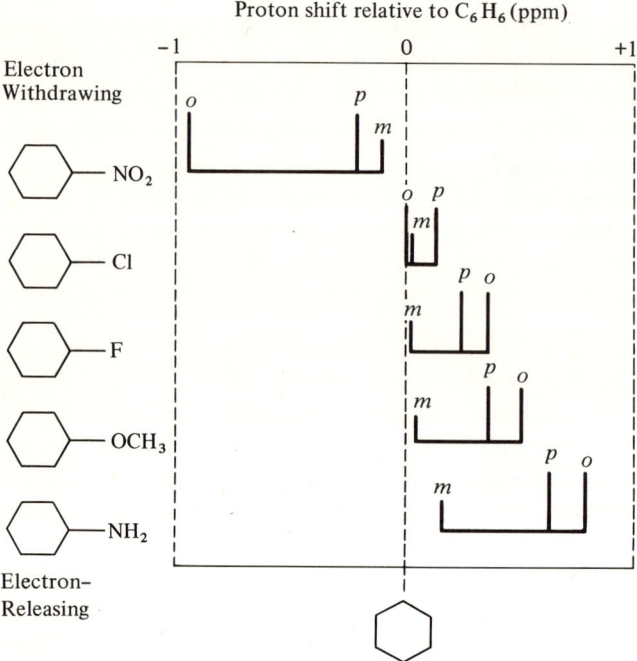

FIG. 6.11.2 Proton shifts in substituted benzene compared to normal benzene. From A. Carrington and A. D. MacLachlan, *Introduction to Magnetic Resonance*, p. 59, Harper & Row, 1967. Reproduced by permission.

the Fermi contact interaction tends to align electronic spins near nucleus A antiparallel to the nuclear spin I_A. This spin polarization of the electronic distribution affects the magnetic field at every other nucleus. It follows that the coupling constant J is sensitive to the extent and nature of the delocalization of electrons between nuclei. Spin–spin coupling occurs mainly through the sigma bonds and tends to decrease as the number of intervening bonds increases.[36] For example, compare the benzene proton–proton spin coupling constants $J_{12} = 7.54$, $J_{13} = 1.37$, and $J_{14} = 0.69$.

[36] J. A. Pople, J. W. McIver, and N. S. Ostlund, *J. Chem. Phys.* **49**, 2965 (1968).

EXPERIMENTAL NMR

The important chemical applications of NMR have involved the observed chemical shifts and spin–spin couplings in the spectra of ^{1}H, ^{19}F, and ^{13}C. Spectra are observed in the pure liquid or in inert solvents. Solid samples yield broad NMR spectra making spin–spin couplings and even chemical shifts difficult to observe. On the other hand, solvent effects due to hydrogen bonding, polar solvents, etc., must be avoided. It is often suggested that the observed shifts should be extrapolated to infinite dilution for reproducibility.

NMR is used extensively in organic structural analysis and conformational studies, for example, boat versus chair cylohexane. It is important to note that if changes occur in the nuclear environment during the NMR experiment, for example, a proton exchange reaction, then the NMR spectrum is a time average over the shifts and couplings which are present. More precisely, we may distinguish two limiting cases called *fast exchange* and *slow exchange*. In the first case the jumping frequency between environments, v_e, is much greater than the separation of two resonance lines affected by the motion, Δv. The result is a *single* resonance line at the average frequency with a width $(\Delta v)^2/v_e$. Thus, in fast exchange an average environment is reported by the resonance experiment. However, in slow exchange $v_e \ll \Delta v$, and now *both* lines are observed but broadened because of the reduced *lifetime* of each environment. This follows from the uncertainty principle: $\Delta\text{time}\, \Delta\text{energy} \geq \hbar/2$ (Section 2.3). A variety of important rate processes have been studied by magnetic resonance, that is, to obtain rate constants. When studied as a function of temperature, activation energies for such processes as internal rotation, electron exchange, proton exchange, and isomerization are obtainable.

REFERENCES

GENERAL REFERENCES

Important reference works in the fields of molecular spectra and group theory have been included as footnotes to the text. The following additional references are of special interest.

MOLECULAR SPECTROSCOPY

Barrow, G., *Introduction to Molecular Spectroscopy*, McGraw-Hill, New York, 1962 (Rotation-Vibration Spectra).

Jaffé, H. H. and Orchin, M., *Theory and Applications of Ultraviolet Spectroscopy*, Wiley, New York, 1962.

Murrell, J. N., *The Theory of the Electronic Spectra of Organic Molecules*, Wiley, New York, 1963.

Sandorfy, C., *Electronic Spectra and Quantum Chemistry*, Prentice-Hall, Englewood Cliffs, New Jersey, 1964.

GROUP THEORY

Cotton, F. A., *Chemical Applications of Group Theory*, Wiley, New York, 1963.
Hammermesh, M., *Group Theory and its Application to Physical Problems*, Addison-Wesley, New York, 1962.
Tinkham, M., *Group Theory and Quantum Mechanics*, McGraw-Hill, New York, 1964.

MAGNETIC RESONANCE SPECTROSCOPY

Jackman, L. M., *Nuclear Magnetic Resonance Spectroscopy*, Pergamon, New York, 1959.
Pople, J. A., Schneider, W. G., and Berstein, H. J., *High Resolution Nuclear Magnetic Resonance*, McGraw-Hill, New York, 1959.

PROBLEMS

1. The separation of rotational lines in the vibration–rotation spectra of CO is 3.85 cm^{-1}. Determine the C—O bond length, R_e(CO).
2. Calculate the (1) reduced mass and (2) moment of inertia of gaseous Na^{35}Cl ($R_e = 2.36$ A). Predict the pure rotational spectrum of this molecule.
3. Calculate the zero-point energies of (1) O$_2$, $\bar{\nu} = 1580.36$ cm^{-1}; (2) H$_2$, $\bar{\nu} = 4395.2$ cm^{-1}; (3) HI, $\bar{\nu} = 2309.5$ cm^{-1}; (4) CO$_2$, ν_1, ν_2, ν_3, see Fig. 7.7.1; (5) H$_2$O, ν_1, ν_2, ν_3 see Fig. 6.7.2. Which of these molecules has a pure vibrational spectrum?
4. The fundamental absorption of H^{81}Br ($v = 0 \to 1$) occurs at 2559 cm^{-1}. It consists of a series of lines approximately equally spaced with an interval of 16.9 cm^{-1}. Plot the potential energy curve of this molecule close to its minimum. Label and scale both axes.
5. The vibrational frequency of HBr is 2559 cm^{-1}. Calculate: (1) the zero-point energy of HBr in kilocalories per mole; (2) The force constant in dynes per centimeter; (3) The vibrational frequency of DBr (deuterium bromide) assuming the same force constant as HBr. Is this a good assumption?
6. Deduce the point group symmetries of the following molecules: (1) H$_2$S, angular triatomic; (2) naphthalene; (3) NCl$_3$, pyramidal; (4) HCN, linear; (5) phenanthrene; (6) CH$_3$Br; (7) cyclohexane, chair, boat and planar; (8) ethylene.
7. List the symmetry elements present in the point groups of problem 6.
8. Consider the elements 1, a, a^2, a^3. Given that $a^3 = 1$ show that these elements satisfy the criteria of a group.
9. List all the allowed transitions between electronic states of H$_2$O, or any molecule of C$_{2v}$ point group, in terms of irreducible representations, and give their polarizations.
10. Given two 1s_H (hydrogen), one 1s_O (oxygen), and one 2p_y (oxygen) atomic functions, find the linear combinations of these functions which transform as irreducible representations of the point group of H$_2$O (i.e., find the symmetry-adapted functions).

6.11 The Chemical Shift and Spin-Spin Coupling in NMR

11. A solution of an organic dye (molecular weight 294) containing 1 g of solute per 100 ml transmits 75% of the incident light at 5000 A in a quartz cell of 1-cm length. (1) What concentration will absorb 50% of the incident light in the same cell? (2) What percent of light is absorbed by a 3 g/100 ml solution of the dye? (3) What cell length at a concentration of 1 gm/100 ml is necessary to absorb 85% of the incident light?
12. Sketch the expected low-resolution proton magnetic resonance spectrum in (1) *n*-propyl alcohol, (2) isopropyl alcohol, (3) *tert*-butyl alcohol, (4) naphthalene. Note the relative peak areas.
13. The limit of continuous absorption of Br_2 gas occurs at 19,750 cm^{-1}. Br_2 photodissociates into an excited atom Br^*, and a normal atom. The atomic transition $Br \rightarrow Br^*$ occurs at 3685 cm^{-1}. Calculate the dissociation energy of Br_2 ($Br_2 \rightarrow 2\ Br$) in kilocalories per mole.
14. Show that the pure rotational transition $J=0$ to $J=2$ is forbidden. Use the wavefunction and transition moment given in Section 6.1.

7
POLYATOMIC MOLECULES

We have been obliged here and there
to take cognizance of the entering wedge
of scientific bolshevism, which we call
the quantum theory . . .

            G. N. LEWIS
      Valence and Structure of Atoms and Molecules, 1923

ALTHOUGH quantum theory was in its infancy, to say the least, at the time the American physical chemist G. N. Lewis wrote his classic book,[1] he exposed, with the help of I. Langmuir, some of the regularities in the nature of chemical bonding in diatomic and polyatomic molecules. In the following sections we often recover a quantum mechanical version of the Lewis electron-pair bond from quantum theory, not as a necessary consequence of the theory, but as a simple localized representation of the electronic structure (see Section 5.14). The quantum mechanical methods of treating the electronic structure of polyatomic molecules are the same as those applied to the diatomic molecule in Chapter 5. Therefore the methods discussed in Sections 5.4, 5.6, 5.7, and 5.15 are equally applicable here. Symmetry is very useful in finding the form of the LCAO–MO in polyatomic molecules without calculations of any kind; in the following sections group theory is applied to that goal.

7.1 THE MOLECULAR SCHRÖDINGER EQUATION

The molecular Schrödinger equation for the electronic wavefunction is, in the Born-Oppenheimer approximation,[2]

$$\hat{H}\Psi = E\Psi \tag{7.1.1}$$

where $\hat{H}$ is the electronic Hamiltonian for a polyatomic molecule containing N electrons and K nuclei of nuclear charge $Z_1, Z_2, \ldots, Z_K$.

$$\hat{H} = \sum_{i=1}^{N}\left[-\tfrac{1}{2}\nabla_i^2 - \sum_{a=1}^{K} Z_a/r_{ia}\right] + \sum_{i>j}^{N} 1/r_{ij} \tag{7.1.2}$$

This Hamiltonian is equivalent to (5.4.4) for the diatomic molecule when $K = 2$. Like equation (5.4.4) equation (7.1.2) can be rewritten

$$\hat{H} = \sum_{i=1}^{N} \hat{h}(i) + \hat{V}(r_{ij})$$

---

[1] G. N. Lewis, *Valence and Structure of Atoms and Molecules*, reprinted by Dover Publications, New York, 1966.
[2] Appendix D.

Again, the presence of interelectronic interactions, $\hat{V}(r_{ij}) = \sum_{i>j} 1/r_{ij}$, makes the Schrödinger equation (7.1.1) intractable. In spite of this, the molecular *Hartree-Fock equation* (5.7.1) permits one to obtain a set of *molecular orbitals* ψ_i, which are the best one-electron wavefunctions for the polyatomic molecule.

$$\hat{H}_{\text{eff}} \psi_i(i) = E_i \psi_i(i) \quad (5.7.1)$$

$\hat{H}_{\text{eff}}$ is defined by equations (5.7.2) to (5.7.4). The physical justification and meaning of the Hartree-Fock method is the replacement of the true interelectronic potential $\hat{V}(r_{ij})$ with an effective potential which accounts for the average (or smeared out) interelectronic interaction.

The *Hartree-Fock-Roothaan* method (Section 5.7) approximates the molecular orbital as a linear combination of atomic orbitals.

$$\psi_i = \sum_{i=1}^{n} c_i \phi_i \quad (7.1.3)$$

The ϕ_i are the atomic orbitals of the component atoms. The variational principle is used to find the best set of coefficients, c_i, satisfying the molecular Hartree-Fock equation when *iterated* to *self-consistency*: Thus *self-consistent-field* or SCF–LCAO–MO. As explained in Section 5.7 [equation (5.7.12)], the ϕ_i, which are usually STO, should have optimized Z_{eff} for the SCF–LCAO–MO to approach the best one-electron wavefunctions for the molecule. Several examples of such MO will be given in the following sections.

The total N-electron closed-shell wavefunction is the antisymmetrized product of the occupied molecular spin orbitals.

$$\Psi(1, 2, \ldots, N) = \hat{A} \psi_1 \alpha(1) \psi_1 \beta(2) \cdots \psi_{N/2} \beta(N) \quad (7.1.4)$$

As explained in Section 5.9 the electronic density ρ and the total energy are invariant to certain linear transformations of the orbitals. Thus, a representation in terms of localized orbitals (LO) is possible, and is usually more chemically intuitive. In the next few sections we will examine the molecular orbitals and localized orbitals for molecules of several symmetry point groups. These representations will be compared in terms of the observable properties of polyatomic molecules such as ionization potential, molecular shape, and bond energy, in Section 7.8.

7.2 LINEAR MOLECULES

The simplest linear polyatomic molecules are the symmetric triatomic molecules (point group $\mathbf{D}_{\infty h}$). For example, CO_2, BeH_2, BO_2, CS_2, $HgCl_2$. We will first examine BeH_2 because its MO and LO display many of the features of polyatomic orbitals in simple form.

The atomic orbitals of Be and the hydrogen atoms form the *basis set* for

7.2 Linear Molecules

the LCAO–MO of BeH_2. If we wish to limit ourselves to a small basis set it is natural to choose the valence shell AO of the component atoms. In that case we need only consider $1s_H$, $2s_{Be}$, and $2p_{Be}$. However, for the greater flexibility needed to approach Hartree-Fock accuracy the basis set should be much larger. It should also include $2s_H$ and $2p_H$ as well as $1s_{Be}$, $3s_{Be}$, $3p_{Be}$, and even $3d_{Be}$, all with optimized Z_{eff}. The use of this extended basis set is a direct application of the mathematical *expansion theorem* [equation (2.5.11)]. The extended basis is not very chemically intuitive, except that it is obvious that many AO will be needed to exactly represent the distortion in the charge density which occurs on bonding. However, the valence shell AO *do* make the largest contributions to the LCAO–MO (Table 7.2.2).

From Fig. 7.2.1 and Section 5.5 the irreducible representations for which the valence shell AO form a basis are easily seen to be σ_g, σ_u, and π_u (Table 7.2.1).

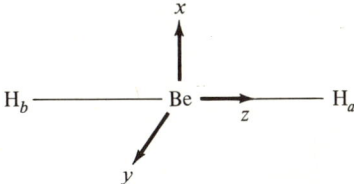

FIG. 7.2.1 Coordinates of BeH_2.

TABLE 7.2.1

AO basis for BeH_2	Irreducible representation
$(1s_a + 1s_b)$, $2s_{Be}$	σ_g
$(1s_a - 1s_b)$, $2p_{z_{Be}}$	σ_u
$\begin{matrix}2p_{x_{Be}}\\2p_{y_{Be}}\end{matrix}$	π_u degenerate

The LCAO–MO follow from Table 7.2.1.

$$\sigma_g = c_1(1s_a + 1s_b) + c_2\, 2s_{Be}$$
$$\sigma_u = c_3(1s_a - 1s_b) + c_4\, 2p_{z_{Be}} \qquad (7.2.1)$$
$$\pi_u = 2p_{x_{Be}} \quad \text{or} \quad 2p_{y_{Be}} \text{ degenerate MO}$$

The application of the Hartree-Fock Roothaan procedure to these LCAO–MO (in an extended basis of STO) yields SCF–LCAO–MO like those of Table 7.2.2. The relative energy ordering and the component AO

TABLE 7.2.2 Some Coefficients of the SCF–LCAO–MO Wavefunction of BeH$_2$[a]

AO	K	MO $1\sigma_g$	$1\sigma_u$	Z_{eff}*
$1s_a \pm 1s_b$	0.00	0.40	0.41	1.065
$1s_{Be}$	0.89	−0.20	—	2.945
$1s'_{Be}$	0.20	−0.02	—	5.748
$2s_{Be}$	0.00	0.17	—	1.785
$2s'_{Be}$	0.00	0.34	—	2.548
$3s_{Be}$	−0.11	−0.02	—	10.89
$2p_{zBe}$	—	—	+0.29	2.092
Orbital energy	−4.678 hartrees	−0.4881 hartree	−0.4469 hartree	

[a] The basis set for this calculation included *several* 1s and 2s beryllium STO (see the examples given in Section 4.10). The basis set also included 2s and 2p hydrogen STO as well as 3d, 4f, and additional 2p beryllium STO. These STO entered the MO with coefficients smaller than 5×10^{-2} and have been omitted from this table for clarity. Unpublished results of J. R. Riter used with permission.
* Z_{eff} as defined in equation (4.9.2).

of the valence shell molecular orbitals are indicated in Fig. 7.2.2. With four valence shell electrons to be accommodated in the MO of Fig. 7.2.2 it is evident that the ground-state MO configuration of BeH$_2$ is

$$\text{BeH}_2 \ K(1\sigma_g)^2(1\sigma_u)^2 \ {}^1\Sigma_g \quad (7.2.2)$$

where $K = (1s_{Be})^2$ is the filled K shell of beryllium.

The forms of the occupied and unoccupied MO of BeH$_2$ are outlined in Fig. 7.2.3. The $1\sigma_g$ and $1\sigma_u$ MO are *bonding* because they pile up electronic density in the Be—H bonding regions. On the other hand $2\sigma_g$ and $2\sigma_u$ are *antibonding* for these MO have nodal planes which bisect the Be—H bonds. Such MO could not, of themselves, lead to stable molecule formation. The degenerate AO, $2p_{xBe}$ and $2p_{yBe}$, form the doubly degenerate nonbonding MO, π_u.

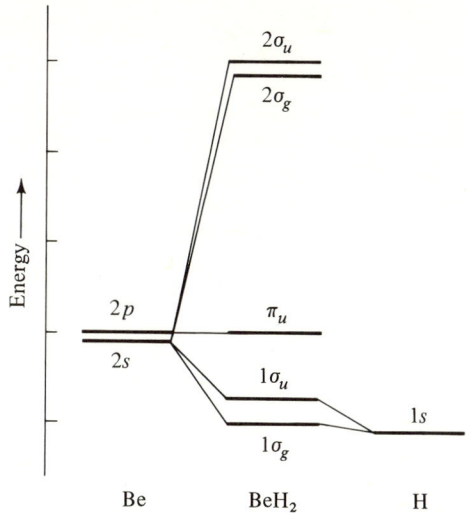

FIG. 7.2.2 Mixing of valence shell AO to form LCAO–MO of BeH_2. The $1s_{Be}$ AO is too low in energy to appear on the same scale.

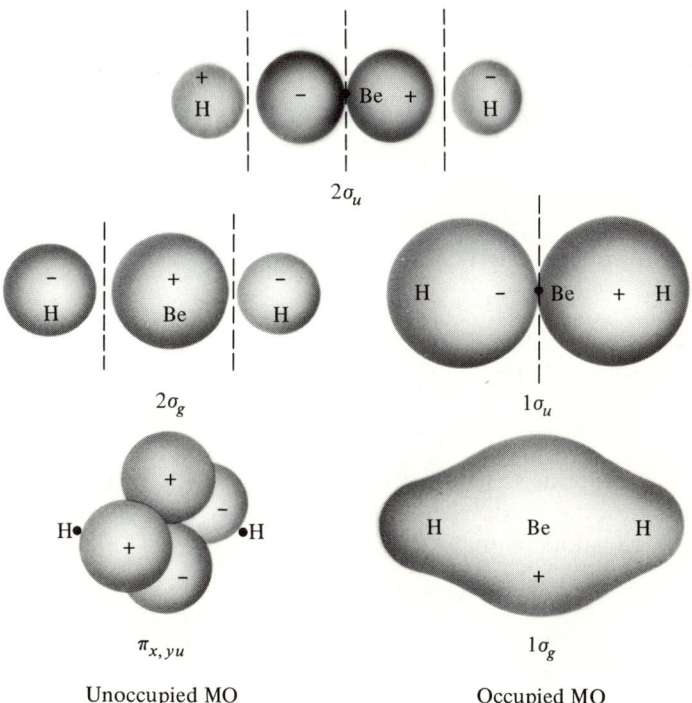

FIG. 7.2.3 Schematic outlines of the BeH_2 molecular orbitals and their relative phases.

BeH$_2$ LO AND sp HYBRIDIZATION

Chemical intuition informs us that the two bonds of BeH$_2$ must be equivalent. Yet, although the bonding orbitals in the MO representation, $1\sigma_g$ and $1\sigma_u$, treat each H-atom equivalently, they are spread out (delocalized) over the whole molecule. Is there a representation which permits independent, distinct, and equivalent bonding orbitals for the equivalent Be—H bonds? Yes, this is the localized orbital representation (Section 5.14). Although, as seen in Section 5.14, the MO may be transformed directly to LO, the easiest and most rapid way to set up a localized orbital representation is through hybridization of $2s$ and $2p$ AO. In the present case hybridization to form the sp hybrids of Fig. 5.14.1 on the Be atom leads immediately to localized bonding orbitals. (See Fig. 7.2.4.)

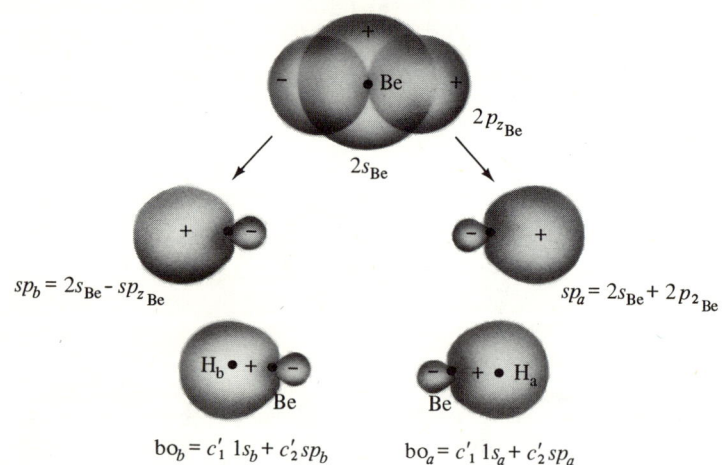

FIG. 7.2.4 The formation of sp hybrids on Be atom and their use in forming localized bonding orbitals.

The sp hybrids, sp_a and sp_b, overlap strongly with $1s_a$ and $1s_b$, respectively, yielding the localized bonding orbitals bo$_a$ and bo$_b$ of Fig. 7.2.4. The localized orbital configuration is

$$\text{BeH}_2 \ K \, (\text{bo}_a)^2 (\text{bo}_b)^2 \tag{7.2.3}$$

Equation (7.2.3) should be compared to (7.2.2), the MO configuration. The advantage of the LO representation is its emphasis on the *equivalent* and *localized* nature of the *bonds* and the traditional assignment of two electrons to each localized bond, as in the *Lewis structure* for BeH$_2$, H:Be:H.

7.2 Linear Molecules

EXERCISE Show that if $c_1 = c_3$ and $c_2 = c_4$ in equation (7.2.1), the linear transformation between LO and MO is simply

$$\begin{pmatrix} bo_a \\ bo_b \end{pmatrix} = \begin{pmatrix} 1 & 1 \\ 1 & -1 \end{pmatrix} \begin{pmatrix} 1\sigma_g \\ 1\sigma_u \end{pmatrix}$$

CO_2 MO

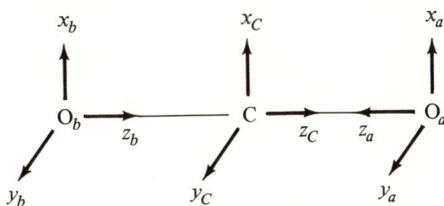

FIG. 7.2.5 Coordinates for CO_2.

CO_2 (a linear symmetric triatomic molecule) also belongs to point group $D_{\infty h}$. Table 7.2.3 summarizes the symmetry properties of the valence shell AO of CO_2 which will be used to form the LCAO–MO. The orbitals are abbreviated $s_a = 2s$ oxygen on nucleus a; $s_C = 2s$ carbon; $x_C = 2p_x$ on the carbon nucleus, etc. etc. Table 7.2.3 may be verified from Fig. 7.2.5, which displays the choice of coordinate axes. The MO of each irreducible representation are

$$\sigma_g = c_1 s_C + c_2(s_a + s_b) + c_3(z_a + z_b)$$

$$\sigma_u = c_4 z_C + c_5(s_a - s_b) + c_6(z_a - z_b)$$

$$\pi_u = \begin{cases} c_7 x_C + c_8(x_a + x_b) \\ c_7 y_C + c_8(y_a + y_b) \end{cases} \quad \text{degenerate MO} \qquad (7.2.4)$$

$$\pi_g = \begin{cases} (x_a - x_b) \\ (y_a - y_b) \end{cases} \quad \text{degenerate MO}$$

TABLE 7.2.3

AO basis for CO_2	Irreducible representation
$(s_a + s_b)$, s_C, $(z_a + z_b)$	σ_g
$(s_a - s_b)$, z_C, $(z_a - z_b)$	σ_u
$(x_a + x_b)$, x_C $(y_a + y_b)$, y_C	π_u degenerate
$(x_a - x_b)$ $(y_a - y_b)$	π_g degenerate

The degenerate π_g MO is clearly nonbonding. The other MO may be bonding or antibonding (or even nonbonding) depending on the values of the coefficients. The coefficients of the occupied SCF–LCAO–MO of CO_2 are listed in Table 7.2.4. These coefficients were determined by the Hartree-

TABLE 7.2.4 The Coefficients of the SCF–LCAO–MO Wavefunction of CO_2 [a,b]

AO	MO				
	$1\sigma_g$	$2\sigma_g$	$1\sigma_u$	$2\sigma_u$	$1\pi_u$
s_C	0.32	0.40	—	—	—
z_C	—	—	0.51	0.65	—
$(s_a \pm s_b)$	0.48	−0.51	0.38	−0.64	—
$(z_a \pm z_b)$	0.21	0.46	0.11	0.37	—
x_C, y_C	—	—	—	—	0.53
$(x_a + x_b)$ $(y_a + y_b)$	—	—	—	—	0.49

[a] STO with unoptimized Z_{eff}. Coefficients refer to the coordinates of Fig. 7.2.5.
[b] Results of J. F. Mulligan, J. Chem. Phys. **19**, 347 (1951). Used by permission.

Fock-Roothaan procedure in one of the earliest applications of this method. The bonding orbitals are $1\sigma_g$, $1\sigma_u$, and $1\pi_u$, with $2\sigma_g$ and $2\sigma_u$ less bonding.

EXERCISE   Write the π_g MO given in (7.2.4) in its normalized form: let $S_{ab} = \langle 2p_a, 2p_b \rangle$.

The resulting ground-state electronic configuration is

$$CO_2 \; KKK \; (1\sigma_g)^2(1\sigma_u)^2(2\sigma_g)^2(2\sigma_u)^2(1\pi_u)^4(1\pi_g)^4 \quad {}^1\Sigma_g \qquad (7.2.5)$$

The nonbonding MO, $1\pi_g$, has the highest orbital energy, in reasonable agreement with the observed ionization potential of CO_2 (13.8 eV). The π MO are shown in Fig. 7.2.6.

The lowest-energy unoccupied MO, $2\pi_u$, may be used to discuss the excited states of CO_2. The lowest-energy transition should be to states of the configuration[3] (σ core) $(1\pi_u)^4(1\pi_g)^3(2\pi_u)^1$, that is, ${}^1\Delta_u$ and ${}^1\Sigma_u$. Likely assignments in the observed spectrum are absorption bands in the region 1700–1400 Å, ${}^1\Sigma_g \to {}^1\Delta_u$, the absorption bands in the region 1390–1240 Å, ${}^1\Sigma_g \to {}^1\Sigma_u$. Both excited states are thought to have *bent* equilibrium nuclear configurations, that is, point group C_{2v} (see Section 7.6).

[3] A. D. Walsh, J. Chem. Soc. **1953**, 2266.

7.2 Linear Molecules

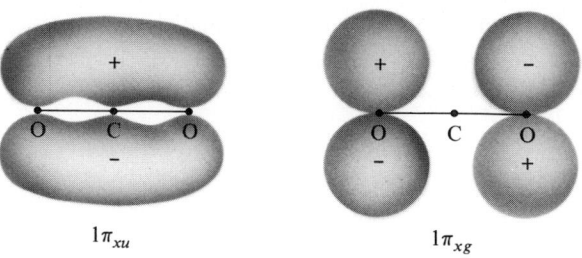

FIG. 7.2.6 The occupied π MO of CO_2.

Although the MO representation gives a quantitative description of the electronic structure and spectra of CO_2, there is an uncertainty in the bonding description due to the delocalization of the MO over all three atoms.

CO_2 LO AND RESONANCE STRUCTURES

A qualitative localized orbital representation of the electronic structure of CO_2 is obtained, without calculation, by invoking hybridization. Form the sp hybrids on carbon and oxygen as in Fig. 7.2.7. The sp hybrids sp_{O_1}

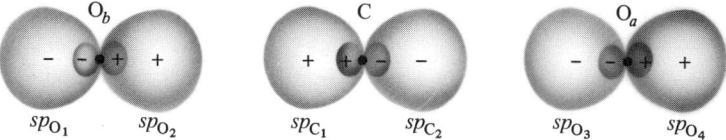

FIG. 7.2.7 sp Hybridization of carbon and oxygen in CO_2.

and sp_{O_4} are nonbonding oxygen orbitals; the following localized bonding orbitals are formed from the remaining overlapping hybrids: $bo_b = c'_1 sp_{C_1} + c'_2 sp_{O_2}$ and $bo_a = c'_1 sp_{C_2} + c'_2 sp_{O_3}$. If we make a distinction at this point between the σ core and the π shell, we may leave the latter undisturbed as MO. The result is a *mixed representation*, with configuration (7.2.6).

$$CO_2 \; KKK \; (lp_a)^2 (lp_b)^2 (bo_a)^2 (bo_b)^2 (1\pi_u)^4 (1\pi_g)^4 \qquad (7.2.6)$$

The mixed representation identifies two equivalent σ-bonding electron pairs and several equivalent nonbonding pairs. The π bonds remain delocalized over the whole molecule. The orbitals $lp_a = sp_{O_4}$ and $lp_b = sp_{O_1}$

are termed "lone-pair" orbitals, lp, because when doubly occupied this electron pair has its density directionally oriented *away* from the molecule.

The bonds may also be localized, but at a high price. Consider the localized π orbitals.

$$\pi_{xa} = c'_7 x_C + c'_8 x_a \qquad \pi_{xb} = c'_7 x_C + c'_8 x_b$$
$$\pi_{ya} = c'_7 y_C + c'_8 y_a \qquad \pi_{yb} = c'_7 y_C + c'_8 y_b$$

The assignment of the π-shell electrons to these localized orbitals and atomic orbitals gives an ambiguous result. There are two degenerate configurations, I and II (Fig. 7.2.8), each corresponding to the classical chemical structure O=C=O, that is, $:\ddot{O}::C::\ddot{O}:$. I and II are called *resonance structures*.

$$\text{I} \quad CO_2 \; KKK (lp_a)^2 (lp_b)^2 (bo_a)^2 (bo_b)^2 (\pi_{xa})^2 (\pi_{yb})^2 (y_a)^2 (x_b)^2 \qquad (7.2.7)$$
$$\text{II} \quad CO_2 \; KKK (lp_a)^2 (lp_b)^2 (bo_a)^2 (bo_b)^2 (\pi_{ya})^2 (\pi_{xb})^2 (x_a)^2 (y_b)^2$$

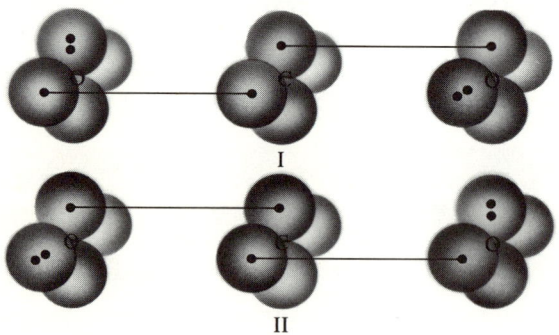

FIG. 7.2.8 Schematic representation of the localization of electron pairs in the resonance structures of CO_2.

The CO_2 molecule is sometimes termed a "resonance hybrid" of resonance structures I and II, or said to be "in resonance" between I and II. Of course, resonance is not something which actually occurs in the molecule (especially not in the coupled oscillator sense of Section 6.9). Resonance between I and II expresses the limitation of the localized orbitals to unambiguously represent the delocalization of the π electrons. In this sense the configurations (7.2.6) and even (7.2.5) are preferable. However, resonance structures need not stop with low-energy nonionic structures such as I and II. Rather, in the spirit of the valence bond method (see Section 5.16), one may mix in other structures on the basis of chemical

7.3 Trigonal Planar Molecules

intuition. For example, according to Pauling[4] the two ionic structures, III $^+$O≡C—O$^-$ and IV $^-$O—C≡O$^+$, contribute to the ground state of CO_2, thereby explaining the observed decrease of the C=O distance in CO_2 compared to that in molecules with only one C=O bond (e.g., H_2CO, 1.21 Å; CO_2, 1.16 Å).

It should be evident from the discussion in Section 5.16 and elsewhere that the inclusion of ionic and other excited configurations together with I and II in the representation of the ground state, for example, $CO_2 = C_1(I + II) + D_1(III + IV) + \cdots$, goes beyond the independent particle model which is the basis of single-configuration representations. The addition of these chemically intuitive structures may, in fact, be a substantial improvement over the independent particle model. However, this is difficult to establish in the absence of extensive calculations. Furthermore, it is not always obvious which structures should be included and which omitted; intuition is not the surest guide.

7.3 TRIGONAL PLANAR MOLECULES

BH$_3$ MO

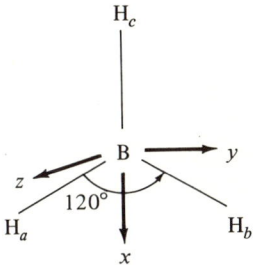

FIG. 7.3.1 The BH_3 molecule is planar and lies in the xy plane.

BH_3 is an example of the molecular point group $\mathbf{D}_{3h}$ (Fig. 7.3.1). the symmetry elements of this group were given in Fig. 6.6.1 and the character table of $\mathbf{D}_{3h}$ is Table 6.6.1. The irreducible representations of the bonding orbitals are easily determined as follows. The three bonds of BH_3 form the basis for the reducible representation with character χ_{red}.

	E	$2C_3$	$3C_2'$	σ_h	$2S_3$	$3\sigma_v$
$\chi_{\text{red}} =$	3	0	1	3	0	1

[4] L. Pauling, *Nature of the Chemical Bond*, Cornell University Press, Ithaca, N.Y., 1939.

EXERCISE   Derive χ_{red} for the three BH bonds of BH_3.

From equation (6.5.3) it follows that χ_{red} is composed of A_1' and E', $\chi_{\text{red}} = A_1' + E'$.

EXERCISE   Demonstrate the last assertion.

Thus we expect bonding MO which form the bases for A_1' and E', respectively. In fact, it was shown in Section 6.6 that the LCAO $1s_a + 1s_b + 1s_c$ forms the basis for A_1' and

$$\begin{bmatrix} 2(1s_a) - 1s_b - 1s_c \\ 2(1s_b) - 1s_a - 1s_c \end{bmatrix}$$

forms the two-dimensional basis for the two-dimensional irreducible representation E' of $\mathbf{D}_{3h}$. Therefore these are the correct linear combinations of hydrogenic AO for use in BH_3 MO. As for the boron AO, we will again limit the basis to valence shell AO with $2s_{\text{Boron}} = s_B$, $2p_{x_{\text{Boron}}} = x_B$, etc. The projection operators show that $\hat{P}^{A_1'} s_B = s_B$ and $\hat{P}^{A_2''} z_B = z_B$, that is, s_B and z_B are bases of A_1' and A_2'', respectively.

EXERCISE   Again, prove the last assertions.

By elimination, x_B and y_B form the basis of the two-dimensional representation E'. The LCAO are summarized in Table 7.3.1.

TABLE 7.3.1

AO basis for BH_3	Irreducible representation[a]
$(1s_a + 1s_b + 1s_c)$, s_B	a_1'
$\left. \begin{array}{l} 2(1s_a) - 1s_b - 1s_c \\ 2(1s_b) - 1s_a - 1s_c \\ x_B \ y_B \end{array} \right\}$	e'
z_B	a_2''

[a] We will indicate MO symmetry by lower case and state symmetry by the upper case symbols of the irreducible representation.

The ground-state MO configuration of BH_3 is thus

$$BH_3 \ K(a_1')^2(e')^4 \quad {}^1A_1' \tag{7.3.1}$$

The nonbonding MO a_2'' is unoccupied. The doubly degenerate MO e' accommodates four electrons. The state symbol ${}^1A_1'$ expresses the fact

7.3 Trigonal Planar Molecules

that the electronic state is a singlet state with totally symmetric spatial function. Again, although chemically it is clear that there are three equivalent BH bonds in BH_3, the MO representation displays this equivalence through the overall symmetry of the electronic wavefunction and charge density. This symmetry is determined by the *direct product* of the irreducible respresentations of the occupied MO. For BeH_2 ground state, as for other *closed-shell* molecular states, the symmetry is A'_1, *totally symmetric*.

BH_3 LO AND sp^2 HYBRIDIZATION

The LO representation of BH_3 will display three equivalent BH bonds. The hybridization required for the central atom will be related to the MO a'_1 and e' (a linear combination of s_B, x_B, and y_B, that is, sp^2 hybridization) The sp^2 hybrids are

$$sp_a^2 = 1/\sqrt{3}\, s_B + \sqrt{2/3}\, x_B$$
$$sp_b^2 = 1/\sqrt{3}\, s_B - 1/\sqrt{6}\, x_B + 1/\sqrt{2}\, y_B \quad (7.3.2)$$
$$sp_c^2 = 1/\sqrt{3}\, s_B - 1/\sqrt{6}\, x_B - 1/\sqrt{2}\, y_B$$

PROOF To prove that (7.3.2) constitute the sp^2 hybrids we may start with sp_a^2. Taking the x axis aligned with the BH_a bond, sp_a^2 will overlap with $1s_a$ if $sp_a^2 = c_{1a} s_B + c_{2a} x_B$. Then the remaining hybrids are $sp_b^2 = c_{1b} s_B + c_{2b} x_B + c_{3b} y_B$ and $sp_c^2 = c_{1c} s_B + c_{2c} x_B + c_{3c} y_B$. We require that the hybrids are orthonormal, as were the original AO, that is, $\langle sp_a^2, sp_a^2 \rangle = c_{1a}^2 + c_{2a}^2 = 1$ and $c_{1b}^2 + c_{2b}^2 + c_{3b}^2 = 1$, $c_{1c}^2 + c_{2c}^2 + c_{3c}^2 = 1$; $\langle sp_a^2, sp_b^2 \rangle = c_{1a} c_{1b} + c_{2a} c_{2b} = 0$, $c_{1a} c_{1c} + c_{2a} c_{2c} = 0$, and $c_{1b} c_{1c} + c_{2b} c_{2c} + c_{3b} c_{3c} = 0$. Finally, it is reasonable that the totally symmetric orbital s_B appear equivalently in all three hybrids with a weight of $\tfrac{1}{3}$ in each, $c_{1a}^2 = c_{1b}^2 = c_{1c}^2 = \tfrac{1}{3}$ ($\tfrac{1}{3}$ "s character" in each hybrid). From the first normalization relation $c_{2a} = \sqrt{\tfrac{2}{3}}$; from the first orthogonality relation $c_{2b} = -c_{1a}^2/c_{2a} = -1/\sqrt{6}$; from the second orthogonality relation $c_{2c} = -1/\sqrt{6}$. The normalization relations yield $c_{3c}^2 = 1 - c_{1c}^2 - c_{2c}^2 = \tfrac{1}{2}$ and $c_{3b}^2 = 1 - c_{1b}^2 - c_{2b}^2 = \tfrac{1}{2}$, but from the final orthogonality relation $c_{3c} = -c_{3b}$. These coefficients are displayed in equation (7.3.2). The sp^2 hybrid orbitals are outlined in Fig. 7.3.2.

EXERCISE Show that the sp^2 hybrids are orthonormal.

The bonding orbitals follow immediately from these hybrids, $bo_a = c'_1 sp_a^2 + c'_2 1s_a$, $bo_b = c'_1 sp_b^2 + c'_2 1s_b$, and $bo_c = c'_1 sp_c^2 + c'_2 1s_c$. The configuration of these LO displays the three equivalent electron-pair bonds of BH_3 as in the Lewis structure H : B : H.
$$\overset{..}{H}$$

$$BH_3 \quad K(bo_a)^2 (bo_b)^2 (bo_c)^2 \quad (7.3.3)$$

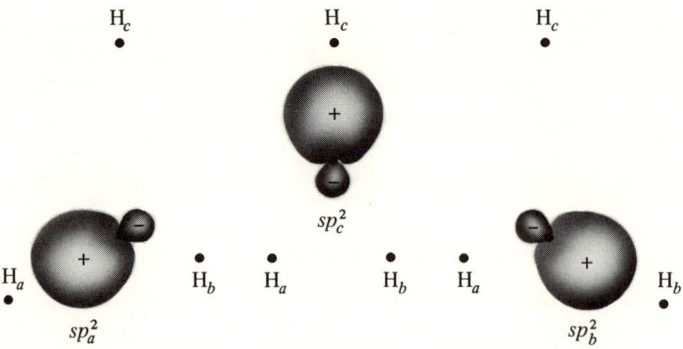

FIG. 7.3.2. The sp^2 hybrids.

7.4 TETRAHEDRAL MOLECULES

The most highly symmetric tetrahedral molecules belong to point group T_d. They form a large class of chemically important species, including CX_4, SiX_4, etc. (X = H, Cl, F, etc.) and ions ClO_4^-, NH_4^+, SO_4^{2-}, etc. Methane, CH_4, is the simplest molecule of the T_d point group.

CH_4 MO

The four bonds of CH_4 form a reducible representation of T_d which decomposes into irreducible representations A_1 and T (T is a three-dimensional representation). Table 7.4.1 summarizes the irreducible components of the valence shell AO.

EXERCISE In the LCAO–MO representation bonding is obtained when AO are combined with the same phase in their region of overlap. Use Fig. 7.4.1. and the phases of s_C, x_C, etc., to show that the carbon AO are combined with the $1s$ hydrogen orbitals of Table 7.4.1 with the same phase in each octant.

The LCAO–MO are formed, as usual, from the AO of Table 7.4.1, with coefficients determined to give the SCF–LCAO–MO.

$$\begin{aligned}
t_x &= c_1 x_C + c_2(1s_a + 1s_c - 1s_b - 1s_d) \\
t_y &= c_1 y_C + c_2(1s_a + 1s_d - 1s_b - 1s_c) \\
t_z &= c_1 z_C + c_2(1s_a + 1s_b - 1s_c - 1s_d) \\
a_1 &= c_3 s_C + c_4(1s_a + 1s_b + 1s_c + 1s_d)
\end{aligned} \qquad (7.4.1)$$

Figure 7.4.2 displays the energies of these MO. The ten electrons of CH_4 are simply accommodated as

$$CH_4 \ K(1a_1)^2(1t)^6 \quad {}^1A_1 \qquad (7.4.2)$$

7.4 Tetrahedral Molecules

TABLE 7.4.1

AO basis for CH_4	Irreducible representation	
$(1s_a + 1s_b + 1s_c + 1s_d), s_C$	a_1	
$(1s_a + 1s_b - 1s_c - 1s_d), z_C$	t_z	
$(1s_a + 1s_d - 1s_b - 1s_c), y_C$	t_y	t degenerate[a]
$(1s_a + 1s_c - 1s_b - 1s_d), x_C$	t_x	

[a] The three degenerate components which form the t basis are labeled t_x, t_y, and t_z.

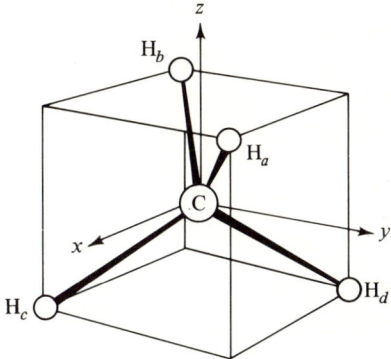

FIG. 7.4.1 Methane. Point group T_d. The four hydrogen atoms are at the corners of a tetrahedron.

CH_4 is one of the best studied polyatomic molecules. Extensive calculations have been performed in extended basis sets. A concise pictorial statement of the electronic structure of CH_4 is given in Fig. 7.4.3 which is a ρ contour map in the plane defined by three nuclei H, C, and H.

Figure 7.4.3 shows the tetrahedral distribution of the charge density in CH_4; however, it reveals no pile-up of charge density in the CH bond regions, but just a general diffuse electronic density of tetrahedral shape. In order to ascertain the net transfer of charge from carbon and hydrogen into the bond regions it is instructive to examine the $\Delta\rho$ map of the molecule (i.e., the density difference contour map $\Delta\rho = \rho_{molecule} - \rho_{atoms}$; see Section 5.12). The $\Delta\rho$ contour map of CH_4 is given by Fig. 7.4.4. It illustrates the net transfer of electronic density from both C and H into the CH bonding regions upon molecule formation. This transfer leads to binding, just as in the case of the diatomic molecule (sections 5.3, 5.12, and 5.13).

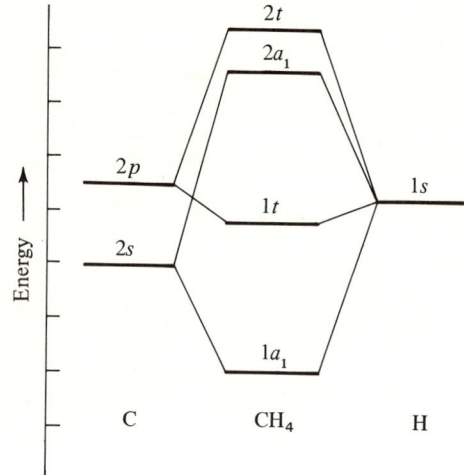

FIG. 7.4.2. Schematic of the orbital energy levels in CH$_4$.

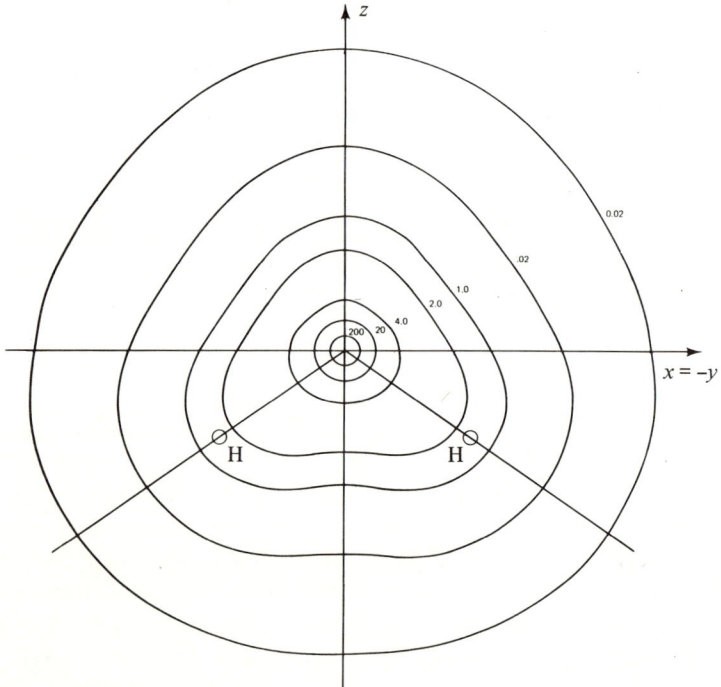

FIG. 7.4.3 Contour map of the electronic density ($4\pi\rho$) in CH$_4$ on a plane defined by three nuclei H, C, and H. From A. G. Turner, A. F. Saturno, P. Hauk, and R. G. Parr, *J. Chem. Phys.* **40**, 1919 (1964). Reproduced by permission.

7.4 Tetrahedral Molecules

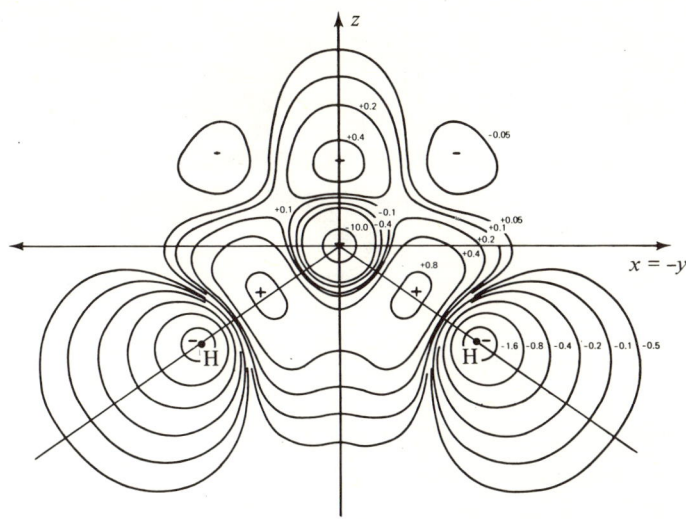

FIG. 7.4.4 The electronic density difference contour map of CH_4. Density difference between CH_4 and its component atoms ($4\pi\Delta\rho$) is plotted in the plane defined by the three nuclei H, C, and H. Regions labeled $+$ have a *greater* electronic density in the molecule. From A. G. Turner, A. F. Saturno, P. Hauk, and R. G. Parr, *J. Chem. Phys.*, **40**, 1919 (1969). Reproduced by permission.

CH_4 LO AND sp^3 HYBRIDIZATION

The angle between the CH bonds of CH_4 is the tetrahedral angle, $109°28'$. The LO representation of CH_4 must display four equivalent bonding orbitals oriented at the tetrahedral angle. The required hybridization is sp^3.

$$sp_a^3 = \tfrac{1}{2}(s_C + x_C + y_C + z_C)$$
$$sp_b^3 = \tfrac{1}{2}(s_C + z_C - x_C - y_C)$$
$$sp_c^3 = \tfrac{1}{2}(s_C + x_C - y_C - z_C)$$
$$sp_d^3 = \tfrac{1}{2}(s_C + y_C - x_C - z_C)$$

(7.4.3)

Equations (7.4.3) are the orthonormal sp^3 hybrids; they are displayed in Fig. 7.4.5. The CH bonding orbitals are constructed equivalently as $bo_i = c_1' sp_i^3 + c_2' 1s_i$ ($i = a, b, c, d$).

$$CH_4 \quad K(bo_a)^2(bo_b)^2(bo_c)^2(bo_d)^2 \qquad (7.4.4)$$

The four equivalent bonding LO are doubly occupied to give an unambiguous classical bonding structure

$$\begin{array}{c} H \\ \ddot{} \\ H:\ddot{C}:H \\ \ddot{} \\ H \end{array}$$

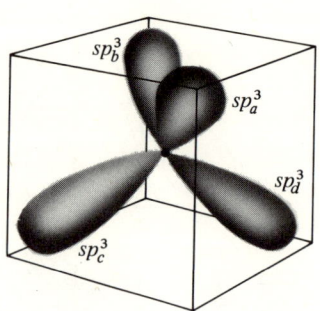

FIG. 7.4.5 The sp^3 hybrids.

We have not demonstrated that the LO representation of four equivalent bonding orbitals may be obtained by transformation of the MO representation (but see the following exercise).

EXERCISE   Show that the LO are obtainable from the MO by the following transformation with the constraint that the MO and LO variational coefficients are related by $c_3 = c_1 = c_1'$ and $c_4 = c_2 = \frac{1}{2}c_2'$, which is almost fulfilled in practice.

$$\begin{pmatrix} bo_a \\ bo_b \\ bo_c \\ bo_d \end{pmatrix} = \begin{pmatrix} \frac{1}{2} & \frac{1}{2} & \frac{1}{2} & \frac{1}{2} \\ \frac{1}{2} & -\frac{1}{2} & -\frac{1}{2} & \frac{1}{2} \\ \frac{1}{2} & \frac{1}{2} & -\frac{1}{2} & -\frac{1}{2} \\ \frac{1}{2} & -\frac{1}{2} & \frac{1}{2} & -\frac{1}{2} \end{pmatrix} \begin{pmatrix} a_1 \\ t_x \\ t_y \\ t_z \end{pmatrix}$$

7.5 ANGULAR TRIATOMIC AND TRIGONAL PYRAMIDAL MOLECULES

Angular triatomic and trigonal pyramidal molecules are two common classes of molecular symmetry. C_{2v} (Fig. 6.3.1 and Table 6.5.1) is the point group of the most prominent angular triatomics, for example, H_2O, SO_2, H_2S, NO_2, O_3. C_{3v} (Fig. 6.4.1) is the point group of the most symmetrical trigonal pyramidal molecules, for example, NX_3, PX_3 (X = H, Cl, F, etc.). The important hydrides H_2O and NH_3 are extremely well studied, and will be the basis of our discussion.

H_2O MO

The symmetry elements of H_2O were illustrated in Fig. 6.3.1. The character table for the irreducible representations of C_{2v} is Table 6.5.1. If we limit ourselves to valence shell atomic orbitals in the construction of LCAO–MO, then the basis set consists of $1s_a$, $1s_b$, s_O, x_O, y_O, and z_O. This set forms the basis of a reducible representation of C_{2v}. Noting that $2p$ orbitals of oxygen transform as the vectors x, y, and z, equation (6.5.2) gives the reducible representation in the basis x_O, y_O, and z_O. Furthermore, the exercise and example following equation (6.5.3) shows that this reducible representation contains the irreducible representations A_1, B_1, and B_2. That is, x_O transforms as B_1, y_O as B_2, and z_O as A_1.

EXERCISE Show from Table 6.5.1 and Fig. 6.3.1 that x_O, y_O, and z_O form the basis of B_1, B_2, and A_1 irreducible representations, respectively.

The $2s$ orbital on oxygen is clearly the basis of an A_1 representation (totally symmetric). The LCAO $(1s_a + 1s_b)$ and $(1s_a - 1s_b)$ are bases of A_1 and B_2 representations, respectively.

PROOF Prove the last assertion.

	Character	
	$(1s_a + 1s_b)$	$(1s_a - 1s_b)$
$\hat{E}(1s_a \pm 1s_b) = (1s_a \pm 1s_b)$	1	1
$\hat{C}_2(1s_a \pm 1s_b) = \pm(1s_a \pm 1s_b)$	1	-1
$\hat{\sigma}_v(1s_a \pm 1s_b) = \pm(1s_a \pm 1s_b)$	1	-1
$\hat{\sigma}'_v(1s_a \pm 1s_b) = (1s_a \pm 1s_b)$	1	1
The characters are those of	A_1 and	B_2

TABLE 7.5.1

AO basis of H_2O	Irreducible representation
$(1s_a + 1a_b)$, s_O, z_O	a_1
$(1s_a - 1s_b)$, y_O	b_2
x_O	b_1

From Table 7.5.1 the LCAO–MO of H_2O are

$$a_1 = c_1(1s_a + 1s_b) + c_2 s_O + c_3 z_O$$
$$b_2 = c_4(1s_a - 1s_b) + c_5 y_O \tag{7.5.1}$$
$$b_1 = x_O$$

The SCF–LCAO–MO calculation[5] yields the orbital energies and the coefficients summarized in Table 7.5.2.

$$H_2O \; K(1a_1)^2(1b_2)^2(2a_1)^2(1b_1)^2 \; {}^1A_1 \tag{7.5.2}$$

[5] F. O. Ellison and H. Shull, *J. Chem. Phys.* **23** 2348 (1955).

TABLE 7.5.2 The Coefficients of the SCF–LCAO–MO Wavefunction of $H_2O^{a,\ b,\ c}$

AO	MO			
	$1a_1$	$2a_1$	$1b_2$	$1b_1$
$(1s_a \pm 1s_b)$	0.15	0.31	0.53	—
s_O	0.82	−0.55	—	—
z_O	0.12	0.76	—	—
y_O	—	—	0.58	—
x_O	—	—	—	1.0
Orbital energies (eV)	−37.43	−14.2	−19.2	−12.8

[a] Results of F. O. Ellison and H. Shull, *J. Chem. Phys.* **23**, 2348 (1955).
[b] Unoptimized Z_{eff}.
[c] Coordinates chosen with oxygen nucleus as the origin and positive z towards the protons, compare Fig. 6.3.1.

The most easily removed electron of H_2O comes from the nonbonding oxygen orbital $1b_1 = x_O$. The negative of the SCF–LCAO–MO orbital energy of $1b_1$ is 12.8 eV, which is in good agreement with the observed ionization potential, 12.6 eV. The $1a_1$ MO is largely s_O (weakly bonding), but the $2a_1$ and the $1b_2$ MO are both bonding.

H_2O has been the object of extensive investigation. SCF–LCAO–MO calculation in an extended basis set[6] gives the total energy −76.005 hartrees (−76.481 hartrees experimental) and dipole moment 2.035 D (1.84 D experimental) as well as the electronic density as a contour map (Fig. 7.5.1).

H_2O LO

The bond angle HOH is 105°, far from the angle between sp^2 hybrid orbitals (120°), but not quite the tetrahedral angle (109°28′). Judging from the bond angle, the bond orbitals of H_2O involve intermediate but nearly tetrahedral hybridization. However, the bond angle does not completely determine the hybridization. There are only a few cases in which the bond angles and the requirement of bond equivalence completely determines the hybridization for the LO, for example, BH_3 and CH_4. In these cases there exist unique transformations between the MO and the LO. The transformation in the case of CH_4 is unique because the four CH bonds form a

[6] S. Aung, R. M. Pitzer, and S. I. Chan, *J. Chem. Phys.* **49**, 2071 (1968).

7.5 Angular Triatomic and Trigonal Pyramidal Molecules

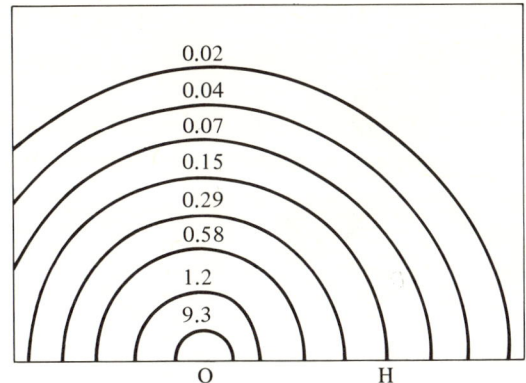
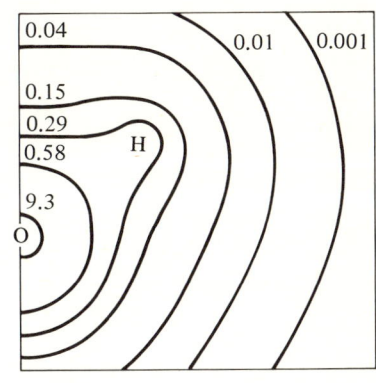

FIG. 7.5.1 (a) Contour map of the electron density in the mirror plane bisecting the HOH angle. In this view the two hydrogen atoms coincide.
(b) Contour map of the electronic density in the plane of the water molecule. Only a half-plane is shown, by symmetry, the other half-plane is identical.
From S. Aung, R. M. Pitzer, and S. I. Chan, *J. Chem. Phys.* **49**, 2071 (1968). Used by permission.

reducible representation decomposable into $A_1 + T$. Therefore the LO are a linear combinations of the a_1 and t MO [see equations (7.4.1) and (7.4.2)]. In H_2O the two OH bonds form a reducible representation decomposable into A_1 and B_2, but now there are *two* bonding valence shell MO of A_1 symmetry, $1a_1$ and $2a_1$, of equation (7.5.2). Any orthonormal combination of these a_1 MO may be used with $1b_2$ to form the LO. As a result, the MO to LO transformation is undetermined to the extent of a parameter which must be fixed by an energetic or localization criterion[7]. The LO of H_2O have been the subject of some research[8], but leaving aside the question of intermediate hybridization for the moment we might first suppose that sp^3 hybridization is a zero-order approximation for all the LO of H_2O. The LO representation is then

$$H_2O \ K(bo_a)^2(bo_b)^2(lp_c)^2(lp_d)^2 \quad (7.5.3)$$

The four localized orbitals of H_2O are the bonding orbitals, $bo_i = c'_1 sp_i^3 + c'_2 1s_i$, and the nonbonding (so-called lone-pair orbitals), $lp_i = sp_i^3$. Figure 7.5.2 gives a schematic illustration of these LO. Comparison of Fig. 7.5.2 to 7.5.1 shows that the former overemphasizes the localization for purposes of illustration, because, of course, the electronic density is the

---
[7] As in the LO determination for diatomic molecules (see Section 5.14).
[8] J. N. Murrell, J. G. Stamper, and N Trinajstic, *J. Chem. Soc.* **A1966**, 1624.

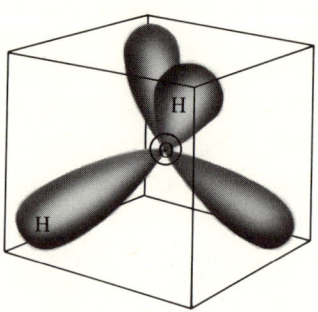

FIG. 7.5.2 Schematic representation of the localized orbitals of H_2O.

same in both MO and LO representations. Furthermore, the sp^3 hybrids of the electron-pair bonds, sp_a^3 and sp_b^3, need not be equivalent to the sp^3 hybrids of the lone-pair orbitals, sp_c^3 and sp_d^3; configuration (7.5.3) would still correspond to the classical Lewis structure H:Ö:H even if they differ. If $2\theta_{cd}$ is the angle between the sp_c^3 and sp_d^3 hybrids, and $2\theta_{ab}$ the corresponding angle between the other sp^3 hybrids, then these angles are related[9] by

$$\cot^2 \theta_{cd} = 1 - \cot^2 \theta_{ab} \qquad (7.5.4)$$

For $\theta_{ab} = \theta_{cd} = \theta$ we have $\cos 2\theta = -\frac{1}{3}$, which defines the regular ($2\theta = 109°28'$) tetrahedral sp^3 hybrids of, say, CH_4. As the angle $2\theta_{ab}$ between sp_a^3 and sp_b^3 *increases* from 90° to 180° the angle $2\theta_{cd}$ between the lone pairs *decreases* from 180° to 90° and one achieves a linear molecule with lone pairs $(x_O)^2$ and $(y_O)^2$. Thus θ_{ab} (or equivalently θ_{cd}) controls the amount of s character in the two sets of hybrids. θ_{ij} ($ij = ab$ or cd) together with the coefficients c_1' and c_2' of the bonding LO bo_i, form a set of three parameters which may be varied to minimize the energy of H_2O. In this way it should be possible to show that $2\theta_{ab} = 105°$, the observed bond angle in H_2O.

NH$_3$ AND THE MOLECULAR DIPOLE MOMENT

The bond angle HNH in NH_3 is 107°, which suggests an increase in s character in the bonding hybrids over H_2O, bringing NH_3 close to regular sp^3 (109°). The LO representation is schematically as in Fig. 7.5.2 with an additional H atom at one of the corners (c or d) of the tetrahedron. The one remaining nonbonded electron pair in NH_3 is the nitrogen lone pair.

---

[9] M. Kotani, K. Ohno and K. Kayama, "Quantum Mechanics of Electronic Structure," *Encyclopedia of Physics*, Vol. XXXVII/2, Springer-Verlag, Berlin/Vienna, 1961.

7.5 Angular Triatomic and Trigonal Pyramidal Molecules

Molecules belonging to a point group containing a *center of symmetry* (Section 6.3.1) have *zero* permanent dipole moment by symmetry, for example, CH_4, CO_2, BH_3, benzene. The dipole moment, when not zero, is an important measurable property of the charge distribution. Since the charge distribution is invariant to transformation of the MO, we may discuss the dipole moment in MO or LO representations. The dipole moment is calculable as the expectation value of the electric dipole operator (Section 5.13).

$$\boldsymbol{\mu} = \boldsymbol{\mu}_{electronic} + \boldsymbol{\mu}_{nuclear}$$

$$\boldsymbol{\mu}_{electronic} = 2\sum_{i=1}^{N}\langle \psi_i, -e\mathbf{r}\psi_i\rangle = 2\sum_{i=1}^{N}\langle \chi_i, -e\mathbf{r}\chi_i\rangle \qquad (7.5.5)$$

Equation (7.5.5) expresses the electronic moment for a configuration of doubly occupied MO (ψ_i) or LO (χ_i). As seen, the molecular moment is the vector sum of electronic and nuclear components, and the electronic component is the vector sum of contributions from each LO. In H_2O, for example, calculating the dipole moment from the oxygen nucleus as origin (the total moment is independent of the choice of origin in neutral species) the total moment will consist of three contributions; M_p the component of the proton moment along the symmetry axis, M_b the contribution of the bonding orbitals to the moment, and M_l the lone-pair contribution. (M_l and M_b are components along the symmetry axis.) These contributions are illustrated in Fig. 7.5.3. The relative magnitudes and signs of the contributions to the total molecular moment of H_2O are also given in Fig. 7.5.3. The lone pairs make an important contribution to the dipole moment. Similarly in NH_3 the single lone pair makes a sizable contribution to the molecular moment ($\mu = 1.46$ D) in comparison to the bond and proton moments ($M_l = 3.6$ D, $M_b = 7.6$ D, $M_p = 5.5$ D)[10].

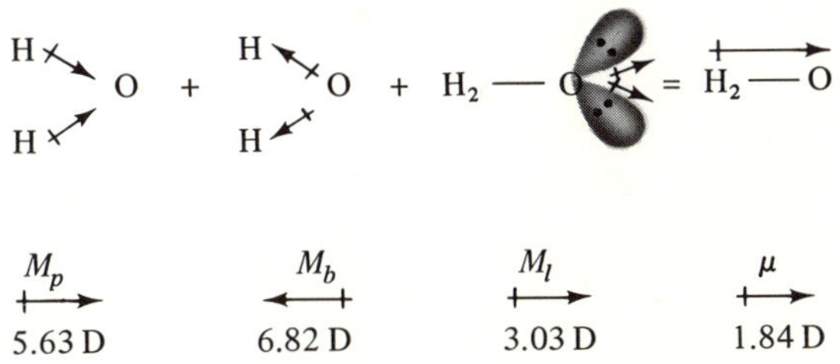

FIG. 7.5.3 Contributions to the dipole moment of H_2O. The positive end of the dipole moment vectors are crossed. Data from A. B. F. Duncan and J. A. Pople, *Trans. Faraday Soc.* **49**, 217 (1953). Used by permission.

---

[10] A. B. F. Duncan and J. A. Pople, *Trans. Faraday Soc.* **49**, 217 (1953).

TABLE 7.5.3 Dipole Moments of Some Angular Triatomic and Trigonal Pyramidal Molecules[a]

Molecule	Molecular dipole moment (D)
H_2O	1.84
H_2S	0.92
SO_2	1.63
NO_2	0.39
O_3	0.52
NH_3	1.46
NF_3	0.23
PH_3	0.55
PF_3	1.03
PCl_3	0.79
AsH_3	0.15
AsF_3	2.82
$AsCl_3$	1.99

[a] From *Tables of Experimental Dipole Moments* by A. L. McClellan. W. H. Freeman and Company, San Francisco, Calif. Copyright © 1963. Used by permission.

7.6 MOLECULAR SHAPE—THE BOND CONCEPT AND MOLECULAR PROPERTIES

In the previous sections of this chapter we have discussed the electronic structure of molecules of a *given* symmetry. We have taken the point group of the molecule as an experimental fact determined by analysis of the infrared spectrum, x-ray diffraction, or some other means. The electronic structure of the ground state was then determined by the use of the variational principal in the Hartree-Fock-Roothaan method using trial LCAO-MO which are symmetry orbitals, that is, form the basis of irreducible representations of the point group. The results are the energy ordering of the MO, the ionization potentials of the molecule, and the electronic configuration. But no explanation has been given of *why* the molecule has this particular shape. The use of hybridization to give localized orbitals which replace the delocalized MO with visualizable bonds and lone pairs also follows *after* the fact of a *known* molecular shape.

Before continuing with this discussion of molecular shape, it should be clear to the student that asking *why* the bond angle in H_2O is 105° instead of 109° or 90° is like asking *why* the equilibrium internuclear distance in

7.6 Molecular Shape—The Bond Concept and Molecular Properties

H_2 is 1.4 A instead of twice the Bohr radius (1.2 A). The answer for bond angles, as for bond distances, is simply that the total molecular energy is a minimum at this nuclear configuration.

Calculations have been carried out at different bond angles and bond distances for most of the molecules discussed in these sections. If the total molecular energy is minimized at each nuclear configuration (e.g., linear H_2O and planar NH_3 as well as bent and pyramidal nuclear configurations) accurate calculations show that the *lowest* total molecular energy corresponds to the experimentally observed nuclear configuration. This is the rather expensive and time-consuming analogy to the calculation of $E(R)$ as a function of R for a diatomic molecule (see Fig. 5.4.2). Often it is instructive (but dangerous) to explain the molecular shape on the basis of minimization of just a *portion* of the total molecular energy, for example, just the orbital energy. An important example of this approach is the interpretation of the sp^3 hybridization representation of the electronic density distribution in the first-row hybrides as an attempt to *minimize the interelectronic repulsion* by placing the electron pairs spatially as far apart as possible.[11]

EXERCISE   On this basis offer an explanation for why the bond angle in H_2S is only 92° compared to 105° in H_2O. Hint: Assume that the interelectronic repulsion of $3p$ electrons is substantially less than that of $2p$ electrons.

The correlation of orbital energies with molecular shape is illustrated by Fig. 7.6.1 which is a "Walsh Diagram"[12] for the simplest case, the triatomic hybride H_2A. Such diagrams are schematic, but we note that the orbital energy orderings are in agreement with those found for H_2O (bent, eight valence electrons) and BeH_2 (linear, four valence electrons). Furthermore, certain *semi-empirical* methods for the calculation of the orbital

EXERCISE CH_2 and BH_2 are bent (C_{2v}) in their ground states; justify these observations on the basis of Fig. 7.6.1.

EXERCISE   Predict the symmetry of the first excited state of BeH_2 C_{2v} or $D_{\infty h}$?

energies of polyatomic molecules (Chapter 8) are not only simply done on modern computers but are quite capable of correctly predicting the orbital energy order and variation with molecular shape.[13]

[11] J. W. Linnett, *Wave Mechanics and Valency*, Wiley, New York, 1960.
[12] A. D. Walsh, *J. Chem. Soc.* **1953**, 2260.
[13] L. C. Allen and D. J. Russell, "Extended Hückel Theory and the Shape of Molecules," *J. Chem. Phys.* **46**, 1029 (1967).

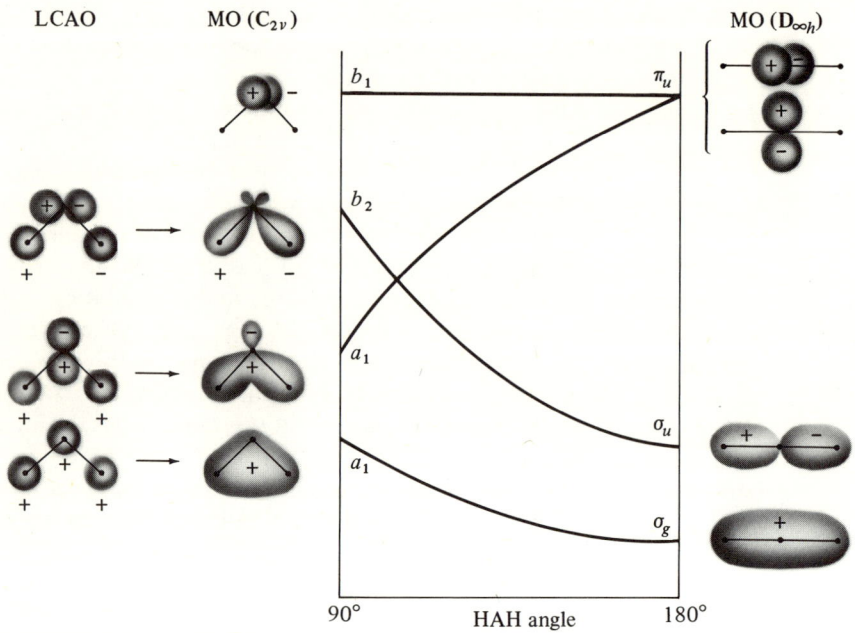

FIG. 7.6.1. The Walsh diagram for H_2A. Energy increases vertically. The most stable orbitals are $a_1(\sigma_g)$ and the least stable orbitals are $b_1(\pi_u)$. Note that the orbital energy orderings are the same as those of H_2O (7.5.2) and BeH_2 (7.2.2).

Alternatively, instead of using total or partial molecular energies, we visualize the positively charged nuclei repelling each other and shaping the electronic charge distribution until the internuclear repulsive forces are neutralized. This is the Hellmann-Feynman (electrostatic) theorem for polyatomic molecules (Section 5.3): *At the equilibrium nuclear configuration the net force on each nucleus is zero.* This implies that changes in the electronic density distribution ρ will of necessity produce changes in molecular shape. Such changes in ρ are produced by excitation to another electronic state, ionization, or perturbation by the close approach of another molecule, as in a chemical reaction. Specifically, a molecule which is linear in the ground state may be bent in an excited state, for example, CO_2, or vice versa. Appreciable changes in ρ are produced not only by adding or removing an electron, for example, $NO_2(132°)$ $NO_2^-(115°)$, but by substituting one atom for another, for example $H_2O(105°)$ $H_2S(92°)$. Such chances are predictable if $\Delta\rho$, the change in electronic density due to substitution, excitation, or ionization is known, but unfortunately this requires extensive calculations of high accuracy.

7.6 Molecular Shape—The Bond Concept and Molecular Properties

We have certainly overemphasized the difficulties by citing cases of small molecules subjected to appreciable changes in ρ. Such molecules will always be best handled by highly accurate calculation. Much more interesting to the chemist are cases of *small perturbation* of bonds in molecules. What happens to a CH bond as CH_4 goes to CH_3X? In these cases nature has been very kind to the chemist. Most of the power and usefulness of quantum chemistry arises from the fact that perturbations due to neighboring bonds will often produce quite small changes in the bond. There are very important exceptions; the C—C bond varies in length between 1.2 and 1.54 A. Of course, these changes in bond length are associated with changes in ρ around the C—C bond, that is, associated with changes in hybridization and delocalization around the C atoms. Next, we list some of the properties of polyatomic molecules which are approximately additive in the bonds and/or atoms which comprise the molecule.

BOND PROPERTIES AND OTHER ADDITIVE PROPERTIES OF MOLECULES

1. Stereochemistry is based on the fact that molecules are composed of groups of bound atoms with bond angles and bond lengths which hardly differ from one molecule to another. The steric property of the group is *transferable* from one molecule to another, and the total molecular dimensions are the sum (to good approximation) of the group dimensions.
2. Bond lengths for a particular pair of bonded atoms are relatively constant from one molecule to another, for example, C—H, 1.08 A; C—C, 1.54 A; C=C, 1.35 A; C—O, 1.43 A. An even more elementary additivity arises from the fact that bond lengths can be expressed as the sum of two characteristic radii of the bonded atoms (see Appendix F Table F.2).
3. The fact that pairs of bonded atoms (and groups of bonded atoms) have characteristic infrared absorption frequencies is used in everyday analysis of IR spectrograms. This accounts for the popularity of IR as an analytical tool. For example, the characteristic absorption of C—H around 3 microns can be used to identify the presence of CH in an unknown molecule and estimate the number of CH bonds.
4. Bond dipole moments and lone-pair moments are approximately additive vectorially to give the total molecular dipole moment. These moments are often transferable from one molecule to another. Magnetic properties (diamagnetic susceptibilities) and optical properties (such as the refractivities) also are approximately additive in bond and atomic contributions.
5. Of principal chemical interest among the additive properties of molecules are the so-called *bond energies*. The bond energy of a diatomic molecule is identical to its dissociation energy measured

relative to the separated atoms in their ground electronic states. However, in a polyatomic molecule such as CH_4 the dissociation energy for $CH_4 \to CH_3 + H$ differs markedly from that for, say, $CH_2 \to CH + H$. Therefore, it is most useful to define an *average* dissociation energy called the bond energy, such that the total binding energy of CH_4 is four times the bond energy of CH. From the bond energy of CH (98.7 kcal/mole) the bond energies of C—C, C=C, etc., may be calculated from the known heats of formation of C_2H_6, C_2H_4, etc. The result is a self-consistent set of bond energies, such as given in Table 7.6.1, which is useful for estimating heats of formation of known and unknown compounds.

TABLE 7.6.1 Bond Energies (kcal/mole)

		H	C	N	O	F
H		104.	98.7	93.	110.	134.6
	Single		83.	73.	85.5	116.
C	Double		145.	147.	170.	—
	Triple		194.	213.	—	—

Properties 1 through 5 and their additivity are empirical facts and not necessarily predictions of quantum theory. They are in agreement with quantum theory in those few cases that permit sufficiently accurate calculation.

EXERCISE Define a bond energy for the CO bond in CO_2. Given $CO_2 \to CO + O$ ($D_0 = 127$ kcal/mole) and $CO \to C + O$ ($D_0 = 256$ kcal/mole).

MO VERSUS LO

In the molecular orbital representation, delocalized symmetry orbitals extend over the whole molecule. A prime example is CH_4, in which two delocalized MO, a_1 and t, comprise the entire valence shell. The MO configuration $K(a_1)^2(t)^6$ neither indicates that CH_4 consists of four equivalent CH bonds, nor does it imply that the bonds possess properties which are transferable from CH_4 to, say, C_2H_6, which belongs to a different point group. Although the MO deal quite effectively with the ionization potentials and spectra, they don't lend themselves to a simple explanation of the observed constancy of CH bond energy, bond length, or force constant for vibration in a variety of hydrocarbons.

We interpret observations 1) to 5) listed above as strong experimental evidence for the concept of independent bonds in molecules. These observations justify the localized orbital representation for polyatomic molecules and, really, the localized orbital approach seems even more valuable and suggestive for polyatomic molecules than it was for diatomics. Although we have not insisted that localized orbitals such as the bonding orbital bo_i in CH_4 be transferable to other hydrocarbons, that is implied by observations 1) to 5). Such transferability is the subject of current research.

The difficulty with localized orbitals is twofold. First, the LO are obtained from the MO on the basis of criteria of localization which leave the total energy and electronic density invariant, but *the LO are not eigenfunctions to the Hartree-Fock problem and consequently the LO energies are not the ionization potentials of the molecule*.[14] Secondly, the use of LO can lead to *ambiguity* in the sense of Section 7.2; that is, if degeneracies arise, resonance structures are required. This has been called *the breakdown of localization*, and such cases encompass a huge class of molecules, for example, aromatic and conjugated molecules. On the whole, the MO and LO representations are complementary and neither can fully satisfy chemical intuition in every case. Both MO and LO (the K shell) have heretofore been used in the same configuration; we will find it often advantageous to use them together.

7.7 FORMALDEHYDE STRUCTURE AND SPECTRA—POPULATION ANALYSIS

There are 16 electrons in H_2CO, a molecule belonging to point group C_{2v}. The coordinate axes in Fig. 7.7.1 have been chosen to coincide with those of H_2O (Fig. 6.3.1) so the vectors x, y, and z form the basis of B_1, B_2, and A_1 irreducible representations.

TABLE 7.7.1

AO basis for H_2CO	Irreducible representation
$(1s_a + 1s_b), s_O, s_C, z_O, z_C$	a_1
$(1s_a - 1s_b), y_O, y_C$	b_2
x_O, x_C	b_1

[14] In another terminology MO and all orbitals obtainable from them by a unitary matrix transformation are called *molecular orbitals* (this includes our LO). The particular molecular orbitals which are eigenfunctions to the Hartree-Fock Hamiltonian are the *canonical* Hartree-Fock orbitals.

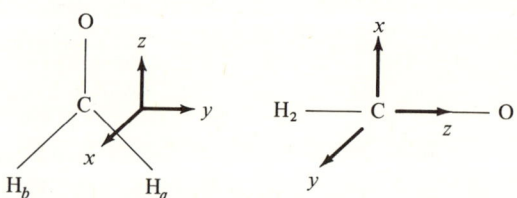

FIG. 7.7.1 Coordinates in H_2CO. Point group C_{2v}.

EXERCISE   Verify Table 7.7.1.

Without the need for SCF–LCAO–MO calculations, but given the equilibrium nuclear configuration of H_2CO, it is evident from chemical intuition that the divalent oxygen and the tetravalent carbon form a double bond. Group theory easily identifies the AO components of the bonding MO. For example, the $\sigma(CO)$ bond is A_1 and therefore largely composed of s_O, s_C, z_O, and z_C. The $\pi(CO)$ bond is B_1 and is composed solely of x_O and x_C. By group theoretical techniques, which we have used previously, we can show that the two CH bonds form the basis of a reducible representation with irreducible components A_1 and B_2. Therefore we expect CH_2 bonding MO of A_1 and B_2 symmetry.

EXERCISE   Show that the two CH bonds form a reducible basis decomposable into $A_1 + B_2$.

The SCF–LCAO–MO results are summarized in the MO configuration[15]

$$H_2CO \quad KK(1a_1)^2(2a_1)^2(1b_2)^2(3a_1)^2(1b_1)^2(2b_2)^2 \; ^1A_1 \qquad (7.7.1)$$

The SCF–LCAO–MO total molecular energy is -113.892 hartrees (-114.562 hartrees experimental). The $1a_1$ MO is largely s_O but, like the CH_2 bonding MO, it is delocalized over all four atoms in the molecule. Thus delocalization, which is inherent in the MO, again defies an unambiguous identification of the bonding MO with specific bonds. Only $1b_1$ which is $c_1 x_O + c_2 x_C$ is a clearly localized bonding MO. In the next few paragraphs we develop an analysis of LCAO–MO wavefunctions (called a population analysis) which permits a quantitative discussion of the electronic density distribution on the component atoms in spite of delocalization.

[15] D. B. Neumann and J. W. Moskowitz, *J. Chem. Phys.* **50** 2216 (1969).

POPULATION ANALYSIS

Mulliken[16] suggested a procedure for quantitatively examining the electronic structure of a molecule through the use of the LCAO–MO wavefunction. This insight is achieved by breaking the total electronic density into *populations* of the component AO in the LCAO–MO or *populations* on the component atoms in the molecule.

Consider the normalized LCAO–MO of AO on atoms 1 and 2.

$$\psi = c_1 \phi_1 + c_2 \phi_2$$

If the MO, ψ, is occupied by N electrons (usually $N = 2$) the N-electron distribution may be considered as divided into three spatial parts given by the following expression:

$$N|\psi|^2 = N(c_1^2 |\phi_1|^2 + 2c_1 c_2 |\phi_1 \phi_2| + c_2^2 |\phi_2|^2)$$

Upon integrating this last expression over all space one obtains the breakdown of the N-electron *population* into three parts

$$N = Nc_1^2 + 2Nc_1 c_2 S_{12} + Nc_2^2$$

where S_{12} is the overlap integral $\langle \phi_1, \phi_2 \rangle$. The overlap population $2c_1 c_2 N S_{12}$ is divided equally between the atoms, leading to the important idea of an *atomic population*, that is, $(Nc_1^2 + Nc_1 c_2 S_{12})$ is the atomic population on atom 1 due to the MO ψ.

To be really useful these definitions must be generalized to the case of a polyatomic molecule with several MO and with several AO on each nucleus. Writing this general MO as $\psi_i = \sum_r \sum_k c_{ir_k} \phi_{r_k}$, then i denotes the MO, r denotes the AO, and k, denotes the atom.

$$N_i |\psi_i|^2 = N_i \sum_k \sum_r c_{ir_k} \phi_{r_k} \sum_l \sum_s c_{is_l} \phi_{s_l}$$

$$N(i; r_k) = N_i \sum_l^{\text{atoms}} \sum_s^{\text{AO}} c_{ir_k} c_{is_l} S_{r_k s_l} \qquad (7.7.2)$$

$N(i; r_k)$ is the *atomic orbital population* of AO ϕ_{r_k} in the MO ψ_i: N_i is the number of electrons occupying ψ_i (usually $N_i = 2$).

$$N(i; k) = \sum_r N(i; r_k) \qquad (7.7.3)$$

$N(i; k)$ is the *partial gross atomic population* on atom k in the MO ψ_i. This population index indicates the total charge population on the kth atom due to the ith MO.

$$N(k) = \sum_i \sum_r N(i; r_k) \qquad (7.7.4)$$

---

[16] R. S. Mulliken, *J. Chem. Phys.* **23**, 1833 (1955).

$N(k)$ is the *gross atomic population* on atom k. This population index indicates the total electronic population on the atom k due to *all* the occupied MO.

EXERCISE $\quad \sum_{k} N(k) = ?$

These definitions are now illustrated with the rather delocalized orbitals of H_2CO. Table 7.7.2 presents these population indices for the MO and AO of H_2CO. It should be noted that the coefficients and the overlap

TABLE 7.7.2 Atomic Populations in H_2CO^a

Nucleus	Atomic orbital[b]	Molecular Orbitals of H_2CO						Gross atomic population
		$1a_1$	$2a_1$	$3a_1$	$1b_1$	$1b_2$	$2b_2$	
Oxygen atom	$2s$	1.369	0.216	0.200	—	—		
	$2p_z$	0.105	0.189	1.147	—	—	—	
	$2p_x$	—	—	—	1.311	—	—	
	$2p_y$	—	—	—		0.606	1.251	
Partial gross atomic population on oxygen		1.474	0.405	1.347	1.311	0.606	1.251	6.394 + K shell = 8.394
Carbon atom	$2s$	0.326	0.791	0.057	—	—	—	
	$2p_z$	0.162	0.243	0.419	—	—	—	
	$2p_x$	—	—	—	0.642	—	—	
	$2p_y$	—	—	—	—	0.841	0.139	
Partial gross atomic population on carbon		0.488	1.034	0.476	0.642	0.841	0.139	3.620 + K shell = 5.620
Two hydrogen atoms	$1s$	0.015	0.536	0.161	—	0.535	0.541	1.788

[a] The table displays the atomic orbital populations in each MO (eq. 7.7.2), the partial gross atomic populations in each MO (eq. 7.7.3), and in the last column the gross atomic population on the atoms (eq. 7.7.4). The data are those of D. B. Neumann and J. W. Moskowitz, *J. Chem. Phys.* **50**, 1, 2216 (1969). Used by permission.

[b] Small contributions from d orbitals on C and O and p orbitals on H have been omitted from the table for clarity. The populations of the K shell have been included in the last column (almost entirely $1s$). The omissions account for the sum of gross atomic populations being 15.8 in a 16-electron molecule!

7.7 Formaldehyde Structure and Spectra—Population Analysis

integrals must be calculated to obtain the energy. Thus population analysis does not require too much additional calculation for the amount of insight into the electronic structure so obtained.

Table 7.7.2 gives quantitative meaning to the statement made earlier that $1a_1$ is largely s_O, that is, the population of s_O in $1a_1$ is 1.369 out of a theoretical maximum of 2.0. In the same way it is evident that $2b_2$ is largely y_O (this is the "oxygen lone pair" orbital). The $1b_1$ MO is the bonding π orbital of H_2CO. Population analysis shows that $1b_1$ has a much greater population on oxygen than on carbon. On the other hand, $1b_2$, the bonding MO of B_2 symmetry, is extremely delocalized over all four atoms with a slightly greater population on carbon than oxygen.

The gross atomic populations display a net migration to oxygen atom (gross population 8.4 compared to neutral atom 8.0) away from carbon atom (gross population 5.6 compared to neutral carbon atom 6.0) and away from the hydrogen atoms too.

MIXED REPRESENTATION

The discussion of the populations in H_2CO shows that it is possible to make a quantitative analysis of the effects of delocalization if the all-electron SCF–LCAO–MO wavefunction is available. For the vast majority of molecules of chemical interest such calculations are not economically feasible. So, with the feeling of going from the sublime to the ridiculous, we now illustrate the use of a *mixed representation* of LO and MO on H_2CO, although the value of such representations lies in the enormous number of large molecules which are not amenable to all-electron calculations. A mixed representation was also given for CO_2 in Section 7.2.

The formaldehyde molecule is planar with bond angles HCH and HCO close to 120°, the angle of sp^2 hybridization. Because of the planarity of H_2CO the π MO, $1b_1$, is of a different irreducible representation from the other MO which constitute the σ core of the molecule. It is useful to give the σ core an LO representation based on sp^2 hybridization of the carbon atom; mathematically, localized linear combinations of the a_1 and b_2 MO are constructed.

$$H_2CO \; KK(s_O)^2(bo_a)^2(bo_b)^2(bo_O)^2(1b_1)^2(y_O)^2 \qquad (7.7.5)$$

bo_a and bo_b are the CH bonding orbitals, $bo_a = c'_1 sp_a^2 + c'_2 1s_a$; bo_O is the CO bonding orbital, $bo_O = c'_3 sp^2 + c'_4 z_O$; s_O and y_O are nonbonding orbitals; finally $1b_1$ is, as before, the π-bonding MO. While in this simple example the $1b_1$ MO is actually localized between C and O, this will not be the case for the large conjugated and aromatic molecules discussed in the next chapter. In these large molecules the bond angles indicate sp or sp^2 hybridization; the σ core may then be given a rapid representation in terms of LO based on hybridized carbon atoms. The π electrons are treated separately and occupy π MO which are delocalized over the whole molecule. *Thus resonance structures need not be invoked to account for π-electron delocalization, and delocalized MO need not be found for localized σ bonds with their roughly similar properties in different molecules.* The implication is that conjugated molecules differ from each other mainly in the delocalized and polarizable π shell, an observation well supported by experimental and theoretical ionization potentials, spectra, and chemical reactivities.

EXERCISE   Justify the configuration (7.7.5) on the basis of the population analysis of Table 7.7.2.

EXERCISE   Define a *total atomic orbital population* (total population of rth AO on kth atom due to occupation of *all* MO). Find the total atomic orbital populations for the AO of Table 7.7.2.

EXERCISE   Which of the population indices are invariant to an MO to LO transformation?

SPECTRA AND PROPERTIES

The dipole moment of H_2CO is 2.33 D with an assumed direction $H_2C^+O^-$. The computed dipole moment[17] (2.82 D $H_2C^+O^-$) is in good agreement. The gross atomic populations (Table 7.7.2) also show that C^+O^- is the net direction of charge migration upon bond formation.

The minimum ionization potential of H_2CO is due to removal of a $2b_2$ "oxygen lone pair" electron (10.86 eV). Excitation of an electron from $2b_2$ is also responsible for the characteristic absorption of H_2CO, and other carbonyl-containing molecules, around 2700–3000 A. The lowest vacant MO is the $2b_1$, an antibonding MO often denoted as simply π^*. If $2b_2$ is denoted n (for nonbonding) then the transition $2b_2 \to 2b_1$ is $n \to \pi^*$. The value of a terminology like $n \to \pi^*$ is that it is transferable from one molecule to another if a carbonyl is present in the molecule, regardless of the point group or energy ordering of the MO.

There are other transitions in H_2CO which lie at higher energy than $n \to \pi^*$; for example, $\pi \to \pi^*$ ($1b_1 \to 2b_1$) and *Rydberg series*. Rydberg series are observed in the high-energy spectra of diatomic and polyatomic molecules. They are characterized by the fact that the excited orbitals are *atomic-like orbitals* of principal quantum number beyond the valence shell. For example, in H_2CO the series of transitions $2b_2 \to 3s, 4s, \ldots, ns$, or $2b_2 \to 3p, 4p, \ldots, np$ form Rydberg series which converge to the ionization potential of the molecule, as in atomic spectra.

The selection rules for polyatomic electronic transitions follows from the required nonvanishing of the transition moment $\langle \psi_i, \boldsymbol{\mu}\psi_j \rangle$ (Section 6.8).

---

[17] D. B. Neumann and J. W. Moskowitz, footnote 15.

7.7 Formaldehyde Structure and Spectra—Population Analysis

This in turn requires that the direct product of the irreducible representations $\psi_i \times \mu \times \psi_j$ contain the totally symmetric representation. The spatial symmetry of filled MO shells is always totally symmetric, so it is only necessary to take the direct product of the irreducible representations of the unfilled MO to determine the spatial symmetry of the state.

EXAMPLE The excited state which results from the $\pi \to \pi^*$ transition in H_2CO has the configuration (core) $(1b_1)(2b_1)$. The direct product $B_1 \times B_1 = A_1$, so the excited state is $^1A_1$ (a totally symmetric singlet state). The ground state is also $^1A_1$ (7.7.1). The transition $\pi \to \pi^*$ ($^1A_1 \to {}^1A_1$) is thus allowed with polarization in the z direction (z transforms as A_1).

EXERCISE Show that the $n \to \pi^*$ transition is electronically forbidden in H_2CO (also see problem 1 at the end of this chapter).

REFERENCES

Gerhard Herzberg, *Molecular Spectra and Molecular Structure*, Vol. III, Electronic Spectra and Electronic Structure of Polyatomic Molecules, D. van Nostrand, New York, 1966.

John C. Slater, *Quantum Theory of Molecules and Solids*, McGraw-Hill, New York, 1963.

PROBLEMS

1. Suggest a mechanism whereby the forbidden $n \to \pi^*$ transition in H_2CO may obtain some intensity.
2. Consider two molecular states or orbitals ψ_1 and ψ_0 which are eigenfunctions of the zero-order Hamiltonian $\hat{H}_0$ with eigenvalues ε_1 and ε_0. The eigenvalues depend on a nuclear parameter R such that at $R = R_a$, $\varepsilon_1(R_a) = \varepsilon_0(R_a)$. From equation (2.7.10) and general group theoretical considerations, what are the conditions under which the energies of these states will actually cross?
3. Derive a mathematical treatment which gives the energetic separation of the states discussed in Problem 2 at all R. Discuss the occurrence of such crossings when several nuclear parameters are available for simultaneous variation as in a polyatomic molecule. [Ref., G. Herzberg and H. Longuet-Higgins, *Discuss. Faraday Soc.* **35**, 77 (1963).]
4. The observed radical NH_2 has seven valence electrons. Discuss the expected symmetry of this species in its ground state; C_{2v} or $D_{\infty h}$? Discuss the symmetry of its first excited state. [Ref., Herzberg, General Reference 1.]
5. Discuss the bond angles in the closed shell hydrides, BeH_2, BH_3, CH_4, NH_3, H_2O, and HF from the point of view of *localized* bonding pairs and lone pairs of electrons and the minimization of interelectronic repulsion. [Ref., J. W. Linnett, *Amer. Scientist* **52**, 459 (1964).]

6. Present an explanation for the observations $\mu(NH_3) = 1.46$ D, but $\mu(NF_3) = 0.23$ D. The FNF angle (103°) does not differ that much from the HNH angle (107°). [Ref., S. R. La Paglia and A. B. F. Duncan, *J. Chem. Phys.* **34**, 1003 (1961)].
7. From the following table of overlap integrals do a population analysis of the H_2O molecule as done for H_2CO in Table 7.7.2. Use the LCAO-MO coefficients of Table 7.5.2. Discuss the results.

	$2s_O$	$2p_{zO}$	$2p_{yO}$	$1s_b$	
$1s_a$	0.4946	0.2118	0.2760	0.3479	Overlap integrals for STO for H_2O

$1s_b$ overlap integrals are the same, except $\langle 1s_b, 2p_{yO}\rangle = -0.2760$. [Ref., R. S. Mulliken, *J. Chem. Phys.* **23**, 1833 (1955)].
8. Give a qualitative LCAO–MO description of the bonding in NO_2, an angular triatomic with 23 electrons. Give a localized orbital description of the same molecule and compare to MO representation. Discuss the spectrum of NO_2.

8
π-ELECTRON THEORY

... the principal purpose of accurate calculations is to assure us that nothing truly significant has been overlooked.

EUGENE WIGNER
Proceedings of the International Conference on Theoretical Physics, Japan, 1953

CALCULATIONS of the electronic structure of small molecules, some of which have been described or summarized in Chapters 5 and 7, have assured us that the molecular Schrödinger equation can be solved to give the properties of these molecules to high accuracy. Of course, chemically accurate dissociation energies, transition energies, transition intensities, etc., require the consideration of electron correlation in solution of the Schrödinger equation (see Sections 4.11 and 5.15). Thus assured of the applicability of the Schrödinger equation to small molecules we must still face the problem of chemically interesting molecules containing large numbers of electrons, which are dealt with inefficiently by the methods of Chapter 5 and 7. Unless we have very large amounts of computer time available we must find approximation methods for the solution of the molecular Schrödinger equation. Very successful, but not infallible, approximation methods have been devised. In this chapter we examine some of the molecules of great importance in organic chemistry through the use of well-established simplifying procedures.

8.1 THE π-ELECTRON APPROXIMATION

The class of *conjugated molecules* is composed of organic molecules whose classical bond formulas are written in terms of alternating single and double carbon–carbon bonds, for example, —C=C—C=C—C=. *Heteroatoms*, such as nitrogen or oxygen, may also be involved in such conjugated systems. This class of molecules is undoubtedly one of the most well studied in all of chemistry and biochemistry.

The typical conjugated molecule contains a large number of electrons. For example, just one double bond, as in ethylene (C_2H_4), is a 16-electron system. In order not to have to treat all the electrons of the molecule in the same way (to the same accuracy) there must be some basis for distinguishing between them. In fact, there are several ways of distinguishing between orbitals of conjugated molecules. First, by symmetry; the conjugated molecule has a carbon skeleton which lies in a plane; the π electrons are so named because they occupy orbitals having a node in the plane of the molecule. Thus, by analogy with the diatomic symmetry designation, these orbitals are denoted of π type. On the other hand, the orbitals which are cylindrically symmetrical about the C—C and C—H bonds are denoted of σ type, again by analogy with the diatomic molecule. There are also the K-shell orbitals. The π, σ, and K electron densities of the simplest organic π system, ethylene, are displayed in Figs. 8.1.1 and 8.1.2. Notice the strong tendency toward spatial differentation of the π, σ, and K densities.

Another means of distinguishing between the electrons of a conjugated molecule is energetic. Consider the orbital energies in ethylene; the K orbitals are lowest in energy, the σ orbitals are substantially higher, and the π orbitals are highest in energy (the least tightly bound). The relation

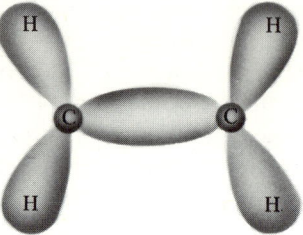

FIG. 8.1.1 Schematic representation of the electronic density in the K shells and the five σ bonds of ethylene.

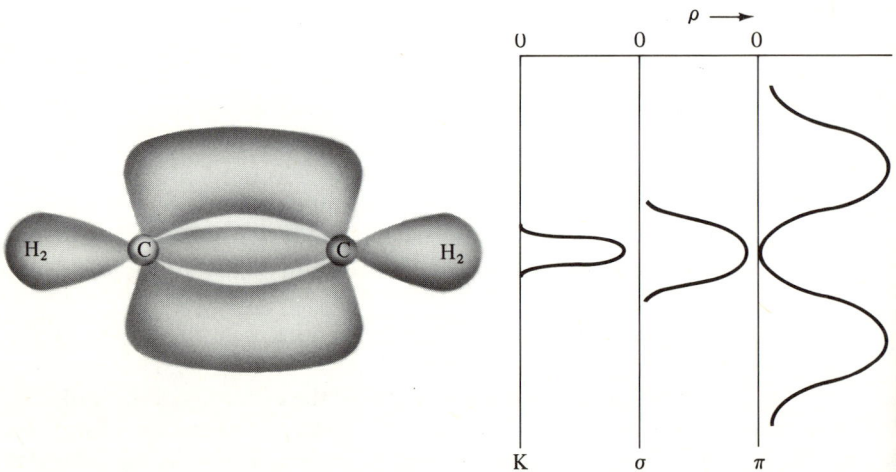

FIG. 8.1.2 Schematic electronic density profiles of the K, σ, and π electronic distributions in ethylene. After Platt, J. R. "The Chemical Bond and the Distribution of Electrons in Molecules," in Vol. XXXVII/2, Springer, Berlin, *Handbuch der Physik*, Bd. 37/2, pp. 173-281; Berlin-Göttingen-Heidelberg: Springer, 1961. Used by permission.

8.1 The π-Electron Approximation

between these energy differences and the localization of the electronic densities is illustrated in Fig. 8.1.3, where the orbital energies are superimposed on the effective potential seen by the electrons. The deep wells are provided by the attraction of the carbon and hydrogen nuclei. The K electrons are unshielded from the carbon nuclei and consequently are tightly bound by the attractive nuclear Coulomb potential, the σ electrons are much less tightly bound, and the π electrons are sufficiently energetic to be delocalized over the whole molecule. The π electrons are the least tightly held, most polarizable, and chemically important electrons of the molecule.

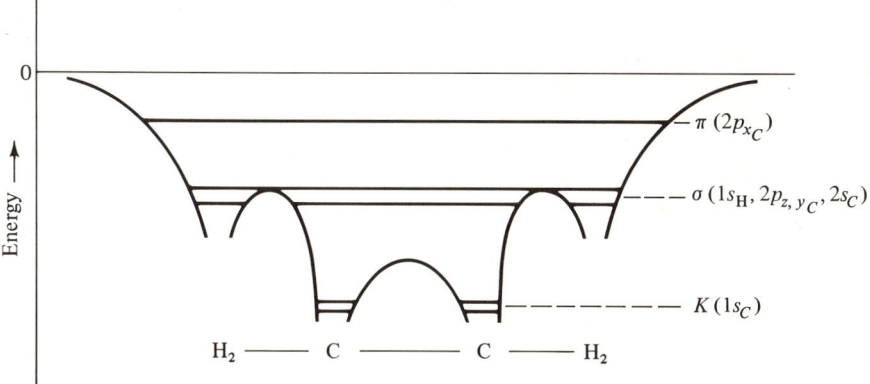

FIG. 8.1.3 Schematic energy relation of the orbitals in ethylene to the potential energy seen by the electrons. After Platt, J. R., "The Chemical Bond and the Distribution of Electrons in Molecules," in *Handbuch der Physik*, Bd. 37/2, pp. 173-281; Berlin-Göttingen-Heidelberg: Springer, 1961. Used by permission.

Having established the physical distinction between the orbitals of a conjugated molecule we next seek a mathematical formulation of the Schrödinger equation which reflects these facts. This is most easily accomplished by postulating a σ-π separability in the following sense. We assume that the total N-electron wave-function Ψ may be written as the product of two separately antisymmetrized wavefunctions Ψ_Π and Ψ_Σ.

$$\Psi = \hat{A}' \Psi_\Pi \Psi_\Sigma \tag{8.1.1}$$

If there are n π electrons in the conjugated molecule then Ψ_Π is the antisymmetrized n-electron wavefunction for these π electrons; consequently Ψ_Σ is the $(N - n)$-electron wavefunction for the electrons of the σ core (i.e., the σ bonds and K shells). Thus, since Ψ_Π and Ψ_Σ are separately antisymmetrized, the antisymmetrizer $\hat{A}'$ need only permute the σ electrons with the π electrons. In simplest form both Ψ_Π and Ψ_Σ are Slater determinants of doubly occupied molecular orbitals.

The wavefunctions are normalized to unity, $\langle\Psi_\Pi, \Psi_\Pi\rangle = 1$ and $\langle\Psi_\Sigma, \Psi_\Sigma\rangle = 1$. In addition, in order to treat the π-electron spectra and structure separately, we require that Ψ_Σ be *unchanged* for all the states of the molecule. These conditions constitute the *π-electron approximation*. Strong experimental evidence has accumulated substantiating the validity of this approach to the spectra, structure, and reactivities of conjugated molecules. We will not argue with success, but remain aware of the approximations involved, and be even more wary of the more severe approximations used later.

The σ–π separability is a useful application of the idea of a mixed representation discussed previously. The Ψ_Σ wavefunction is composed of $1s$ carbon orbitals and localized σ-bond orbitals constructed from sp^2 hybrids, while the Ψ_Π wavefunction is composed of π molecular orbitals which are linear combinations of $2p_{x_C}$ atomic orbitals. The σ core is thus of trivial interest in conjugated molecules (but see Section 8.4) and it only remains to find the π MO.

We now write the Hamiltonian (7.1.2) in a manner to express σ–π separability.

$$\hat{H} = \hat{H}_\sigma^0 + \hat{H}_\pi \tag{8.1.2}$$

$$\hat{H}_\pi = \sum_{i=1}^{n}\left(-\tfrac{1}{2}\nabla_i^2 - \sum_{a=1}^{K} Z_a/r_{ia}\right) + \sum_{i=j+1}^{N}\sum_{j=1}^{n} 1/r_{ij} \tag{8.1.3}$$

$$\hat{H}_\sigma^0 = \sum_{i=n+1}^{N}\left(-\tfrac{1}{2}\nabla_i^2 - \sum_{a=1}^{K} Z_a/r_{ia}\right) + \sum_{i>j>n}^{N} 1/r_{ij} \tag{8.1.4}$$

The π-electron Hamiltonian $\hat{H}_\Pi$ includes a sum over the n π electrons numbered 1 to n. $\hat{H}_\pi$ also includes the interelectronic repulsion of the n π electrons with *all* N electrons in the molecule. $\hat{H}_\sigma^0$ is limited to σ–σ repulsions and the one-electron operators for the σ electrons (kinetic energy and electron-nuclear attraction). A further simplification is achieved by replacing the true σ–π repulsions in $\hat{H}_\Pi$ with a smeared out, or averaged, σ–π potential $\hat{V}_{\text{aver}}(i)$.

$$\hat{H}_\pi = \sum_{i=1}^{n}\left(-\tfrac{1}{2}\nabla_i^2 - \sum_{a=1}^{K} Z_a/r_{ia} + \hat{V}_{\text{aver}}(i)\right) + \sum_{i>j}^{n} 1/r_{ij}$$

where (see Section 5.7)

$$\hat{V}_{\text{aver}}(i) = \sum_{k=n+1}^{N}\left\langle \psi_k(k), \left(\frac{(1-\hat{P}_{ik})}{r_{ik}}\right)\psi_k(k)\right\rangle_k$$

$$\hat{H}_\pi = \sum_{i=1}^{n} \hat{H}_{\text{core}}(i) + \sum_{i>j}^{n} 1/r_{ij} \tag{8.1.5}$$

8.2 The Hückel Molecular Orbital Method

The Hamiltonian $\hat{H}_{\text{core}}$ includes the π-electron kinetic energy and electron–nuclear Coulomb potential, as well as all interactions with the σ-core. It can now be shown from (8.1.4), (8.1.5), and 8.1.1 that

$$\langle \Psi, \hat{H}\Psi \rangle = E_\sigma^0 + E_\pi \tag{8.1.6}$$

$$E_\sigma^0 = \langle \Psi_\Sigma, \hat{H}_\sigma^0 \Psi_\Sigma \rangle \tag{8.1.7}$$

$$E_\pi = \langle \Psi_\Pi, \hat{H}_\pi \Psi_\Pi \rangle \tag{8.1.8}$$

EXERCISE   Obtain equation (8.1.2) by comparison to equation (7.1.2).

EXERCISE   Derive equations (8.1.6)–(8.1.8).

E_σ^0 is the constant contribution to the total energy due to the σ core, independent of Ψ_Π. Thus the energy is minimized by application of the variational principle to the π-electron wavefunction, that is, let $W_\pi = \langle \tilde{\Psi}_\Pi, \hat{H}_\pi \tilde{\Psi}_\Pi \rangle / \langle \tilde{\Psi}_\Pi, \tilde{\Psi}_\Pi \rangle$ be the variational energy with a trial π electronic wavefunction, $\tilde{\Psi}_\Pi$, then $W_\pi \geq E_\pi$.

This is as far as the π-electron approximation will take us by itself. We still have a formidable problem involving the variational solution to the many-π-electron Schrödinger equation, $\hat{H}_\pi \tilde{\Psi}_\Pi = E_\pi \tilde{\Psi}_\Pi$, including the interelectronic repulsions (see Section 7.1.5). However, in the framework of the LCAO–MO method, it is possible to make several drastic approximations which allow us to proceed quite simply to find Ψ_Π.

8.2 THE HÜCKEL MOLECULAR ORBITAL METHOD

The Hückel molecular orbital method is the most widely used computational procedure employed by chemists. It owes its popularity to its simplicity and its ability to correlate the properties of homologous series of molecules. To step from the π-electron approximation to the Hückel method we *assume* that $\hat{H}_\pi$ can be written

$$\hat{H}_\pi = \sum_{i=1}^{n} \hat{h}_{\text{eff}}^\pi (i) \tag{8.2.1}$$

where $\hat{h}_{\text{eff}}^\pi$ is an effective Hamiltonian which includes the effect of the interelectronic repulsions $\sum_{i>j} 1/r_{ij}$ in *some* average manner. With (8.2.1) for the Hamiltonian the π Schrödinger equation immediately separates.

$$\hat{H}_\pi \Psi_\Pi = E_\pi \Psi_\Pi \tag{8.2.2}$$

$$\Psi_\Pi = \hat{A} \psi_1 \alpha(1) \psi_1 \beta(2) \psi_2 \alpha(3) \cdots \tag{8.2.3}$$

$$E_\pi = \sum_{i=1}^{n} E_i \tag{8.2.4}$$

$$\hat{h}_{\text{eff}}^\pi \psi_i = E_i \psi_i \tag{8.2.5}$$

The key equation is the one-electron Schrödinger equation, (8.2.5), which resembles the Hartree-Fock equation used in previous chapters. However, comparison of (8.2.4) with (5.7.5) shows that the analogy is not complete because (8.2.1) is *not* assumed in the Hartree-Fock procedure.

To obtain the solution of the one-electron Schrödinger equation we expand the MO as a LCAO and apply the variational principle.

$$\psi_i = \sum_{j=1}^{n} c_{ij} \phi_j \tag{8.2.6}$$

The ϕ_j are $2p_{x_C}$ orbitals located on atom j; the sum extends over all the carbon atoms in the conjugated system. The c_{ij} are unknown coefficients. The minimization of the energy with respect to the coefficients, $W = \langle \psi_i, \hat{h}^\pi_{\text{eff}} \psi_i \rangle / \langle \psi_i, \psi_i \rangle$, $W \geq E_i$, yields the secular equation.

$$\det |\mathbf{H} - \mathbf{S}W| = 0 \tag{8.2.7}$$

The secular equation has for its roots the set of molecular orbital energies $E_1, E_2, \ldots, E_n$ obtainable from LCAO–MO of the form (8.2.6). At this point the student might want to review the variational procedure in the LCAO method (Sections 5.6, 5.2, and Appendix A).

The important distinguishing features of the Hückel MO (HMO) method are embodied in the following drastic simplifications of the secular determinant.

1. The matrix elements H_{ii} of $\mathbf{H}$ are the same for each carbon atom. H_{ii} is usually denoted α and called the "*Coulomb integral.*"
2. All the overlap integrals are set equal to zero and the orbitals are normalized. Thus $S_{ij} = 0$ for $i \neq j$ and $S_{ii} = 1$ ($S_{ij} = \delta_{ij}$).
3. The $H_{ij} (i \neq j)$ matrix elements of $\mathbf{H}$ take the following values for pairs of carbon atoms; $H_{ij} = \beta$, i and j are directly bonded; $H_{ij} = 0$, i and j are not directly bonded. β is called the "*resonance integral.*"

Two instructive examples of the HMO method now follow.

ETHYLENE

$H_2C{=}CH_2$ is not a conjugated molecule, but it is the simplest example of α, β parametization of the secular determinant. The HMO is $\psi_i = c_{i1}\phi_1 + c_{i2}\phi_2$, where $\phi_1 = 2p_x$ on carbon 1 and $\phi_2 = 2p_x$ on carbon 2. The resulting secular equation is

$$\begin{vmatrix} H_{11} - S_{11}W & H_{12} = S_{12}W \\ H_{21} - S_{21}W & H_{22} - S_{22}W \end{vmatrix} = 0$$

The HMO method simplifications yield

$$\begin{vmatrix} \alpha - W & \beta \\ \beta & \alpha - W \end{vmatrix} = 0 = (\alpha - W)^2 - \beta^2$$

8.2 The Hückel Molecular Orbital Method

EXERCISE Justify the Hückel simplifications of the secular determinant for ethylene. Use the symmetry and simplicity of the molecule.

The roots of the ethylene HMO secular equation are $E_1 = \alpha + \beta$ and $E_2 = \alpha - \beta$ (bonding and antibonding HMO). Ethylene is, practically speaking, a diatomic π system in the HMO method. The two π electrons accomodate themselves in the lowest-energy HMO, ψ_1, to give a π-electron energy, $E_\pi = 2\alpha + 2\beta$. (Note that α and β, since they represent energy integrals, are negative quantities; see further.)

EXERCISE How might one relate α to the ionization potential of carbon?

EXERCISE Find the HMO coefficients for the ethylene molecule, c_{11}, c_{12}, c_{21}, and c_{22}.

BUTADIENE

Butadiene is a simple example of conjugation. The classical bond formula is

$$H_2C = C - C = CH_2$$
$$H H$$

The molecule is planar in its equilibrium nuclear configuration. *cis*- and *trans*-Butadiene have the same secular equation in the HMO method (which should alert us to the severity of the approximations). Consequently we picture the molecule in its trans conformation, Fig. 8.2.1.

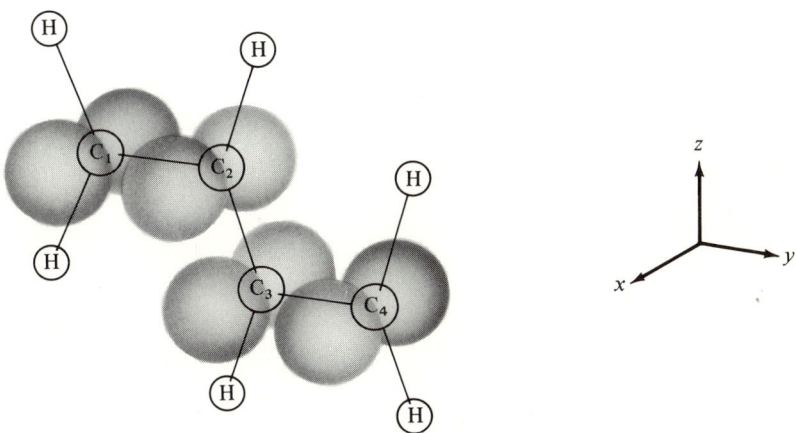

FIG. **8.2.1** $2p_x$ carbon AO in *trans*-butadiene.

EXERCISE Examine the secular determinants for *cis*- and *trans*-butadiene. Why are they the same in the HMO method but, in general, different?

The HMO have the following form: $\psi_i = \sum_{j=1}^{4} c_{ij}\phi_j$. The resulting secular determinant (see Section 5.6) looks like

$$\begin{vmatrix} H_{11} - S_{11}W & H_{12} - S_{12}W & H_{13} - S_{13}W & H_{14} - S_{14}W \\ H_{21} - S_{21}W & H_{22} - S_{22}W & H_{23} - S_{23}W & H_{24} - S_{24}W \\ H_{31} - S_{31}W & H_{32} - S_{32}W & H_{33} - S_{33}W & H_{34} - S_{34}W \\ H_{41} - S_{41}W & H_{42} - S_{42}W & H_{43} - S_{43}W & H_{44} - S_{44}W \end{vmatrix} = 0 \quad (8.2.8)$$

The HMO simplifications of the secular determinant yield

$$\begin{vmatrix} \alpha - W & \beta & 0 & 0 \\ \beta & \alpha - W & \beta & 0 \\ 0 & \beta & \alpha - W & \beta \\ 0 & 0 & \beta & \alpha - W \end{vmatrix} = 0$$

The secular equation is not modified by the change of variable $x = (\alpha - W/\beta)$, which is obtained by dividing through by β.

$$\begin{vmatrix} x & 1 & 0 & 0 \\ 1 & x & 1 & 0 \\ 0 & 1 & x & 1 \\ 0 & 0 & 1 & x \end{vmatrix} = 0 \quad (8.2.9)$$

From Appendix A, equation (A.2.6), the determinant is evaluated and the secular equation is found to be $x^4 - 3x^2 + 1 = 0$.

PROOF The first step in the evaluation of the determinant is

$$x \cdot \begin{vmatrix} x & 1 & 0 \\ 1 & x & 1 \\ 0 & 1 & x \end{vmatrix} - 1 \cdot \begin{vmatrix} 1 & 1 & 0 \\ 0 & x & 1 \\ 0 & 1 & x \end{vmatrix} + 0 \cdot \begin{vmatrix} 1 & x & 0 \\ 0 & 1 & 1 \\ 0 & 0 & x \end{vmatrix} - 0 \cdot \begin{vmatrix} 1 & x & 1 \\ 0 & 1 & x \\ 0 & 0 & 1 \end{vmatrix} = 0$$

Further reduction yields

$$x \left(x \cdot \begin{vmatrix} x & 1 \\ 1 & x \end{vmatrix} - 1 \cdot \begin{vmatrix} 1 & 1 \\ 0 & x \end{vmatrix} + 0 \cdot \begin{vmatrix} 1 & x \\ 0 & 1 \end{vmatrix} \right)$$

$$- 1 \left(1 \cdot \begin{vmatrix} x & 1 \\ 1 & x \end{vmatrix} - 1 \cdot \begin{vmatrix} 0 & 1 \\ 0 & x \end{vmatrix} + 0 \cdot \begin{vmatrix} 0 & x \\ 0 & 1 \end{vmatrix} \right) = 0$$

so finally, $x[x(x^2 - 1) - x] - (x^2 - 1) = 0$

$$x^4 - 3x^2 + 1 = 0$$

8.2 The Hückel Molecular Orbital Method

The equation is quadratic, the substitution $y = x^2$ yielding

$$y^2 - 3y + 1 = 0 \quad \text{or} \quad y = 3 \pm \frac{\sqrt{9-4}}{2} \quad \text{and} \quad x = \pm \left(\frac{3 \pm \sqrt{9-4}}{2}\right)^{1/2}$$

$$x = \pm 1.61804, \pm 0.61804 \qquad (8.2.10)$$

From $x = (\alpha - W)/\beta$ the roots of the secular equation give the π-orbital energy levels of butadiene in terms of the *parameters* α and β (Fig. 8.2.2).

The quantities α and β, the Coulomb and resonance integrals, are negative, which accounts for the orbital energy order of Fig. 8.2.2. The four π-electrons of butadiene are found in the ground-state configuration $(\psi_1)^2(\psi_2)^2$. The π electronic energy is $E_\pi = 4\alpha + 4.472\beta$. The occupied orbital energies are estimates of the ionization potentials of the molecule.

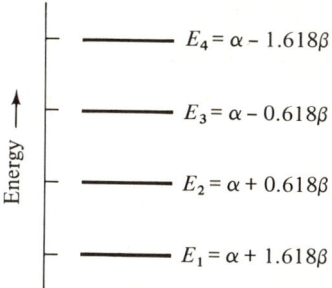

FIG. 8.2.2 π-Orbital energy levels of butadiene in terms of the Hückel parameters.

DELOCALIZATION ENERGY

The delocalization energy (or resonance energy), DE_π, of a conjugated molecule is the difference between $E_\pi(\text{HMO})$ and the π-electron energy of the classical bond structure. Because $H_2C=CH-CH=CH_2$ indicates two independent ethylenic linkages in the molecule the classical π energy would be twice that of ethylene, or $4\alpha + 4\beta$. Thus the delocalization energy of butadiene is $DE_\pi = (4\alpha + 4.472\beta) - (4\alpha + 4\beta) = 0.472\beta$. This is the stabilizing energy which results from delocalization of the π electrons over the entire molecule. DE_π may be compared to resonance energies estimated from experiment,[1] or compared in homologous series of molecules (structurally related compounds) to gauge their relative stabilities. The delocalization of the π electrons is demonstrated next by calculating the coefficients in the HMO.

---

[1] See the discussion of the HMO parameters. Also, A. Streitwieser, Jr., *MO Theory for Organic Chemists*, Wiley, New York, 1961.

HÜCKEL MO COEFFICIENTS, CHARGES, AND BOND ORDERS

In Appendix A (eq. (A.2.16)) it is shown that if the roots of the secular equation are known, the ratio of the coefficients for a given root is determined by the ratio of cofactors, $c_i/c_k = A^{ji}/A^{jk}$. If $|A|$ is the secular determinant, A^{ji} is the cofactor of the element A_{ji}, that is, $A^{ji} = (-1)^{i+j}$ times the determinant which is formed by striking out the ith column and the jth row of the secular determinant. It is simplest to take $j = 1$, the first row, so $c_i/c_k = A^{1i}/A^{1k}$. It further simplifies the process to take the coefficient ratios relative to the same coefficient, say, c_1.

$$c_i/c_1 = A^{1i}/A^{11} \tag{8.2.11}$$

For butadiene we then have

$$\frac{c_2}{c_1} = -\frac{\begin{vmatrix} 1 & 1 & 0 \\ 0 & x & 1 \\ 0 & 1 & x \end{vmatrix}}{\begin{vmatrix} x & 1 & 0 \\ 1 & x & 1 \\ 0 & 1 & x \end{vmatrix}} = \frac{-(x^2-1)}{x^3-2x} \qquad \frac{c_3}{c_1} = \frac{\begin{vmatrix} 1 & x & 0 \\ 0 & 1 & 1 \\ 0 & 0 & x \end{vmatrix}}{x^3-2x} = \frac{x}{x^3-2x}$$

$$\frac{c_4}{c_1} = -\frac{\begin{vmatrix} 1 & x & 1 \\ 0 & 1 & x \\ 0 & 0 & 1 \end{vmatrix}}{x^3-2x} = \frac{-1}{x^3-2x} \tag{8.2.12}$$

EXERCISE   Verify the cofactors in (8.2.12) from (8.2.9).

Table 8.2.1 summarizes the results for the lowest-energy root, $x = -1.618$.

TABLE 8.2.1[a]

i	c_{1i}/c_{11}	$(c_{1i}/c_{11})^2$	$\Sigma_i (c_{1i}/c_{11})^2$	c_{1i}
1	1.0	1.0		0.3717
2	1.618	2.618	7.236	0.6015
3	1.618	2.618		0.6015
4	1.0	1.0		0.3717

[a] From eq. (8.2.12) with $x = -1.618$.

From the normalization condition $\langle \psi_1, \psi_1 \rangle = 1$, and the Hückel zero-overlap assumption, $S_{ij} = 0$ ($i \neq j$), it follows that $c_{11}^2 + c_{12}^2 + c_{13}^2 + c_{14}^2 = 1$, or $(c_{11}/c_{11})^2 + (c_{12}/c_{11})^2 + (c_{13}/c_{11})^2 + (c_{14}/c_{11})^2 = 1/c_{11}^2$. This allows

8.2 The Hückel Molecular Orbital Method

us to find c_{11} from the sum in column 3 of Table 8.2.1, $c_{11} = \sqrt{1/7.236} = 0.3717$. The remaining coefficients of this HMO then follow from the quotients in column 2 and are displayed in the last column. (The positive sign is chosen for c_{11} because multiplication of all the coefficients by -1 does not change the wavefunction.)

EXERCISE Find the coefficients c_{3i} in the HMO ψ_3 ($x = 0.618$) and those of ψ_4 ($x = 1.618$).

For the HMO ψ_2 we have the root $x = -0.618$, which leads, by the same process illustrated in Table 8.2.1, to the coefficients c_{2i}. The two occupied HMO are

$$\psi_1 = 0.3717(\phi_1 + \phi_4) + 0.6015(\phi_2 + \phi_3)$$
$$\psi_2 = 0.6015(\phi_1 - \phi_4) + 0.3717(\phi_2 - \phi_3)$$
(8.2.13)

The HMO are diagrammed in Fig. 8.2.3.

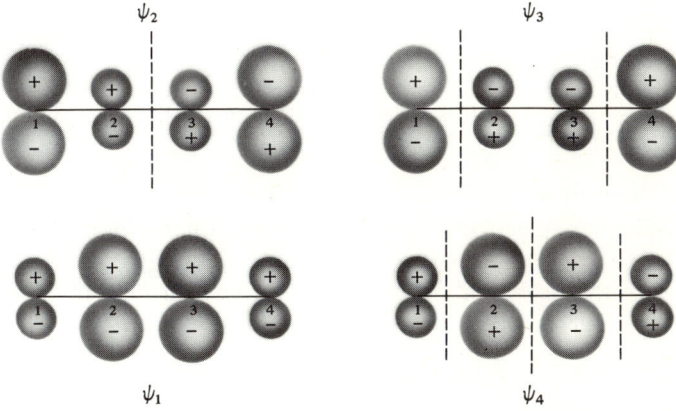

FIG. 8.2.3 HMO nodal patterns in butadiene.

The lowest occupied MO of butadiene shows the delocalization of π-electronic density over the whole molelule. This delocalization may also be described by the free-electron method of Chapter 3. Delocalization of π density in conjugated molecules stabilizes them relative to hypothetical molecules having isolated ethylenic linkages, that is, classical bond for-formulas $-HC=CH-CH=CH-CH=CH-$.

EXERCISE Compare the nodal patterns of occupied and unoccupied HMO to those of the free-electron model of butadiene (Section 3.7). Discuss the similarities and differences between these two methods.

EXERCISE   Solve the Hückel secular equation for hexatriene $H_2C=CH-CH=CH-CH=CH_2$. Find the resonance energy.

The amount of π bonding between adjacent carbon atoms is related to the π-electron density in the bond region (Chapter 5). Thus, the occupied HMO indicate π bonding exists between carbon atoms 2 and 3 as well as atoms 1 and 2, and 3 and 4. This delocalization of π-electron density across the central bond in butadiene is in sharp contrast to the classical bond formula of the compound. A quantitative index of the π bonding between a pair of adjacent carbon atoms is given by the π-*bond order*, P_{ij}.

$$P_{ij} = \sum_{k}^{occ.\,MO} N_k c_{ki} c_{kj} \qquad \pi\text{-BOND ORDER BETWEEN ADJACENT CARBON ATOMS } i \text{ AND } j \qquad (8.2.14)$$

The sum in (8.2.14) is over all the occupied HMO, and N_k is the occupation of the HMO (usually $N_k = 2$).

An index of the π-electron density on the ith carbon atom is given by the π-*electron charge density*, Q_i.

$$Q_i = \sum_{k}^{occ.\,MO} N_k c_{ki}^2 \qquad \pi\text{-CHARGE DENSITY ON } i\text{th ATOM} \qquad (8.2.15)$$

EXERCISE   Show that Q_i corresponds to the π-*electron population* [eq. (7.7.4)] on the ith atom in the zero-overlap approximation used in the Hückel method.

π-electron charge densities are sometimes used as indices of the probability of electrophilic attack in a reaction mechanism. A less useful index is the so-called *free valency index*, F_i, of the ith atom. F_i seems to correlate with the probability of attack by a free radical, a quantitative equivalent of Thiele's idea of residual partial valence (see Chapter 1)

$$F_i = \sqrt{3} - \sum_j P_{ij} \qquad \text{FREE VALENCE OF } i\text{th ATOM} \qquad (8.2.16)$$

where $\sqrt{3}$ is the maximum value which may be taken by the sum of bond orders to a single carbon atom. $\sum_j P_{ij}$ is the actual sum of all bond orders to the ith atom.

EXAMPLE   The HMO indices for butadiene are, following eqs. (8.2.14)–(8.2.16),

$$P_{12} = 2(0.3717)(0.6015) + 2(0.6015)(0.3717) = 0.894$$
$$P_{23} = 2(0.6015)^2 - 2(0.3717)^2 = 0.448$$

$P_{23} = 0.448$ (instead of zero as in the classical bond formula) indicates partial π bonding between carbon atoms 2 and 3.

$$Q_1 = 2(0.3717)^2 + 2(0.6015)^2 = 1.0$$
$$Q_2 = 2(0.6015)^2 + 2(0.3717)^2 = 1.0$$

All $Q_i = 1$, indicating a neutral molecule of zero dipole moment.

$$F_1 = 1.732 - 0.894 = 0.838$$
$$F_2 = 1.732 - 0.448 - 0.894 = 0.39$$

Of course, by symmetry, $F_1 = F_4$, $Q_1 = Q_4$, $P_{12} = P_{14}$, etc.

The complete HMO description of the π-electron structure of butadiene is summarized by the HMO indices in a diagram as follows:

$$\begin{array}{cccc}
1.0 & 1.0 & 1.0 & 1.0 \\
\text{CH}_2 \overset{0.894}{\text{---}} \text{CH} \overset{0.448}{\text{---}} \text{CH} \overset{0.894}{\text{---}} \text{CH}_2 \\
\downarrow & \downarrow & \downarrow & \downarrow \\
0.838 & 0.39 & 0.39 & 0.838
\end{array}$$

EXERCISE Discuss the first excited state of butadiene. Give the HMO configuration, excitation energy, and the HMO indices (P_{ij}, Q_i, and F_i) of the excited state. Compare excited-state to ground-state indices.

ALTERNANT AND NONALTERNANT SYSTEMS

Consider the carbon skeleton of the conjugated molecule. If the carbon atoms can be divided into two sets such that the atoms in one set are formally bonded only to atoms in the other set the molecule is said to be an *alternant* hydrocarbon. In practice one draws a diagram of the molecule and stars alternate atoms, if no two starred atoms, or no two unstarred atoms, are bonded, the hydrocarbon is alternant.

Alternant   Non alternant

It is necessary to distinguish between alternant and nonalternant molecules because there are several theorems which apply to only one class. Notice in Fig. 8.2.2, the orbital energies of butadiene come in pairs, $E_i = \alpha \pm m\beta$. Butadiene, like all conjugated polyenes, is an alternant hydrocarbon.

The HMO pairing theorem states that for every HMO energy $\alpha + m\beta$ in an alternant hydrocarbon there exists another HMO energy $\alpha - m\beta$. A special case arises when the number of carbon atoms in the alternant system is odd. Then there is one root $m = 0$ corresponding to a nonbonding orbital (example: allyl radical C=C—C·).

Another important property of alternant hydrocarbons containing N π electrons and N carbon atoms is that all the Q_i are equal to unity, that is, the molecule is nonpolar. In a nonalternant molecule the π-electron charges vary from carbon to carbon and the molecule may be quite polar. Compare naphthalene and azulene, both of which contain 10 carbon atoms and 10 π electrons.

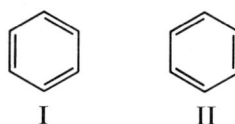

Because the HMO method is based on a constant Coulomb integral for each carbon atom it will not be as reliable for a nonalternant molecule because such a molecule presents different environments for different carbon atoms. Procedures exist for adjusting α to make it consistent with the calculated Q_i.[2]

Cyclic conjugated systems are dealt with in the HMO method in the same way as linear systems. The Hückel procedure, being a MO method, is particularly well suited to the extensive delocalization of the aromatic π-electron systems.

BENZENE

Benzene is the prime example of a molecule having an ambiguous localized bond structure. Benzene, like other aromatic molecules, cannot be represented by a single valence structure, but partakes equally of both Kekulé resonance structures.

There are also contributions from the much higher energy Dewar structures.

---

[2] Refer to footnote 1 of this chapter.

8.2 The Hückel Molecular Orbital Method

III IV V

The superposition of all these resonance structures represents the delocalization of the π electrons of benzene, delocalization which is inherent in the simple Hückel treatment of benzene.

To solve the HMO benzene problem set up the secular determinant as in (8.2.7); with the Hückel simplifications and the substitution $x = (\alpha - W)/\beta$ for each diagonal element and 1 for each pair of bonded atoms we obtain

$$\begin{vmatrix} x & 1 & 0 & 0 & 0 & 1 \\ 1 & x & 1 & 0 & 0 & 0 \\ 0 & 0 & x & 1 & 0 & 0 \\ 0 & 0 & 1 & x & 1 & 0 \\ 0 & 0 & 0 & 1 & x & 1 \\ 1 & 0 & 0 & 0 & 1 & x \end{vmatrix} = 0$$

Expansion of the secular determinant yields the equation

$$x^6 - 6x^4 + 9x^2 - 4 = 0$$

or

$$(x+2)(x-2)(x+1)^2(x-1)^2 = 0$$

Thus the six roots yield HMO energies $E_1 = \alpha + 2\beta$, $E_2 = E_3 = \alpha + \beta$, $E_4 = E_5 = \alpha - \beta$, and $E_6 = \alpha - 2\beta$, which occur in *pairs* because benzene is an alternant hydrocarbon (see Fig. 8.2.4). Figure 8.2.5 illustrates the HMO of benzene through their nodal patterns and phases.

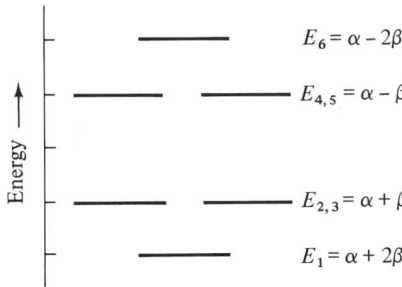

FIG. 8.2.4 π-Orbital energy levels of benzene in terms of the Hückel parameters.

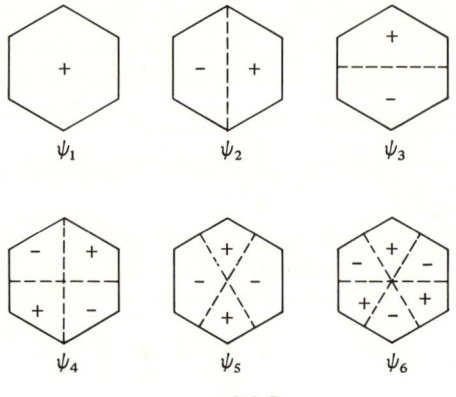

FIG. 8.2.5

Typically, the number of nodes in the HMO increases with the increasing energy of the MO and so does the antibonding character of the orbital. The energy of three isolated double bonds is $3(2\alpha + 2\beta)$ but E_π(benzene) = $2(\alpha + 2\beta) + 4(\alpha + \beta) = 6\alpha + 8\beta$. Thus the delocalization energy of the benzene ring is 2β. HMO calculations on other cyclic and polycylic systems yield somewhat similar resonance energies (substantially larger than for linear conjugated molecules). For example, the bicyclic molecules of 10 carbon atoms, naphthalene and azulene, have delocalization energies $DE_\pi = 3.68\beta$ and 3.36β, respectively.

THE HMO METHOD AND ITS PARAMETERS

The characteristic feature of the Hückel method is its expression in terms of the energy parameters α and β, which may not always be constant. In nonalternant hydrocarbons the asymmetric distribution of π-electron density on the various carbon atoms requires different α for different atoms. A high electronic density on the ith atom serves to decrease the attraction around that atom toward other π electrons. The net result is to make α_i less negative. This is simply accomplished by the Wheland-Mann formula $\alpha_i = \alpha + \omega(1 - Q_i)\beta$ (the ω technique), where Q_i is the π charge density on the ith atom. It has been suggested that $\omega = 1.4$ gives the best aggreement with experiment.[3]

An initial calculation with α yields the Q_i, which in turn yields the α_i; a calculation with α_i yields a new set of Q_i, etc., until self-consistency is obtained.

---

[3] See Streitweiser, footnote 1.

8.2 The Hückel Molecular Orbital Method

If *heteroatoms* such as oxygen or nitrogen have $2p_x$ orbitals involved in the conjugated system, the HMO method may proceed as before but different values of α and β are required for the heteroatoms. Relative to α and β chosen for, say, benzene we may express the heteroatom parameters as $\alpha_Z = \alpha + k_Z \beta$ and $\beta_{C-Z} = k_{Z-C} \beta$ (values of k_Z and k_{Z-C} have been tabulated in Streitwieser's book).[3]

In the absence of heteroatoms all the excitation and delocalization energies in the HMO method are given in units of the parameter β. This should be evident for the HMO energies of ethylene, butadiene, and benzene (Figs. 8.2.2. and 8.2.3), as well as their DE_π. Now β, like α, has been formally defined as an energy integral. However, not only is the operator $\hat{h}_{\text{eff}}^\pi$ ill-defined, but there are so many simplifications in the Hückel procedure that the only feasible numerical procedure is to regard β as a parameter to be determined by experiment, that is, a *semiempirical parameter*. An alternative, nonnumerical procedure, is to express all energies in units of β; this is especially useful for comparisons among a homologous series which should have nearly identical parameters.

Attempts have been made to define universal numerical values of β for the aromatic hydrocarbons[3] and other classes of conjugated molecules. One can correlate the value of β with the observed $\pi \to \pi^*$ transition energies of the conjugated molecules. For example, the observed $\pi \to \pi^*$ transition energy of butadiene is 5.9eV and the HMO transition energy is -1.24β (see Fig. 8.2.2)[4]. The value of β may also be identified by correlating the resonance energy DE_π with observed resonance energies. Because of the extensive π-electron delocalization in conjugate systems the bond orders are variable and the molecular dissociation energy is *not* additive in bond energies as it is to high accuracy in saturated hydrocarbons. Thus DE_π is related to the *difference* between the sum of the bond energies and the experimental dissociation energy of the molecule. (The experimental delocalization energy is taken as the difference between the observed heat of combustion of the hydrocarbon and the heat of combustion predicted on the basis of additivity of the bond energies in the classical bond diagram.[5]) Such a correlation for a large number of aromatic hydrocarbons gives [3]

$$\beta = -16 \text{ kcal/mole } (-0.69 \text{ eV}).$$

This value of β permits one to estimate the dissociation energy of an aromatic hydrocarbon to within a few kilocalories per mole. *Unfortunately*

---

[4] Because of bond length alternation in butadiene and other linear polyenes the use of two values of β has been suggested.

[5] L. Salem, *Molecular Orbital Theory of Conjugated Molecules*, pp. 98–101, Benjamin, New York, 1966.

it differs from the value of β which correlates the $\pi \to \pi^*$ spectra of the same molecules[6]. The value of β which best predicts the frequency of the first strong absorption band in the aromatic hydrocarbons is

$$\beta = -21{,}900 \text{ cm}^{-1} \, (-2.71 \text{ eV})$$

Actually, we should not be surprised at these variations in the "value" of β with the property under investigation. Molecular energies such as ionization potentials, excitation energies, total molecular dissociation energies, and electron affinities are not entirely expressible in terms of one-electron integrals even in the Hartree-Fock method (see Sections 4.10 and 5.7). Interelectronic repulsion integrals are involved in each of these molecular energies. In fact, β represents a *different* (if overlapping) set of energy integrals for each molecular energy.

An important extension of the HMO procedure is the inclusion of the interelectronic integrals as additional *parameters* while maintaining the other Hückel approximations, including the zero-overlap approximation. This is known as the Pariser-Parr-Pople[7] method (PPP Method). It is closer in mathematical form to the π-electron approximation, equations (8.1.5)–(8.1.7), is and therefore a true attempt to parametize the π-electron Schrödinger equation.[8]

One might question the seemingly drastic zero-overlap approximation of these semiempirical theories, because the overlap between $2p_x$ orbitals on bonded carbon atoms is about $\frac{1}{4}$, which is far from zero. However, the parameters do not *need* to refer to a given basis of atomic orbitals; the MO method could just as logically be set up in a basis of *orthogonalized* atomic orbitals, that is, $S_{ij} = 0$ $(i \neq j)$ is no longer an assumption but is characteristic of the basis set. From comparative calculations in orthogonalized and ordinary basis sets[7,9] one concludes that the zero-overlap approximation is not at all bad in π electron calculations.

8.3 TWO APPLICATIONS OF SYMMETRY

THE APPLICATION OF GROUP THEORY TO THE SIMPLIFICATION OF THE HMO METHOD

As the number of carbon atoms in the conjugated molecule under investigation becomes larger and larger the secular determinant soon becomes unwieldy. However, because the secular determinant contains matrix

---

[6] E. Heilbronner and J. N. Murrell, *J. Chem. Soc.* **1962**, 2611.

[7] Robert G. Parr, *Quantum Theory of Molecular Electronic Structure*, Benjamin, New York, 1963.

[8] Semiempirical values for the interelectronic integrals have been deduced; see M. K. Orloff and O. Sinanoğlu, *J. Chem. Phys.* **43**, 49 (1965).

[9] P.-O. Löwdin, *J. Chem. Phys.* **18**, 365 (1950).

8.3 Two Applications of Symmetry

elements $H_{ij} = \langle \phi_i, \hat{h}^\pi_{\text{eff}} \phi_j \rangle$ it is evident from the discussion of Section 6.6 that if the basis set were chosen to form the basis of irreducible representations of the molecular point group, many of the H_{ij} would vanish by symmetry. Specifically, all matrix elements between orbitals forming the basis of *different* irreducible representations are zero. The net result of using *symmetry-adapted orbitals* (Section 6.6) is to *partition* the secular determinant into smaller determinants, one for each irreducible representation contained in the complete $2p_x$ basis set. We now illustrate the use of symmetry in the two examples previously discussed.

Ethylene. The HMO are completely determined by symmetry in this case. Previously the HMO of ethylene were written $\psi_i = c_{i1}\phi_1 + c_{i2}\phi_2$, where $\phi_1 = 2p_x$ on carbon 1 and $\phi_2 = 2p_x$ on carbon 2. The point group of ethylene is $\mathbf{D}_{2h}$ but one symmetry element is sufficient to distinguish between the HMO of this molecule, rotation about the twofold axis, C_2.

$$C_2$$
$$H_2C \stackrel{(+)\,(+)}{\underset{(-)\,(-)}{=}} CH_2$$

The operator relation $\hat{C}_2 \phi_1 = \phi_2$ suggests that $\pi_u = (1/\sqrt{2})(\phi_1 + \phi_2)$ and $\pi_g = (1/\sqrt{2})(\phi_1 - \phi_2)$ with eigenvalues $+1$ and -1 are the proper normalized ($S_{12} = 0$) symmetry orbitals ($\hat{C}_2 \pi_u = \pi_u$ and $\hat{C}_2 \pi_g = -\pi_g$) as in the diatomic molecule. From group theory, $\langle \pi_g, \hat{h}^\pi_{\text{eff}} \pi_u \rangle = 0$ and $\langle \pi_g, \pi_u \rangle = 0$. The symmetry-adapted basis π_g and π_u thus gives the ethylene secular determinant

$$\begin{vmatrix} H_{gg} - S_{gg}W & 0 \\ 0 & H_{uu} - S_{uu}W \end{vmatrix} = 0$$

$S_{uu} = S_{gg} = 1$, so $(H_{uu} - W)(H_{gg} - W) = 0$. The roots of the secular equation are $E_g = H_{gg}$ and $E_u = H_{uu}$, or $E_u = \frac{1}{2}\langle(\phi_1 + \phi_2), \hat{h}^\pi_{\text{eff}}(\phi_1 + \phi_2)\rangle = \frac{1}{2}(2\alpha + 2\beta) = \alpha + \beta$, and $E_g = \alpha - \beta$ as before.

trans-Butadiene. Again the C_2 axis perpendicular to the plane of the molecule is sufficient to distinguish between the irreducible representations for which the orbitals form a basis. $\hat{C}_2 \phi_1 = \phi_4$ and $\hat{C}_2 \phi_2 = \phi_3$ suggest that ϕ_1 and ϕ_4 must appear in the HMO with the same coefficients, likewise for ϕ_2 and ϕ_3. Thus $X_1 = (1/\sqrt{2})(\phi_1 + \phi_2)$, $X_2 = (1/\sqrt{2})(\phi_2 + \phi_3)$, $X_3 = (1/\sqrt{2})(\phi_1 - \phi_4)$ and $X_4 = (1/\sqrt{2})(\phi_2 - \phi_3)$ is our choice of symmetry-adapted orbitals. (Alternatively, the character table of the full point group of butadiene may be used to find the LCAO which form the basis of irreducible representations.) Noting that $\hat{C}_2 X_1 = X_1$ and $\hat{C}_2 X_2 = X_2$ have eigenvalue $+1$, and $\hat{C}_2 X_3 = -X_3$ and $\hat{C}_2 X_4 = -X_4$ have eigenvalue -1, it follows that the secular determinant of butadiene in the symmetry-adapted basis looks like

$$\begin{vmatrix} H'_{11} - S'_{11}W & H'_{12} & 0 & 0 \\ H'_{21} & H'_{22} - S'_{22}W & 0 & 0 \\ 0 & 0 & H'_{33} - S'_{33}W & H'_{34} \\ 0 & 0 & H'_{43} & H'_{44} - S'_{44}W \end{vmatrix} = 0$$

where $H'_{ij} = \langle X_i, \hat{h}^{\pi}_{\text{eff}} X_j \rangle$. With the total determinant reduced to blocks connected only by zero elements, the determinant has been factored, that is, it can be written as the product of two 2×2 determinants. Thus each of the reduced determinants must separately vanish.

$$\begin{vmatrix} H'_{11} - S'_{11}W & H'_{12} \\ H'_{21} & H'_{22} - S'_{22}W \end{vmatrix} = 0 \qquad (8.3.1)$$

$$\begin{vmatrix} H'_{33} - S'_{33}W & H'_{34} \\ H'_{43} & H'_{44} - S'_{44}W \end{vmatrix} = 0 \qquad (8.3.2)$$

The four roots are found from solution of the two quadratic secular equations (8.3.1) and (8.3.2). For example,

$$(H'_{11} - W)(H'_{22} - W) - H'_{12} H'_{21} = 0$$

or

$$(\alpha - W)(\alpha - W) - \beta^2 = 0$$

and

$$W = \alpha + \beta \left(\frac{1 \pm \sqrt{5}}{2} \right)$$

Thus the two roots of (8.3.1) are the orbital energies $E_1 = \alpha + 1.618 \beta$ and $E_3 = \alpha - 0.618 \beta$ of Fig. 8.2.2.

EXERCISE Obtain the roots of (8.3.2). Find the coefficients of the symmetry orbitals X_1 and X_2 in the HMO corresponding to the root E_1 and show that the resulting HMO is equivalent to ψ_1 of (8.2.13).

EXERCISE In a previous exercise the Hückel method has been applied to hexatriene. Repeat this exercise in a basis of symmetry-adapted orbitals.

Group theory is a powerful method of simplifying a secular determinant by partitioning it into determinants of lower order. However, in practice,

8.3 *Two Applications of Symmetry*

the organic chemist does not solve the secular equation by himself, but by means of a high-speed digital computer. HMO computer programs are available for handling most chemically interesting conjugated molecules by direct LCAO–MO without the use of symmetry. These programs yield DE_π, Q_i, and P_{ij} for alternant and nonalternant hydrocarbons. Further developments occur in this area from year to year, making these methods available to larger numbers of chemists.

AN APPLICATION OF ORBITAL SYMMETRY TO THE STUDY OF CONCERTED REACTIONS

We ask the student to distinguish between *concerted* reactions, which proceed directly through a transition state from reactants to products, and nonconcerted reactions, which proceed through a diradical intermediate. Woodward and Hoffman[10] found that the characteristic of concerted reactions is that in certain well-defined circumstances it is possible to transform continuously the molecular orbitals of the reactants into those of the products in such a way as to preserve the bonding character of all occupied molecular orbitals at all stages of the reaction. They designate these concerted reactions as "symmetry-allowed." If there is such a pathway for the reaction, then no occupied level moves to higher energy in the transition state for the concerted reaction (i.e., no occupied bonding MO becomes nonbonding or antibonding). Thus a relatively low activation energy for the reaction is assured.

To establish if a given process is symmetry-allowed Woodward and Hoffman suggest one begin with the orbitals of the reactants or products and mentally follow them through the reaction. All that is needed is a simple appreciation of molecular symmetry and the ability to keep track of the sign of the orbital phases. *No quantitative calculation is involved.* This is regarded as the greatest strength of the resulting predictions. Specifically, they conclude that *the highest-energy occupied molecular orbitals, and their tendency to rise or fall in energy under the molecular distortion leading to reaction, determine the course of the reaction.* The important role of the highest occupied molecular orbitals may be justified by thinking of them as containing the least tightly held, most easily polarizable electrons of the molecule; consequently, the electrons most easily perturbed in the initial stages of reaction. (This was one of the reasons for focusing our attention on the π electrons in conjugated systems.)

As an example of symmetry-allowed and symmetry-forbidden concerted mechanisms we examine the formation of cyclobutene from *cis*butadiene. The MO of each species are illustrated in Fig. 8.3.1. Figure 8.3.2 illustrates

---

[10] R. B. Woodward and R. Hoffmann, *J. Amer. Chem. Soc.* **87**, 395, 2511 (1965); R. Hoffmann and R. B. Woodward, *J. Amer. Chem. Soc.* **87**, 2046, 4388, 4389 (1965).

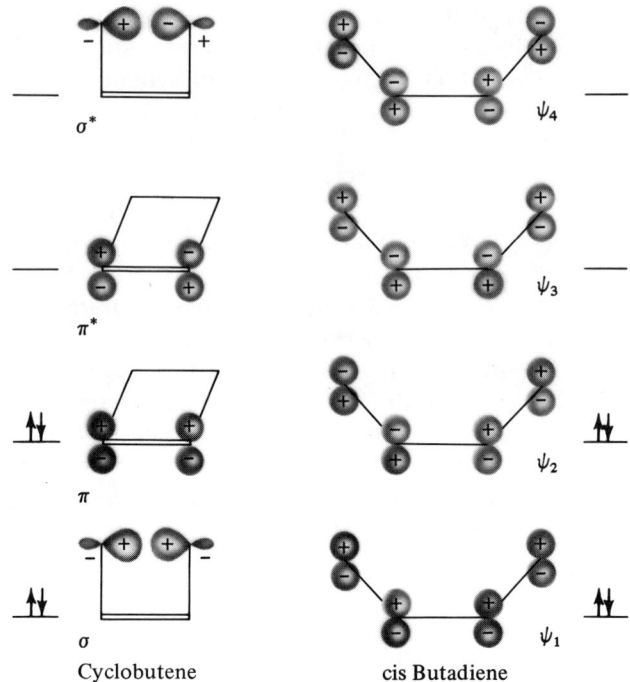

FIG. 8.3.1 Molecular orbitals of cyclobutene and *cis*-butadiene.

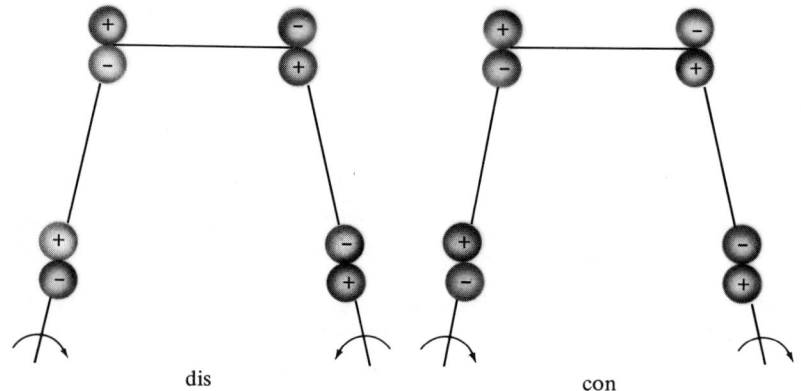

FIG. 8.3.2 Disrotatory and conrotatory processes in the MO ψ_2 of *cis*-butadiene.

the two processes which could be taken as the initial step in the formation of cyclobutene. The *disrotatory* process brings about the overlap of two orbitals of different phase sign. The result is antibonding, destabilizing, and repulsive. The *conrotatory* process overlaps orbitals of the same phase sign, leading to a bonding, stabilizing, and attractive interaction. (In the diatomic case the resulting charge distributions were found to be *binding*.) The result of the conrotatory process is the possible formation of a σ bond. Note that the highest occupied MO, ψ_2, of *cis*-butadiene is bonding, as is the resulting σ MO of cyclobutene.

The conrotatory process also transforms the bonding π MO, ψ_1, of *cis*-butadiene into the bonding π MO of cyclobutene. Thus we have achieved a correlation of the bonding orbitals of the reactant with those of the product while conserving MO symmetry. The conrotatory process is "symmetry-allowed" and the reaction (*cis*-butadiene $\rightleftarrows$ cyclobutene) is energetically feasible in the sense of having a predicted activation energy which is low.[11] The stereochemical consequences of many concerted reactions have been studied in this way. The MO symmetry analysis has been applied to thermal and photochemical ring closure, isomerizations and tautomerizations, elimination and addition reactions, and so on.

8.4 EXTENSIONS OF THE HÜCKEL METHOD

Some features of the Hückel π-electron methodology may be extended to the treatment of saturated molecules (only σ orbitals) or to the treatment of *all* the valence orbitals (σ and π) in a conjugated molecule.[12,13] Harking back to equation (8.1.1) and Fig. 8.1.3, the total wavefunction is rewritten $\Psi = \hat{A}' \Psi_K \Psi_V$ where Ψ_K is the antisymmetrized wavefunction of the K-shell electrons of carbon and Ψ_V is the antisymmetrized wavefunction of the valence electrons. Instead of the π-electron approximation we have a *valence shell* approximation which is again strongly substantiated by the experimental spectra and chemistry of hydrocarbons. The valence shell wavefunction is an antisymmetrized product of molecular orbitals. The MO are linear combinations of atomic orbitals, specifically, for $C_n H_m$, STO (Slater type orbitals, Section 4.9) $n(2s_C)$, $3n(2p_C)$, and $m(1s_H)$ in number for a total of $4n + m$ AO. Again, *assuming* an effective Hamiltonian analogous to (8.2.1) for M valence electrons, $\hat{H}_V = \sum_{i=1}^{M} \hat{h}_{eff}(i)$, we seek the variational solution to the valence shell Schrödinger equation $\hat{H}_V \Psi_V = E_V \Psi_V$. Like (8.2.2) this many-electron Schrödinger equation separates

---

[11] No comment is intended about either the overall free energy of a given reaction or its rate.
[12] R. Hoffmann, *J. Chem. Phys.* **39**, 1397 (1963); J. A. Pople and D. P. Santry, *Mol. Phys.* **7**, 269 (1964).

into M one-electron equations, $\hat{h}_{\text{eff}} \psi_i = E_i \psi_i$. The variational principle determines the coefficients c_{ij} and the roots E_i of the LCAO–MO, $\psi_i = \sum_{j=1}^{4n+m} c_{ij} \phi_j$ ($\phi_j = 2s_C, 2p_C$ or $1s_H$). The resulting secular determinant is of order $4n + m$; $|\mathbf{H} - S W| = 0$. Assumptions must now be made about the numerical value of the matrix elements of $\mathbf{H}$, H_{ij} and $\mathbf{S}$, S_{ij}.

In contrast to the Hückel π-electron method, none of the matrix elements is arbitrarily chosen to be zero. For sp^3 hybridized carbon the H_{ii} matrix elements are taken as the negative of the ionization potentials of $2s_C$ and $2p_C$ in this configuration.[13] $H_{2s2s} = -21.4$ eV, $H_{2p2p} = -11.4$ eV, $H_{1s1s} = -13.6$ eV (negative of the H-atom ionization potential).

EXERCISE   Examine the integrals H_{ii} and justify this choice of numerical values.

An approximate form for the matrix elements H_{ij} ($i \neq j$), which was first used by Wolfsberg and Helmholtz,[14] is usually assumed.

$$H_{ij} = K \frac{(H_{ii} + H_{jj})}{2} S_{ij} \qquad (8.4.1)$$

EXERCISE   Equation (8.4.1) is an overlap-weighted arithmetic mean approximation for H_{ij} in terms of the diagonal matrix elements. Postulate a similar approximation for H_{ij} utilizing a *geometric mean*.

K is a scaling parameter with the value of 1.75. All the overlap integrals in the secular determinant are determined analytically by integration over the STO. This requires that the distance between the atoms be known. If the distances are not known, or even if they are, the calculation can be repeated at many internuclear distances. The result is an estimate of the variation of total energy as a function of the nuclear configuration; stable conformers can be determined and barriers to internal rotation can be estimated.

One of the most interesting recent developments in quantum chemistry has been the established utility and success of this extended Hückel method in predicting molecular geometries.[15] For almost all the hydrocarbons investigated[16] extended Hückel calculations carried out at various nuclear configurations give a minimum total energy not far from the correct experimental geometry of the molecule.

---

[13] H. O. Pritchard and H. A. Skinner, *Chem. Rev.* **55**, 745 (1955); also G. Pilcher and H. A. Skinner, *J. Inorg. Nucl. Chem.* **24**, 937 (1962).
[14] M. Wolfsberg and L. Helmholtz, *J. Chem. Phys.* **20**, 837 (1952).
[15] L. C. Allen and J. D. Russell, *J. Chem. Phys.* **46**, 1029 (1967); G. Blyholder and C. A. Coulson, *Theor. Chim. Acta* **10**, 316 (1968).
[16] See R. Hoffmann, *J. Chem. Phys.* **39**, 1397 (1963).

8.4 Extensions of the Hückel Method

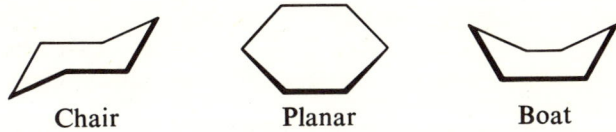

Chair   Planar   Boat

For example, of the three illustrated geometries of cyclohexane, extended HMO calculations predict that the "chair" form is preferred and the "boat" form has the next higher energy, in agreement with experiment. If heteroatoms are present in the molecule, the system is best handled analogously to the iteration on atomic charges in the Hückel method.[17]

Fundamentally, extended Hückel theory, because it is concerned with an effective one-electron Schrödinger equation $\hat{h}_{\text{eff}} \psi_i = E_i \psi_i$, is an attempt to reproduce the results of the Hartree-Fock method. Its advantage over the ordinary Hückel method for conjugated molecules is the removal of the assumption of σ–π separability. The explicit consideration of σ electrons makes it possible to discuss the geometry of the molecule, which is largely due to the σ core, that is, sp, sp^2, or sp^3 hybridization.

REFERENCES

Daudel, R., Le Febvre, R., and Mośer, C., *Quantum Chemistry, Methods and Applications*, Interscience, New York, 1959.

Murrell, J. N., Kettle, S. F. A., and Tedder, J. M., *Valence Theory*, Wiley, New York, 1965. Chapters 16 and 17 on the electronic structure and chemical reactivity of organic molecules are highly recommended.

Roberts, J. D., *Notes on Molecular Orbital Calculations*, Benjamin, New York, 1962. Especially good for its exercises and worked examples.

Salem, L., *The Molecular Orbital Theory of Conjugated Systems*, Benjamin, New York, 1966. A useful and complete compendium of topics in this field.

PROBLEMS

1. Apply the HMO methodology to the MO of H_3, H_3^-, and H_3^+ written as LCAO of $1s_H$ AO. Specifically, determine whether the linear or equilateral-triangular nuclear configuration is the most stable for each of these species. Use the nearest-neighbor and zero-overlap approximations.

2. Find the HMO coefficients for the occupied MO of benzene. Calculate the π-electron charge and bond order indices, Q_i and P_{ij}, for benzene.

3. Make the following assumptions about the parameters of the heteroatoms N and O. $\beta_{ZZ} = \beta_{CC}$.

	C	N	O
Coulomb integral, α_Z	α	$\alpha + \beta$	$\alpha + 2\beta$
Resonance integral, β_{ZC}	β	β	$\sqrt{2}\beta$

---

[17] R. Rein *et al.*, *J. Chem. Phys.* **45**, 4743 (1966).

Compare the HMO calculations of the following molecules to the butadiene results, that is, E_π, Q_i, P_{ij}.
(1) C=C—C=O, acrolein
(2) N=C—C=N
(3) O=C—C=O
(4) C=O—O=C
(5) C=N—N=C

4. Apply the Hückel method to the nonalternant hydrocarbon, fulvene. Determine the π-electron charge density on each atom, Q_i.
5. What do you expect is the relation (if any) between the π-bond order P_{ij} and the bond distance between the ith and jth carbon atoms?
6. Find the orbital energies of the cyclopropenyl (C_3H_3) radical in Huckel theory and compare them to the orbital energies of

 C_6H_4

Compare the delocalization energy of this molecule with that of two cyclopropenyl radicals. Do you expect the molecule to be stable?

9
LIGAND FIELD THEORY

The goal of the structural investigation
of a system is the description of the system
in terms of simpler entities . . . it is
usually convenient to resolve it first
into the next simpler parts, rather than
into its ultimate constituents . . .

LINUS PAULING
The Nature of the Chemical Bond, 1938

THE TRANSITION metal atoms and ions have incomplete shells of d electrons. There are three groups of transition metal ions, beginning with ions of the elements Ti, Zr, and Hf. The first series, Ti, V, Cr, Mn, Fe, Co, Ni (and the ions of Cu and Zn), comprises the main subject of this chapter. We are concerned with the spectra, bonding, and magnetic properties of their coordination compounds, ionic complexes, and crystals.[1]

Typical geometries found in the chemistry of the transition metal complexes are illustrated in Fig. 9.1.1. In general, the transition metal ion is surrounded by four, five, or six *ligands*, that is, it has a *coordination number*[2] of 4, 5, or 6. Ligands are the neutral molecules, atoms, or ions bound to the central ion (atom). Figure 9.1.2 illustrates a common crystal structure involving octahedral coordination of a transition metal. Octahedral coordination occurs quite commonly among ionic crystals and transition metal complexes. Our discussion will be mainly concerned with this coordination. Figure 9.1.3 presents the relation between the cube and octahedral and tetrahedral coordination. The octahedral group O_h arises by placing the ligands at the corners of an octahedron, just as the tetrahedral point group T_d arises by placing the ligands at the corners of a tetrahedron.

According to the variational principle, and the methods based on this principle which we developed in Chapter 5 and 7, we should now undertake the solution of the Hartree-Fock equations for *all* the electrons of the transition metal and its ligands in a typical complex. This is simply not feasible at the present time. As in the last chapter, we must rely on semiempirical methods and models. Fortunately there are several useful approximations which can be applied to the elucidation of the spectra and magnetic properties of these systems in the absence of accurate calculations.

There are three approximate theoretical approaches to the structure (or the spectra) of the transition metal and its ligands. All three focus attention

---

[1] The rare earths of the lanthanide and actinide series have incomplete shells of f electrons. The approach developed in this chapter for the d electrons of the transition metals is also applicable to the rare earths.

[2] The coordination numbers found in ionic crystals and the dependence of the coordination number on the cationic and anionic radii are discussed in L. Pauling, *Nature of the Chemical Bond*, 3rd ed., Cornell Univ. Press, Ithaca, N.Y., 1960.

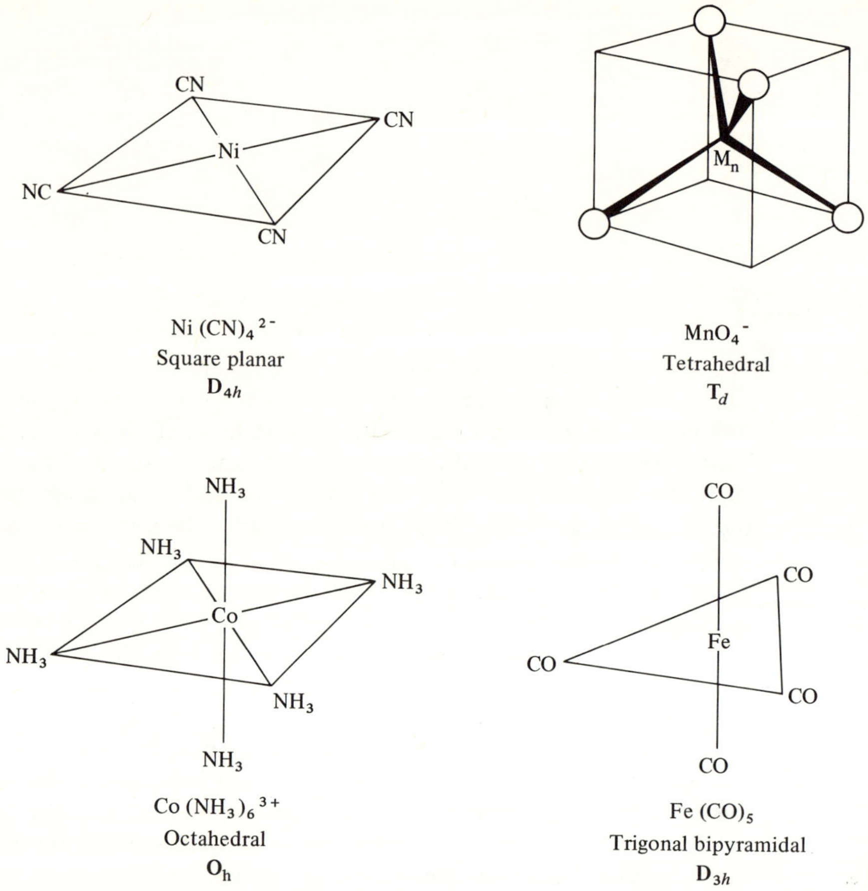

CN
CN
Ni
NC
CN
Mn
Ni (CN)$_4$$^{2-}$
Square planar
D_{4h}
MnO$_4$$^-$
Tetrahedral
T_d
NH$_3$
NH$_3$
NH$_3$
Co
NH$_3$
NH$_3$
NH$_3$
CO
CO
Fe
CO
CO
CO
Co (NH$_3$)$_6$$^{3+}$
Octahedral
O_h
Fe (CO)$_5$
Trigonal bipyramidal
D_{3h}

FIG. 9.1.1 Some examples of point groups which occur in ligand field theory.

FIG. 9.1.2 The perovskite structure ($KNiF_3$, $KMgF_3$, etc.). A commonly occurring crystal structure first observed in the mineral perovskite, $CaTiO_3$. The large ions form a close-packed array with the large cations at the corners of the cube. The small cation (Ni^{2+}, Mg^{2+}, etc.) is at the center of the cube surrounded by an octahedral array of anions (shaded spheres).

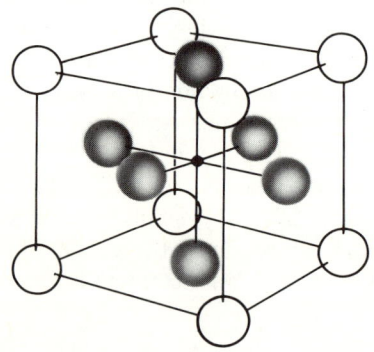

9.1 Crystal Field Theory

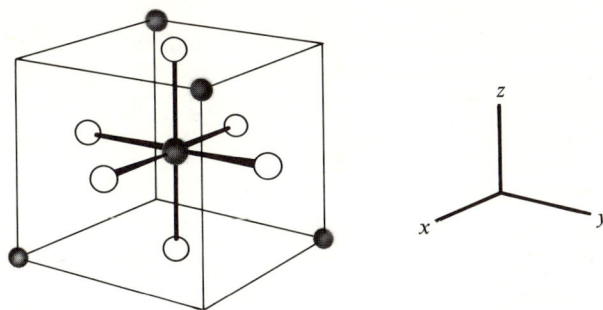

FIG. 9.1.3 Octahedral and tetrahedral point groups. The open circles form the octahedral array.

on the importance of the d orbitals. The earliest effort applied the valence bond approach based on hybridization of the d, s, and p orbitals.[3] Pauling suggested either $d^2sp^3(3d^24s4p^3)$ or $sp^3d^2(4s4p^34d^2)$ hybridization for octahedral coordination. We give only this brief mention to valence bond theory because, although it satisfactorily explains many features of coordination bonding, it cannot explain the electronic spectra. Molecular orbital theory gives a satisfactory treatment of the bonding and spectra, but there is no facile MO method, like the Hückel theory, for the coordination complexes. Another theory which concerns itself with the behavior of the d orbitals has helped explain the spectral and magnetic properties of the transition metal complexes and ionic crystals. Known as the *crystal field theory*, it succeeds by resolving the complex system into its next simpler parts.

9.1 CRYSTAL FIELD THEORY

The crystal field theory takes its name from its concentration on the behavior of the d electrons and their states under the influence of the electrostatic field produced at the central metal atom by the surrounding ligands.[4] We begin with a simple one-electron example. Consider a single $3d$ electron as in Ti^{3+}. In the isolated (free ion) the d electron could occupy any of the five d orbitals, d_{xy}, d_{xz}, d_{yz}, $d_{x^2-y^2}$, d_{z^2}, which were discussed in Section 4.1. All five d orbitals have the same energy in the free ion. We now represent the octahedral electrostatic field by placing negative electric charges at $\pm x$, $\pm y$, and $\pm z$ at equal distances around the Ti^{3+} ion (see Fig. 9.1.3). An electron in a particular d orbital is more or less repelled depending on the orientation of the d orbital relative to the charges. This is to say that the five-fold degeneracy of the d orbitals is *reduced* in the octahedral field. From Figs. 4.1.3, 4.1.5, and 9.1.3 we deduce that d_{xy}, d_{xz}, and

---
[3] L. Pauling, cited in footnote 2.
[4] H. A. Bethe, *Ann. Phys.* **3**, 133 (1929).

d_{yz} remain degenerate. They are called the t_{2g} orbitals (i.e., d_{xy}, d_{xz}, and d_{yz} from the basis of the triply degenerate irreducible representation T_{2g} in the octahedral point group O_h); $d_{x^2-y^2}$ and d_{z^2} remain degenerate and are called the e_g orbitals (i.e., $d_{x^2-y^2}$ and d_{z^2} form the basis of the doubly degenerate irreducible representation E_g in the octahedral point group).

Figures 4.1.4, 4.1.5, and 9.1.3 show that the t_{2g} orbitals are oriented *away* from the point charges making up the octahedral electrostatic field, but the e_g orbitals are not. Thus the t_{2g} orbitals are *lower* in energy than the e_g orbitals.

The group theoretical argument which demonstrates the transformation properties (irreducible representations) of the d orbitals in point group O_h should be understandable from the ideas developed in Chapter 6. Examine the character table of O_h, Table 9.1.1.

TABLE 9.1.1. Character Table of O_h

	E	$3C_2$	$6C_2'$	$8C_3$	$6C_4$	I	$6S_4$	$3\sigma_h$	$6\sigma_d$	$8S_6$
A_{1g}	1	1	1	1	1	1	1	1	1	1
A_{1u}	1	1	1	1	1	−1	−1	−1	−1	−1
A_{2g}	1	1	−1	1	−1	1	−1	1	−1	1
A_{2u}	1	1	−1	1	−1	−1	1	−1	1	−1
E_g	2	2	0	−1	0	2	0	2	0	−1
E_u	2	2	0	−1	0	−2	0	−2	0	1
T_{1g}	3	−1	−1	0	1	3	1	−1	−1	0
T_{1u}	3	−1	−1	0	1	−3	−1	1	1	0
T_{2g}	3	−1	1	0	−1	3	−1	−1	1	0
T_{2u}	3	−1	1	0	−1	−3	1	1	−1	0

The d orbitals are five-fold degenerate but O_h shows no irreducible representations more than three-dimensional (three-fold degenerate). Clearly some of the degeneracy of the d levels is removed in octahedral symmetry, that is, the d orbitals form the basis of a reducible representation in O_h.

The general form of the d orbitals was given in Chapter 4. The radial wavefunction $R(r)$ has no angular dependence. It is invariant to the symmetry operations of O_h and will not concern us. $\Theta(\theta)$ depends on the angle θ, but we will choose all rotations about the z axis in obtaining the character of the reducible representation. Consequently $\Theta(\theta)$ will be unchanged. By elimination, we need only consider the alteration of the $\Phi(\phi)$ functions by the operations of O_h.

$$\Phi(\phi) = \exp im\phi \qquad m = 0, \pm 1, \pm 2$$

9.1 Crystal Field Theory

The effect of a rotation through an angle γ on $\exp im\phi$ is $\exp im(\phi + \gamma)$. Thus for the set of d orbitals

$$\hat{C}_\gamma \begin{pmatrix} \exp(2i\phi) \\ \exp(i\phi) \\ 1 \\ \exp(-i\phi) \\ \exp(-2i\phi) \end{pmatrix} = \begin{pmatrix} \exp[2i(\phi+\gamma)] \\ \exp[i(\phi+\gamma)] \\ 1 \\ \exp[-i(\phi+\gamma)] \\ \exp[-2i(\phi+\gamma)] \end{pmatrix}$$

The matrix representation of the operation $\hat{C}_\gamma$ in the d basis set is

$$\hat{C}_\gamma = \begin{pmatrix} \exp(2i\gamma) & 0 & 0 & 0 & 0 \\ 0 & \exp(i\gamma) & 0 & 0 & 0 \\ 0 & 0 & 1 & 0 & 0 \\ 0 & 0 & 0 & \exp(-i\gamma) & 0 \\ 0 & 0 & 0 & 0 & \exp(-2i\gamma) \end{pmatrix}$$

EXERCISE Show that the trace (character) of $\hat{C}_\gamma$ is $\chi(\gamma) = \dfrac{\sin 5\gamma/2}{\sin \gamma/2}$. Hint: Sum the geometric series.

The characters of the reducible representation follow as

$$\chi(\hat{C}_2) = \chi(\gamma = \pi) = \frac{\sin 5\pi/2}{\sin \pi/2} = 1$$

$$\chi(\hat{C}_3) = \chi(\gamma = 2\pi/3) = \frac{\sin 5\pi/3}{\sin \pi/3} = -1$$

$$\chi(\hat{C}_4) = \chi(\gamma = \pi/2) = \frac{\sin 5\pi/4}{\sin \pi/4} = -1$$

Of course, $\chi(\hat{E}) = 5$. From these characters and Table 9.1.1 the methods of Chapter 6 yield the irreducible components E and T_2.

EXERCISE Show that the representation for which the five d orbitals are the basis is reducible to $E + T_2$ in $\mathbf{O}_h$.

Since we have not considered inversion or improper rotations we have not distinguished between g and u representations. However, the d orbitals are inherently of g symmetry (see Fig. 4.1.5). Consequently, E_g and T_{2g} are the irreducible representations for the d orbitals in $\mathbf{O}_h$. The same method may be applied to determine the splitting of a D term of the free ion in an octahedral field (with the same result, $D = E_g + T_{2g}$) or the splitting of other orbital degeneracies in octahedral or other symmetries. The group theoretical analysis of the irreducible representations spanned by the s, p, and d orbitals in environments of nonspherical symmetry is summarized in Table 9.1.2. The angular dependence of the real orbitals is given for s through d orbitals because it may be useful to the student (see Table 4.1.1 and 4.1.2).

TABLE 9.1.2 Irreducible Representations[a] of Orbitals in Various Symmetries

Real orbitals	O_h	T_d	D_{4h}	Spherical harmonics
s	a_{1g}	a_1	a_{1g}	1
p_x	$\big\}t_{1u}$	$\big\}t_2$	$\big\}e_u$	$\tfrac{1}{2}(Y_{11}+Y_{1-1})$
p_y				$-\tfrac{1}{2}i(Y_{11}-Y_{1-1})$
p_z			a_{2u}	Y_{10}
d_{z^2}	$\big\}e_g$	$\big\}e$	a_{1g}	Y_{20}
$d_{x^2-y^2}$			b_{1g}	$\tfrac{1}{2}(Y_{22}+Y_{2-2})$
d_{xy}	$\big\}t_{2g}$	$\big\}t_2$	b_{2g}	$-\tfrac{1}{2}i(Y_{22}-Y_{2-2})$
d_{yz}			$\big\}e_g$	$-\tfrac{1}{2}i(Y_{21}-Y_{2-1})$
d_{xz}				$\tfrac{1}{2}(Y_{21}+Y_{2-1})$

[a] a and b representations are nondegenerate; e representations are doubly degenerate; and the t representations are triply degenerate.

EXERCISE   By the same method just applied to the d orbitals show that the orbitals p_0, p_{-1} and p_{+1} remain degenerate in O_h and form the basis of a T_{1u} irreducible representation.

Note that the triply degenerate p orbitals *remain* degenerate in O_h, because with the usual choice of Cartesian coordinates (see Fig. 9.1.3) they have equivalent relations to the octahedral field. According to Table 9.1.2 p_x, p_y, and p_z form the basis of the three-dimensional irreducible representation T_{1u} in O_h.

Examination of the d orbitals in O_h and T_d symmetries (Table 9.1.2) shows the same partitioning of the orbitals into two crystal field components, but an examination of Fig. 9.1.3 shows that while the t_{2g} orbitals are lower in energy than e_g in O_h, in T_d the e orbitals are lowest in energy. [The $t_2(d_{xy}, d_{xz}, d_{yz})$ orbitals have lobes oriented towards the point charges in T_d symmetry while the e ($d_{z^2}, d_{x^2-y^2}$) orbitals are oriented away from the charges.] The inversion of the d-shell splitting under a change in crystal field symmetry has important consequences. Next, we examine the possibility of giving a quantitative treatment to the d-shell splitting induced by these crystal fields.

9.2 PARAMETIZATION OF THE CRYSTAL FIELD

The potential at the transition metal due to the presence of a point charge Ze located a distance R away is Ze/R, in electrostatic units. The potential due to an *arbitrary array* of charges may be expanded in spherical harmonics.[5] For the octahedral array of charges (Fig. 9.1.3) each charge is located a distance R from the central ion and the expansion takes the form

$$V(r, \theta, \phi) = \sum_{l=0}^{\infty} V_l \left(\frac{Ze}{R} \left(\frac{r}{R} \right)^l, Y_{lm} \right) \qquad R > r \qquad (9.2.1)$$

$V(r, \theta, \phi)$ is the potential due to the electrostatic crystal field at the position of the d electron (r, θ, ϕ), if $R > r$.

The first term in the sum (9.2.1) which can remove the degeneracy of the d orbitals is (in Cartesian coordinates)

$$V_4 = \frac{35Ze}{4R^5}(x^4 + y^4 + z^4 - \tfrac{3}{5}r^4) \qquad (9.2.2)$$

V_4 is the largest part of the electrostatic perturbation which lifts the d-level degeneracy. The first-order perturbation energies of the t_{2g} and e_g orbitals are $\langle e V_4 \rangle$, where e is the electronic charge.

$$\varepsilon_4^{(1)}(t_{2g}) = \langle d_{xy}, eV_4\, d_{xy} \rangle = \iiint (d_{xy})^2 eV_4\, r^2\, dr \sin\theta\, d\theta\, d\phi$$

$$\varepsilon_4^{(1)}(e_g) = \langle d_{z^2}\, eV_4\, d_{z^2} \rangle = \iiint (d_{z^2})^2 eV_4\, r^2\, dr \sin\theta\, d\theta\, d\phi \qquad (9.2.3)$$

Performing the indicated integrations over the angles and substituting D for the constant factor $35\, Ze/4R^5$ we get the first-order splitting in an octahedral field.

$$\left. \begin{array}{l} \varepsilon_4^{(1)}(t_{2g}) = -4Dq \\ \varepsilon_4^{(1)}(e_g) = 6Dq \end{array} \right\} 10Dq \qquad (9.2.4)$$

where

$$q = \frac{2e}{105} \int [R_{3d}(r)]^2 r^4 r^2\, dr = \frac{2e}{105} \langle r^4 \rangle$$

---

[5] J. O. Hirschfelder, C. F. Curtiss, and R. B. Bird, *Molecular Theory of Gases and Liquids*, Wiley, New York, 1954. The one-center (or von Neumann) expansion for the potential due to an aribtrary array of charges, equation (12.1-26a).

$R_{3d}(r)$ is the normalized radial $3d$ function of the free metal atom or ion. The radial wavefunction is the same for all the d functions (see Chapter 4); consequently the splitting, $10\,Dq$, is determined by the angular integrations. The magnitude of the splitting depends on the radial factor q, which is essentially the expectation value of r^4. Because of the absence of accurate radial wavefunctions, q is not available from calculation. Furthermore, because the crystal field theory is only a rough approximation to reality, there is no reason to believe that a purely theoretical q would be advantageous to the chemist.[6] We are led to the definition of the *crystal field parameter*, Δ.

$$\Delta = 10\,Dq \begin{cases} 6Dq = \tfrac{3}{5}\Delta \\ -4Dq = -\tfrac{2}{5}\Delta \end{cases} \tag{9.2.5}$$

Δ is treated as an empirical quantity, obtainable from spectroscopic experiment for a particular ion and ligand.

In equations (9.2.3) and (9.2.4) $\varepsilon_4^{(1)}$ is the largest part of the crystal field electrostatic energy which *splits* the d levels. The other contributions to the first-order energy are much greater than $\varepsilon_4^{(1)}$ and affect all the d orbitals uniformly, producing the uniform shift illustrated in Fig. 9.2.1. While Δ is of the order of only a few electron volts the uniform shift is 20 to 40 eV. By thus focusing attention on the energy splittings rather than the overall energy of formation of the molecule, the crystal field theory addresses itself, mainly, to the spectral and magnetic properties of the molecule. However, crystal field theory has also distinguished itself by explaining certain irregularities in the thermodynamic properties of the transition metal coordination complexes. Consider that the t_{2g} orbital is *stabilized* by $-\tfrac{2}{5}\Delta$ in an octahedral field relative to the uniform (spherical) field which would have the same effect on all the d levels (Fig. 9.2.1). The e_g orbital is *destabilized* by an amount $\tfrac{3}{5}\Delta$. We define a *crystal field stabilization energy* (CFSE) which is the total energy of the electronic configuration of d orbitals relative to the uniform field limit. Table 9.2.1 gives the CFSE.

An example[7] of the use of CFSE is in the oxidation potentials, $M^{2+} \to M^{3+} + e$, of the first series of transition metal ions. The oxidation potential should parallel the variation in the third ionization potential I_3 of the free metal, M. However, the oxidation potentials are measured in acqueous solution and the heat of the reaction $M_{aq}^{2+} + H_{aq}^{+} \to M_{aq}^{3+} + \tfrac{1}{2}H_2$, varies erratically through the transition series because the crystal field stabilization energy makes various contributions to the heats of hydration of the

---

[6] A situation analogous to the semiempirical α and β parameters of the Hückel method.

[7] P. George and D. McClure, *Progr. Inorg. Chem.* **1**, 381 (1959), present several examples of the effect of the crystal field on the thermodynamic properties of transition metal crystals and coordination complexes.

9.2 Parametization of the Crystal Field

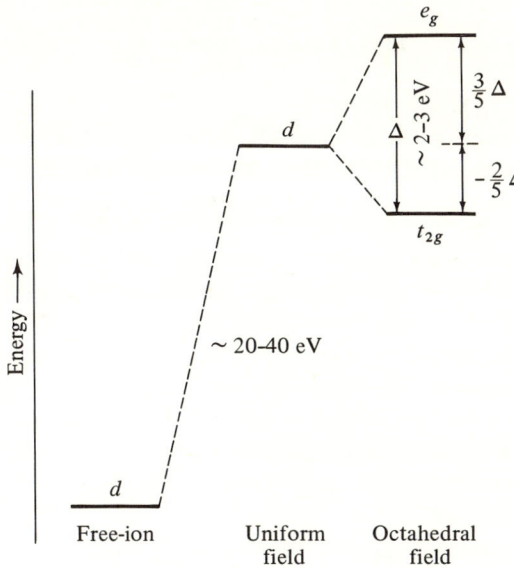

FIG. 9.2.1 Energy of the d orbitals in a uniform (spherical) electrostatic field and in an octahedral electrostatic field. Δ (or 10 D_q) is the crystal field parameter.

ions M^{2+} and M^{3+}. If the thermodynamic energies are corrected by the CFSE of the ions (Δ being determined from the spectrum, see further), then the oxidation potentials are found to parallel I_3 as expected.

TABLE 9.2.1 Crystal Field Stabilization Energies

Number of d electrons	Configuration	CFSE
1	(t_{2g})	$-\frac{2}{5}\Delta$
2	$(t_{2g})^2$	$-\frac{4}{5}\Delta$
3	$(t_{2g})^3$	$-\frac{6}{5}\Delta$
4	$(t_{2g})^3(e_g)$ or $(t_{2g})^4$	$(-\frac{3}{5}$ or $-\frac{8}{5})\Delta$
5	$(t_{2g})^3(e_g)^2$ or $(t_{2g})^5$	$(0$ or $-\frac{10}{5})\Delta$
6	$(t_{2g})^4(e_g)^2$ or $(t_{2g})^6$	$(-\frac{2}{5}$ or $-\frac{12}{5})\Delta$
7	$(t_{2g})^5(e_g)^2$ or $(t_{2g})^6(e_g)$	$(-\frac{4}{5}$ or $-\frac{9}{5})\Delta$
8	$(t_{2g})^6(e_g)^2$	$-\frac{6}{5}\Delta$
9	$(t_{2g})^6(e_g)^3$	$-\frac{3}{5}\Delta$
10	$(t_{2g})^6(e_g)^4$	0

9.3 Δ, THE CRYSTAL FIELD PARAMETER

FACTORS INFLUENCING Δ

By the same procedure which led to equations (9.2.4) we may determine the splitting of the d levels in tetrahedral symmetry. The result is

$$\Delta_{tet} = -\tfrac{4}{9}\Delta_{oct} \tag{9.3.1}$$

where the minus sign reminds us that e lies *lower* than t_2 in $\mathbf{T}_d$ symmetry. This relation is approximately confirmed by experiment. The value of Δ depends on several other factors, as well as geometry: Δ depends on the particular transition metal, the charge on the metal ion, and on the nature of the ligands. In regard to the latter factor, the ligands are ordered according to their ability to split the d levels. Interestingly, it is found that the order is the same regardless of which transition metal ion is used. Ordering the ligands according to the magnitude of Δ we obtain the *spectrochemical series*.

$$CO, CN^- > NO_2^- > NH_3 > H_2O > OH^-,$$
$$F^- > SCN^- > Cl^- > Br^- > I^-$$

THE SPECTROCHEMICAL SERIES

The name of the series derives from the fact that Δ is observed spectroscopically as the transition energy $t_{2g} \to e_g$ in octahedral symmetry. For example, continuing with our one-electron example, $Ti(H_2O)_6^{3+}$ has a transition with maximum intensity at 20,300 cm^{-1}, consequently, $\Delta = $ 20,300 cm^{-1}. Similarly, TiF_6^{3-} yields $\Delta = 17,000$ cm^{-1}, in agreement with the spectrochemical series. In general Δ is about 8000–14,000 cm^{-1} for divalent and 16,000–24,000 cm^{-1} for trivalent ions of the first transition series. For example, $Co(H_2O)_6^{2+}$, $\Delta = 9300$ cm^{-1}; $Co(H_2O)_6^{3+}$, $\Delta = 18,200$ cm^{-1}. The value of Δ is larger for the higher transition series, $5d > 4d > 3d$. For example, $Ir(NH_3)_6^{3+}$, $Rh(NH_3)_6^{3+}$, and $Co(NH_3)_6^{3+}$ have Δ values of about 40,000, 34,000, and 23,000 cm^{-1}, respectively.

SPECTROSCOPIC DETERMINATION OF Δ

A characteristic feature of the first series of transition metal complexes is their color. They undergo transitions near the low-energy (visible) portion of the spectrum (4500–8000 A, or 22,000–13,000 cm^{-1}). The low energy and low intensity of these transitions identify them as d–d transitions between d levels split by the octahedral field. Their low intensity is a consequence of their forbidden character (see Section 9.6).

Observations of a weak low-energy transition in a complex in which the central ion has configuration $(d)^1$, that is, $(t_{2g})^1$ gives the value of Δ directly. The spectroscopic term of the free-ion of configuration $(d)^1$ is ^2D (see Chapter 4). Figure 9.3.1 shows the splitting of the term ^2D in an octahedral

9.3 Δ, The Crystal Field Parameter

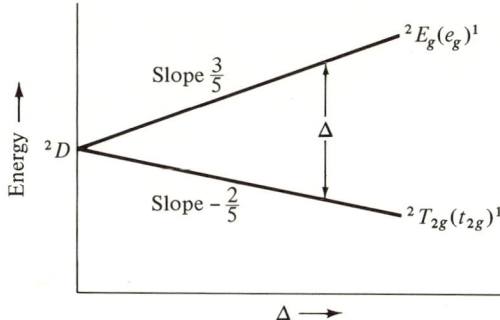

FIG. 9.3.1 Splitting of the ^2D term, $(d)^1$, in an octahedral field.

field. Consequently, the spectrum of the $(d)^1$ complex consists of a single band at Δ cm^{-1}. The single absorption band of Ti(H$_2$O)$_6^{3+}$ at 20,300 cm^{-1} was mentioned previously. This transition quantitatively identifies Δ for this ion–ligand complex and, coincidentally, gives solutions of this ion a reddish violet color.

To proceed with the application of crystal field theory to the spectra of other transition metal complexes we must consider configurations $(d)^2$, $(d)^3$, $(d)^4$, etc. In Chapter 4 it was shown that these configurations result in *several spectroscopic terms*, only one of which is associated with the ground state. For example, $(d)^2$ gives rise to the terms (see Table 4.7.2) ^3F, ^3P, ^1G, ^1D, and ^1S. ^3F is the ground-state term. However, an additional complexity results from the fact that each term is $(2S + 1)(2L + 1)$-fold degenerate. This degeneracy is removed (partially) in the crystal field.

Russell-Saunders coupling (Chapter 4) was used to find the spectroscopic terms arising from a given configuration. The electron-nuclear attraction was considered paramount in determining the energy, followed by the interelectronic repulsions which split the energy into that of the spectroscopic terms. Finally, spin-orbit coupling was considered a small perturbation which splits the terms into their component J states (see Fig. 4.7.3). In dealing with the atom or ion in a crystal field we proceed in the same way with Russell-Saunders coupling, but two distinct cases arise depending on the magnitude of Δ.

HIGH-SPIN COMPLEXES, LOW-SPIN COMPLEXES, AND THE MAGNITUDE OF Δ

For the first series of transition metals the crystal field perturbation may be considered either *before* the interelectronic repulsions because Δ is large, in which case we have a *strong field* or *low-spin* complex, or *after*

TABLE 9.3.1 High-Spin (Weak Field) and Low-Spin (Strong Field) d-Electron Configurations

Total number of d electrons		1	2	3	4	5	6	7	8	9
Weak Field	e_g	— —	— —	— —	↑ —	↑ ↑	↑ ↑	↑ ↑	↑ ↑	↑↓ ↑
	t_{2g}	↑ — —	↑ ↑ —	↑ ↑ ↑	↑ ↑ ↑	↑ ↑ ↑	↑↓ ↑ ↑	↑↓ ↑↓ ↑	↑↓ ↑↓ ↑↓	↑↓ ↑↓ ↑↓
Number of unpaired electrons		1	2	3	4	5	4	3	2	1
Multiplicity		2	3	4	5	6	5	4	3	1
Strong field	e_g	— —	— —	— —	— —	— —	— —	↑ —	↑ ↑	↑↓ ↑
	t_{2g}	↑ — —	↑ ↑ —	↑ ↑ ↑	↑↓ ↑ ↑	↑↓ ↑↓ ↑	↑↓ ↑↓ ↑↓	↑↓ ↑↓ ↑↓	↑↓ ↑↓ ↑↓	↑↓ ↑↓ ↑↓
Number of unpaired electrons		1	2	3	2	1	0	1	2	1
Multiplicity		2	3	4	3	2	1	2	3	2

the interelectronic repulsions and *before* spin-orbit coupling because Δ is smaller, in which case we have a *weak field* or *high-spin* complex.[8] The two limiting field strength cases become associated with spin because the magnitude of Δ controls the order in which the orbitals t_{2g} and e_g are filled. In an octahedral field the t_{2g} spin-orbitals are filled first, but because of interelectronic repulsion, the electrons prefer not to be placed in the same spatial orbitals if it can be avoided. Energetically, by placing electrons in the same space orbital instead of different space orbitals, we lose some stabilizing exchange energy and gain some destabilizing Coulomb repulsion. Thus, if Δ is small enough, e_g spin-orbitals are filled before the t_{2g} orbitals are doubly occupied. Consequently, the weak field leads to high spin. The maximum multiplicity, $2S + 1 = 6$, is reached for the $(d)^5$ configuration in the weak field case. On the other hand, if Δ is large the crystal field energy dominates and determines the electronic configuration. That is, *all* the t_{2g} spin-orbitals must be occupied before an e_g level can be occupied. The two limits are contrasted in Table 9.3.1. Note that it is configurations $(d)^4$, $(d)^5$, $(d)^6$, and $(d)^7$ which offer a choice between low-spin and high-spin complexes. The different CFSE for these were given in Table 9.2.1.

9.4 WEAK FIELD COMPLEXES

The Russell-Saunders coupling scheme (Chapter 4) gives the terms which arise for $(d)^n$ configurations (Table 4.7.2). The effect of a weak octahedral field on these terms is summarized in Table 9.4.1. Because the field is

TABLE 9.4.1 Splitting of $(d)^n$ Terms in an Octahedral Field, O_h

Term	O_h Irreducible representations
S	A_{1g}
P	T_{1g}
D	$E_g + T_{2g}$
F	$A_{2g} + T_{1g} + T_{2g}$
G	$A_{1g} + E_g + T_{1g} + T_{2g}$
H	$E_g + T_{1g} + T_{1g} + T_{2g}$
I	$A_{1g} + A_{2g} + E_g + T_{1g} + T_{2g} + T_{2g}$

[8] The crystal field perturbation is always greater than the spin-orbit coupling for first series transition metals in even the weakest crystal fields. However, this is no longer necessarily true for the second and third series of transition metals.

electrostatic, it does not interact directly with the spin. However the, $(2L + 1)$-fold orbital angular momentum degeneracy is split. For example, ^3F is the ground-state term of $(d)^2$ configuration. Application of the octahedral field splits ^3F into $^3A_{2g}$, $^3T_{1g}$, and $^3T_{2g}$ components. The magnitude of the splitting is obtainable in terms of Δ by evaluating the expectation value of the two-electron eigenfunctions of the ^3F term over the perturbation eV_4, as was done in Section 9.2 for the one-electron case.[9] This gives the splitting to first order in the energy, which is sufficient in a weak field. The resultant splittings are (relative to an arbitrary zero) $^3T_{1g}(-\frac{3}{5}\Delta)$, $^3T_{2g}(\frac{1}{5}\Delta)$, and $^3A_{2g}(\frac{6}{5}\Delta)$. Consequently, $^3T_{1g}$ is the ground state of the ion in an octahedral field. Table 9.4.2 summarizes the crystal field splittings of the ground-state terms of the free ions placed in a weak octahedral field.

TABLE 9.4.2 Splittings of the Ground-State Terms of $(d)^n$ in a Weak Octahedral Field[a,b]

Configuration	Term	Crystal field energy
$(d)^1$	^2D	$^2E_g(\frac{3}{5}\Delta)$, $^2T_{2g}(-\frac{2}{5}\Delta)$
$(d)^2$	^3F	$^3A_{2g}(\frac{6}{5}\Delta)$, $^3T_{2g}(\frac{1}{5}\Delta)$, $^3T_{1g}(-\frac{3}{5}\Delta)$
$(d)^3$	^4F	$^4T_{1g}(\frac{3}{5}\Delta)$, $^4T_{2g}(-\frac{1}{5}\Delta)$, $^4A_{2g}(-\frac{6}{5}\Delta)$
$(d)^4$	^5D	$^5T_{2g}(\frac{2}{5}\Delta)$, $^5E_g(-\frac{3}{5}\Delta)$
$(d)^5$	^6S	$^6A_{1g}(0)$

[a] Note that the splittings of $(d)^{10-n}$ are obtained from the above $(d)^n$ splittings by changing the signs.
[b] ^6S is unaffected by the crystal field because it is not orbitally degenerate.

EXERCISE   Compare the ground-state crystal field energies of configurations $(d)^n$ ($n = 1, 2, \ldots, 10$) given in Table 9.4.2 to the CFSE for the weak field configurations found in Table 9.2.1. To what can you ascribe the discrepancies? Hint: What is taken account of in Table 9.4.2 but ignored in Table 9.2.1?

Because configuration $(d)^{10-n}$ corresponds to $(d)^n$ with n positive holes replacing the n electrons, the splittings are merely *inverted* for $(d)^{10-n}$ relative to $(d)^n$, for example, the ground state of $(d)^8$ is $^3A_{2g}(-\frac{6}{5}\Delta)$.

[9] An additional complication in such calculations is that the eigenfunctions of ^3F are mixed under the influence of V_4.

9.5 STRONG FIELD COMPLEXES

In the strong field case Δ is sufficiently large so that the ground state of the complex is not necessarily derived from the ground-state term of the free ion; it may have a different parentage. Recall that in the strong field case the crystal field is more important than the interelectronic interaction in determining the ground-state configuration. Consequently, *all* the states of the free ion which arise from the $(d)^n$ configuration must be considered. If perturbation theory is used to calculate the effect of the crystal field on the energy of the terms, then the second-order perturbation energy is also important.

We will take the $(d)^2$ ion as an example of the strong field complex, although the results are not essentially different in weak or strong fields for this configuration (see Table 9.3.1 and further). The three possible configurations of $(d)^2$ in octahedral symmetry are, in order of increasing energy,

$$(t_{2g})^2 \quad (t_{2g})(e_g) \quad (e_g)^2$$

The crystal field energy difference between each configuration and the next higher is Δ. Now the "perturbation" due to the interelectronic interaction must be added. Just as $(d)^2$ gave rise to ^3F, ^1D, ^3P, etc., the above three configurations will give rise to octahedral states. These are found by reduction of the direction product representation, for example, for $(t_{2g})^2$ we have $t_{2g} \times t_{2g} = A_{1g} + E_g + T_{1g} + T_{2g}$. The multiplicities can be either singlet or triplet for this two-electron case, but they must be chosen in accordance with the Pauli exclusion principle. The states resulting from the three octahedral two-electron configurations are summarized in Table 9.5.1.

The ground state is $^3T_{1g}$, which is the same result obtained from weak field considerations. Table 9.5.1 is analogous in octahedral symmetry to that part of Table 4.7.2 which pertains to $(d)^2$ in spherical symmetry.

TABLE 9.5.1 States Arising from the Two-Electron Configurations in Octahedral Symmetry

Configuration	States in O_h
$(t_{2g})^2$	$^3T_{1g}, {}^1T_{2g}, {}^1E_g, {}^1A_{1g}$
$(t_{2g})(e_g)$	$^3T_{2g}, {}^3T_{1g}, {}^1T_{2g}, {}^1T_{1g}$
$(e_g)^2$	$^3A_{2g}, {}^1E_g, {}^1A_{1g}$

EXERCISE   From Table 9.4.1 and the $(d)^2$ terms ^3F, ^1D, ^3P, ^1G, and ^1S show that all the predicted states in octahedral symmetry are accounted for in Table 9.5.1.

From the above exercise it appears that the ground state of the $(d)^2$ complex, $^3T_{1g}$, is the same as that arising from the ground-state term ^3F in the weak field case. It is, except that the term ^3P also gives rise to a $^3T_{1g}$ state. The two levels interact, because they have the same symmetry, and $\langle ^3T_{1g}(^3P), eV_4^3 T_{1g}(^3F)\rangle = \tfrac{2}{5}\Delta$. This interaction appreciably stabilizes the ground state in the strong field. Figure 9.5.1 illustrates the Δ-dependent interaction between the $^3T_{1g}$ states. Note that as $\Delta \to 0$ the limiting slopes of the $^3T_{1g}$ states are those of first-order perturbation theory (weak field limit).

Diagrams of the behavior of the states of $(d)^n$ in a crystal field as a function of Δ, from $\Delta = 0$ to the strong field limit, are called *Tanabe-Sugano* or *Orgel diagrams*. Figures 9.3.1 and 9.5.1 are simple examples of such diagrams.[10]

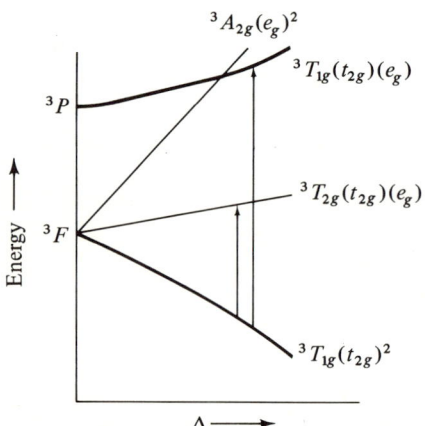

FIG. 9.5.1 States of the $(d)^2$ configuration in an octahedral field. Interaction between the $^3T_{1g}(^3P)$ and $^3T_{1g}(^3F)$ states leads to the indicated curvature at large Δ. At small Δ the limiting slopes of the lines as $\Delta \to 0$ are 0 and $-3/5$, respectively (see Table 9.4.2). The arrows indicate expected transitions. The states arising from other terms of $(d)^2$ have been omitted for clarity.

---

[10] Complete diagrams of the states of $(d)^n$ can be found in treatises and papers on ligand field theory: J. S. Griffith, *The Theory of Transition Metal Ions*, Cambridge Univ. Press, London, 1961; Y. Tanabe and S. Sugano, *J. Phys. Soc. (Japan)* **9**, 753, 766 (1954); L. E. Orgel, *J. Chem. Phys.* **23**, 1004, 1819, 1824 (1955).

9.5 Strong Field Complexes

COMPARISON OF STRONG FIELD AND WEAK FIELD LIMITS

In discussing Table 9.3.1 we mentioned that the magnitude of Δ was critical for determining the properties of $(d)^n$ ($n = 4, 5, 6$, and 7). However, for $(d)^n$ ($n = 1, 2, 3, 8$, and 9) the ground state in octahedral symmetry is derived from the lowest term of the free ion for all values of Δ. Therefore the number of unpaired electrons is the same for all Δ. We have just observed this for $(d)^2$ (Fig. 9.5.1).

In configurations $(d)^n$ ($n = 4, 5, 6$, and 7) the ground state in octahedral symmetry is that derived from the ground-state term of the free ion *only up to some critical value of* Δ. Beyond this value of Δ a state of lower multiplicity, with another free-ion term parentage, falls lower in energy and becomes the ground state. This is then the strong field (low-spin) limit displayed in Table 9.3.1. For example, for small Δ the ground state of $(d)^5$ in an octahedral field is $^6A_{1g}$, which is the only state arising from the configuration $(t_{2g})^3(e_g)^2$. The parentage of $^6A_{1g}$ is ^6S, the ground-state term of $(d)^5$. However, as Δ is increased, $^6A_{1g}$ rises in energy relative to a state $^2T_{2g}$ of configuration $(t_{2g})^5$. Eventually, at sufficiently large Δ, $^2T_{2g}$ becomes the ground state. The parentage of $^2T_{2g}$ is ^2I of $(d)^5$. Examples are the coordination complexes of Mn^{2+} and Fe^{3+}. What has just been stated in words is illustrated in Fig. 9.5.2 with a portion of the Tanabe–Sugano diagram for $(d)^5$.

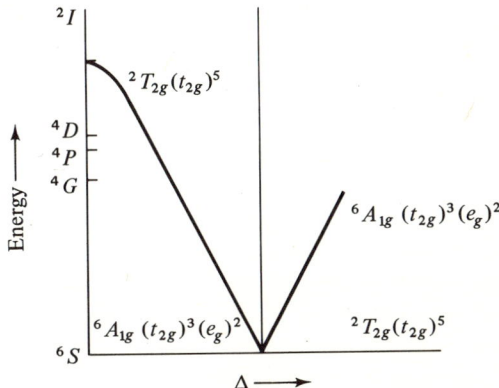

FIG. 9.5.2 A small portion of the Tanabe–Sugano diagram for the states of $(d)^5$ in an octahedral crystal field. The diagram is drawn so that the energy of the ground state is always taken as the zero of energy. Where the ground state changes (at the vertical line) there is a sharp change in slope. This is an artifact of the diagram and does not represent a discontinuity in the energy. For clarity, the states arising from the other terms of $(d)^5$ have been omitted.

The crystal field theory predicts quite different magnetic properties for $(d)^n$ ($n = 4, 5, 6,$ and 7) configurations in the two limiting octahedral fields. Predictions of this kind are in excellent agreement with experimental observations. The $(d)^n$ configurations of tetrahedral complexes could also be examined in weak field and strong field limits. Practically, however, this would be of little value since Δ_{tet} never seems to get large enough to give a strong field complex [see equation (9.3.1)]. Consequently, the tetrahedral complexes are all weak field (high-spin) complexes.

9.6 ELECTRONIC SPECTRA

The arrows in Fig. 9.5.1 indicate the expected transitions in the $(d)^2$ complex. Both transitions are *forbidden*.

$$^3T_{1g}(t_{2g})^2 \to {}^3T_{1g}(t_{2g})(e_g)$$
$$^3T_{1g}(t_{2g})^2 \to {}^3T_{2g}(t_{2g})(e_g)$$

Each of these transitions is multiplicity-allowed $\Delta S = 0$, and a one-electron transition, $t_{2g} \to e_g$. [Two electron-transitions such as $^3T_{1g}(t_{2g})^2 \to {}^3A_{2g}(e_g)^2$ are highly forbidden.[11]] The forbidden character of the crystal field transitions arises from their origin as *d–d* transitions.[12] In octahedral symmetry the *d* orbitals and the states are all of *g* symmetry (Table 9.4.1). But the nonvanishing transition moment[13] requires the selection rule: *g* states combine only with *u* states. How can $^3T_{1g} \to {}^3T_{1g}$ be observed? It is generally accepted that the mechanism involved is *vibronic coupling*. That is, if the transition occurs from the vibrationless ground state (g), the excited state to which the transition occurs is vibrationally excited with an odd number of quanta of a *u* symmetry vibrational mode. The overall vibronic symmetry of the excited state is thus $g \times u = u$ and the transition is *vibronically allowed*. Alternatively, the vibrational distortion of *u* symmetry removes the center of symmetry, thereby allowing the crystal field transition to occur. The overall diminution in the intensity due to the forbidden character of the *d–d* electronic transitions is such that the observed transitions are 10^{-3} weaker than an ordinary allowed electronic transition.

The spectrum of $Ti(H_2O)_6^{3+}$ has been discussed. It is shown in Fig. 9.6.1 and consists of a single weak peak at 20,300 cm^{-1} assigned to the

---

[11] Made allowed by configuration interaction (Section 4.11).
[12] *d–d* transitions are forbidden in spherical symmetry due to the selection rule $\Delta l = \pm 1$ (Section 4.8).
[13] Allowed electric dipole radiation (Section 6.8 and Appendix H). Note that the vector components x, y, and z transform as u in O_h.

9.6 Electronic Spectra

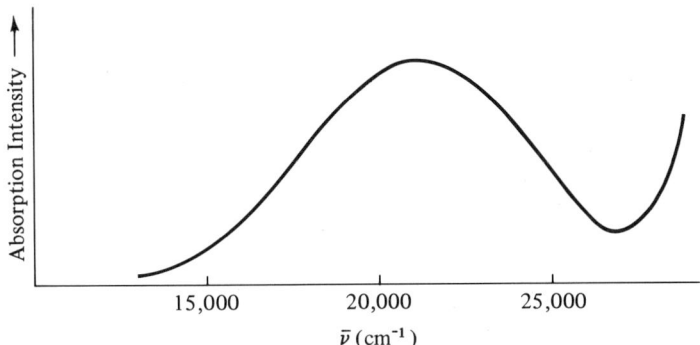

FIG. 9.6.1 Absorption spectrum of Ti(H$_2$O)$_6^{3+}$).

transition $^2T_{2g}(t_{2g}) \to {}^2E_g(e_g)$ which gives the value of Δ directly (Fig. 9.3.1).

Among ions which have the $(3d)^2$ configuration V^{3+} has received the greatest experimental attention, although Ti^{2+} and Cr^{4+} are known. The absorption spectrum of V(H$_2$O)$_6^{3+}$ shows two weak peaks at 17,100 cm^{-1} and 25,500 cm^{-1}. These are assigned to the $^3T_{1g} \to {}^3T_{2g}$ and $^3T_{1g} \to {}^3T_{1g}$ transitions, respectively (see Fig. 9.5.1). In the crystal, corundum the V^{3+} spectrum shows corresponding transitions at 17,400 and 25,200 cm^{-1}, respectively.

The ground-state term of $(d)^3$ is ^4F. In both weak field and strong field limits the ground state in octahedral symmetry is $^4A_{2g}$ (see Table 9.4.2 and Fig. 9.6.2). The observed spectrum of V(H$_2$O)$_6^{2+}$ correlates well with the multiplicity-allowed transitions suggested by Fig. 9.6.2. The spectrum is illustrated in Fig. 9.6.3. The assignments are

$$^4A_{2g} \to {}^4T_{2g} \qquad 12{,}300 \text{ cm}^{-1}$$
$$^4A_{2g} \to {}^4T_{1g}({}^4F) \qquad 18{,}500 \text{ cm}^{-1}$$
$$^4A_{2g} \to {}^4T_{1g}({}^4P) \qquad 27{,}900 \text{ cm}^{-1}$$

EXERCISE From Table 9.4.2 and the above assignments in the V^{2+} spectrum find the value of Δ. Predict the transition energy $^4A_{2g} \to {}^4T_{1g}({}^4F)$ and compare to experiment.

EXERCISE Given the energy difference for V^{2+}, $E({}^4F - {}^4P) = 11{,}500$ cm^{-1}, predict the transition energy $^4A_{2g} \to {}^4T_{1g}({}^4P)$ and compare to experiment. To what do you attribute deviations between experiment and prediction? Fill in the strong field portion of Fig. 9.6.2. Hint: $^4T_{1g}({}^4P)$ and $^4T_{1g}({}^4F)$ have the same symmetry.

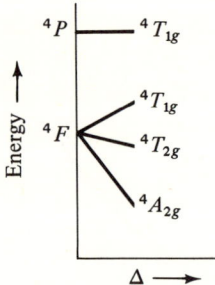

FIG. 9.6.2 Low-energy and weak field portion of the $(d)^3$ crystal field diagram. States arising from other terms are omitted.

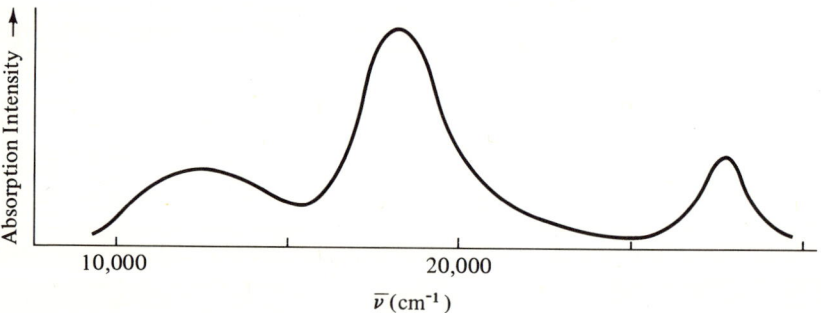

FIG. 9.6.3 Absorption spectrum of $V(H_2O)_6^{2+}$.

Table 9.1.2 shows that in tetrahedral symmetry the p orbitals are of t_2 symmetry, as are some of the d orbitals. This offers the opportunity of mixing p with d orbitals, an opportunity which is absent in $\mathbf{O}_h$. An observable consequence of the mixing is that the crystal field transitions are much more intense in $\mathbf{T}_d$ than in $\mathbf{O}_h$ complexes. The transitions are no longer pure d–d and therefore $g \leftrightarrow g$ electronically forbidden. Permanganate ion, MnO_4^-, is a commonly known example of a $\mathbf{T}_d$ complex with an intense absorption.

OTHER TRANSITIONS

In Section 4.7 we discussed spin-orbit coupling in atomic spectra. We noted that in the presence of spin-orbit coupling, in the limiting case, S ceases to be a good quantum number so $\Delta S \neq 0$ (multiplicity-forbidden) transitions may occur. Even though spin-orbit coupling is always weaker than the crystal field interaction in the first series of transition metals, multiplicity-forbidden transitions do occur. However, these transitions are generally 100 times weaker than ordinary crystal field transitions.

A much more important component of the spectra of the coordination complexes are the broad, intense transitions which generally occur at higher energies than the *d–d* transitions we have discussed. These transitions involve the *direct* participation of the ligand orbitals. In MO language, the transitions take place from MO having predominantly ligand character to those having largely *d* character or vice versa. These transitions are called *charge transfer transitions.* Although the intuitive meaning of the name is clear, it should not be interpreted too literally in the absence of more accurate information about the wavefunctions involved in these transitions, or the change in charge densities occurring on excitation. In general, the intensities of charge transfer bands are 10^2 or 10^3 greater than those of the *d–d* class. To obtain a better understanding of the structure of the transition metal compounds and the role of the ligands, we now introduce the molecular orbital method, which allows the direct participation of the ligand orbitals in our discussion. We also are able to consider the nature of the covalent bonding between ligand and transition metal through the use of the MO method.

9.7 THE MOLECULAR ORBITAL THEORY OF THE OCTAHEDRAL COMPLEX

Crystal field theory ignores the overlap of ligand orbitals with the orbitals of the transition metal. Molecular orbital theory provides a more realistic view of the transition metal complex by providing for delocalization of electrons between ligand and metal. We will show that the essential features of the crystal field theory are not lost in the MO method and that the division into weak field and strong field complexes is still valid. Furthermore, the bonding and antibonding MO describe the covalent bonding in the complex and permit one to discuss the charge transfer bands in the spectrum.

The approach to the MO in coordination complexes is essentially the same as in large organic molecules. In both cases the molecule is too big for a Hartree-Fock calculation, but can be treated by semiempirical parametization of the energy integrals. An important simplification is that we have to consider the molecular orbitals in only a few point groups, mainly $\mathbf{O}_h$, while in organic chemistry a variety of point groups occur. As a result, we can treat the octahedral complex very abstractly, giving all the results in terms of the type of ligand orbitals involved and the irreducible representations of $\mathbf{O}_h$.

First, we choose the valence shell atomic orbitals to use in forming the LCAO–MO. Secondly, we find the irreducible representations of which the AO from the basis. Linear combinations of AO forming the basis of the same irreducible representation are combined together to form a MO. Next, if numerical answers are needed, the secular determinant is set up and the matrix elements are parametized, usually by using the Wolfsberg-Helmholtz approximation [see eq. (8.4.1)]. This requires only the calculation of overlap integrals and the knowledge of the appropriate ionization potentials.[14] Finally, the orbital energies are found by solution of the secular equation and the LCAO-MO coefficients are obtained.[15] This can be followed by a population analysis (Section 7.7) in order to observe the distribution of valence shell electronic charge in the complex.

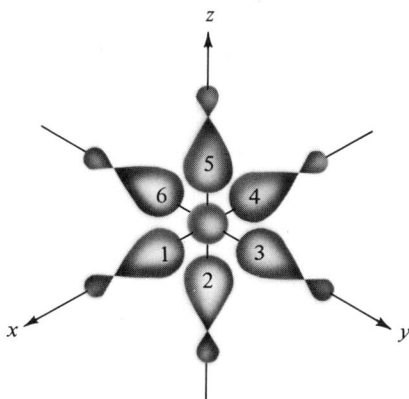

FIG. 9.7.1 Ligand σ orbitals surrounding the transition metal ion in an octahedral complex.

The valence shell AO consist of the $4s$, $4p$, and $3d$ orbitals of the first series of transition metal ions and the six σ ligand orbitals shown in Fig. 9.7.1. We need not specify the exact s–p character of the ligand σ orbital because it is different for different ligands, and even differs from the free-molecule orbital. For H_2O and NH_3 ligands, σ is derived from the nonbonding lone-pair orbital (Section 7.5).

The reducible representation for which the six σ ligand orbitals form the basis has the following character in $\mathbf{O}_h$.

---

[14] C. J. Ballhausen and H. B. Gray, *Molecular Orbital Theory*, Chap. 8, Benjamin, New York, 1964.

[15] The procedure for solution of secular equations was summarized in Section 5.6. In practice, existing computer programs are utilized.

9.7 The Molecular Orbital Theory of the Octahedral Complex

E	$3C_2$	$6C_2'$	$8C_3$	$6C_4$	I	$6S_4$	$3\sigma_h$	$6\sigma_d$	$8S_6$
6	2	0	0	2	0	0	4	2	0

Using the method of Chapter 6 we find this representation is reducible to $A_{1g} + E_g + T_{1u}$.

EXERCISE Verify that the sum of the characters of these irreducible representations gives the reducible character of the σ ligand orbitals (Table 9.1.1).

Projection operators (Section 6.6) may be used to project the a_{1g}, e_g, and t_{1u} orbitals from the σ basis, but it is both easier and quicker in this case to note that the transformation properties of the transition metal orbitals are

$$\begin{array}{ll} s & a_{1g} \\ \left.\begin{array}{l} d_{x^2-y^2} \\ d_{z^2} \end{array}\right\} e_g & \left.\begin{array}{l} p_x \\ p_y \\ p_z \end{array}\right\} t_{1u} \end{array} \qquad (9.7.1)$$

The σ ligand orbitals (group orbitals) of each irreducible representation and degenerate member of that irreducible representation are found by combining σ orbitals with the central ion orbitals having the same phase sign in that octant. For example, from Fig. 9.7.1 it is evident that $\sigma_5-\sigma_2$ transforms like p_z (i.e., this particular component of t_{1u}). From Fig. 4.1.5 and Fig. 9.7.1 we obtain

$$\left.\begin{array}{l}(\sigma_5 - \sigma_2)_z \\ (\sigma_3 - \sigma_6)_y \\ (\sigma_1 - \sigma_4)_x\end{array}\right\} t_{1u} \qquad \left.\begin{array}{l}(\sigma_1 + \sigma_4 - \sigma_3 - \sigma_6)_{x^2-y^2} \\ (2\sigma_5 + 2\sigma_2 - \sigma_1 - \sigma_3 - \sigma_4 - \sigma_6)_{z^2}\end{array}\right\} e_g$$

$$(\sigma_1 + \sigma_2 + \sigma_3 + \sigma_4 + \sigma_5 + \sigma_6) \qquad a_{1g}$$

(9.7.2)

Combination of (9.7.1) and (9.7.2) yields the trial LCAO–MO, for example, $\psi_z = c_1(\sigma_5 - \sigma_2)_z + c_2 p_z$. The MO are illustrated in Fig. 9.7.2.

We have accounted for the LCAO–MO formed from the σ ligand orbitals, but the t_{2g} d orbitals are left. The t_{2g} orbitals are properly disposed for π bonding with the ligands. π Bonding occurs if the ligands have available filled (or empty) π AO (or π MO), for example, F^-, CN^-, Cl^-. If the π MO of the ligand are filled they lead to *ligand-to-metal* π bonding which delocalized the electronic charge of the filled π orbitals towards the metal. But if the π MO of the ligand are empty they lead to *metal-to-ligand* π bonding, or *back donation*. F^- is an example of a ligand-to-metal π-bonding ligand. CN^-, which has a low-lying empty antibonding π MO, is an example of a ligand capable of back donation.

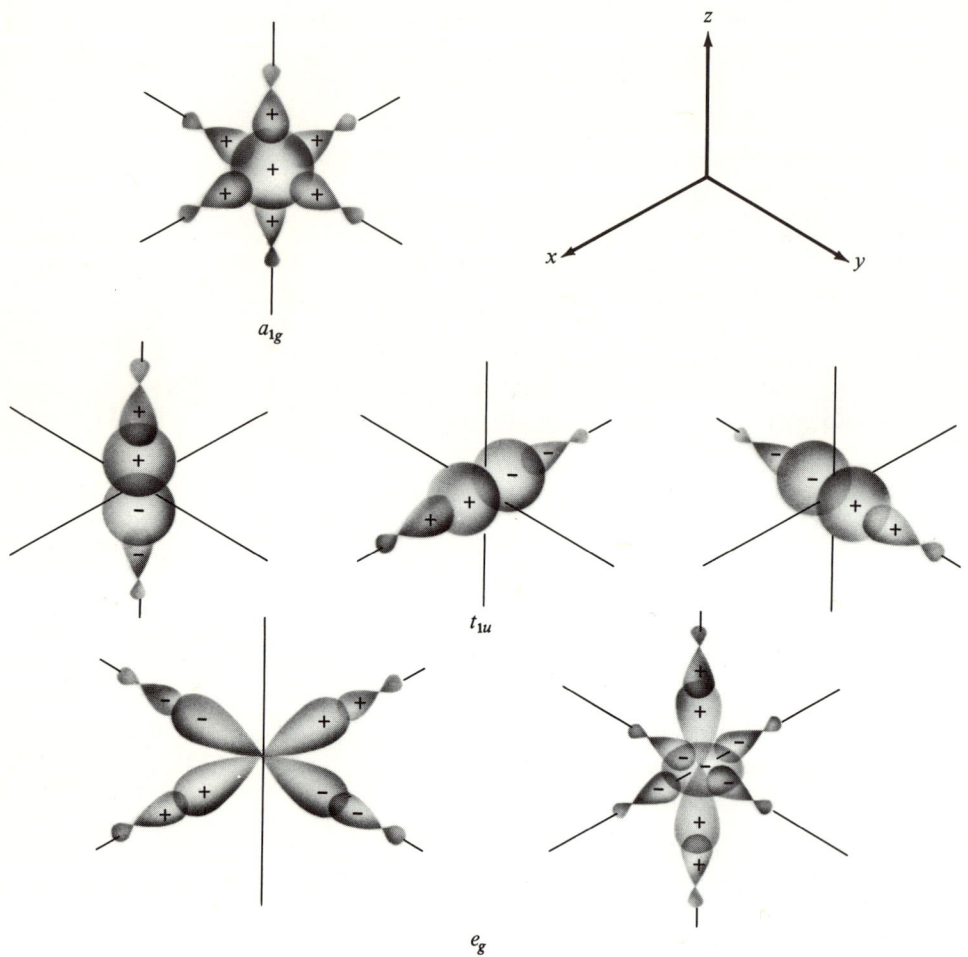

FIG. 9.7.2 Molecular orbitals formed from σ ligand orbitals.

If we limit our considerations to σ-donating ligands, for example, H_2O, NH_3, the t_{2g} d orbitals remain nonbonding. In a qualitative, rather than a quantitative fashion, the energy levels of the metal, complex, and ligands are related as in Fig. 9.7.3. The MO energy level diagram is constructed from average known ionization potentials of the transition metal ions and the ligands, together with estimates of the metal–ligand interaction energies. Figure 9.7.3 emphasizes only the salient features of the molecular orbital theory of the octahedral complexes by omitting π, π^*, and σ^* ligand orbitals.

9.7 The Molecular Orbital Theory of the Octahedral Complex

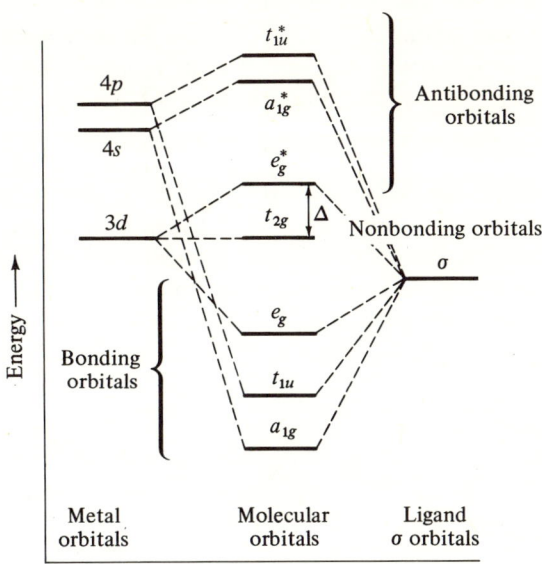

FIG. 9.7.3 Schematic orbital energy levels arising from σ ligand orbitals in an octahedral transition metal complex. Starred orbitals are antibonding.

The nonbonding t_{2g} orbitals and the higher e_g^* levels are analogous to ordinary crystal field theory. It is expected that the bonding MO consist mostly of the more stable ligand orbitals, while the antibonding MO consist mostly of the metal orbitals, in the case of e_g^*, d_{z^2} and $d_{x^2-y^2}$. Consequently the energy separation $t_{2g} \to e_g^*$ in Fig. 9.7.3 is labeled Δ. This assumption is supported by the comparison of molecular orbital and crystal field theories. For example, consider the $(d)^1$ complex $Ti(H_2O)_6^{3+}$. The Ti^{3+} ion is bonded to six H_2O molecules. Each of the six valence shell MO of the ligand involved in the coordinate-covalent bonding is doubly occupied. This gives a total of 13 valence shell electrons in the complex. The 13 electrons are accomodated in the MO of Fig. 9.7.3, giving the ground-state configuration of $Ti(H_2O)_6^{3+}$.

$$(a_{1g})^2(t_{1u})^6(e_g)^4(t_{2g})\ ^2T_{2g} \tag{9.7.3}$$

The ground state $^2T_{2g}$ is the same as that predicted from crystal field theory. The $^2T_{2g} \to\ ^2E_g$ transition, Fig. 9.3.1 in crystal field theory, consists of the one-electron MO excitation $t_{2g} \to e_g^*$. The analogy with crystal field theory holds in other complexes in addition to $Ti(H_2O)_6^{3+}$, because in octahedral coordination complexes the 12 electrons donated by the six ligands fully occupy the bonding a_{1g}, t_{1u}, and e_g MO. The remaining electrons are the 3d electrons of the transition metal. These must be distributed among t_{2g} and e_g^*. This is entirely analogous to crystal field theory where the d electrons are distributed among t_{2g} and e_g orbitals. However, the antibonding character of the e_g orbital is not explicit in cyrstal field theory.

In practical applications of the MO theory the magnitude of Δ is still a parameter; this semiempirical approach to the spectra and orbitals of the transition metal complex is called *ligand field theory*. The magnitude of Δ still controls the distribution of electrons among the t_{2g} and e_g^* orbitals; consequently, the weak and strong field limits also exist in MO theory and the discussions of Sections 9.3 to 9.6 are applicable.

REFERENCES

Ballhausen, C. J., *Ligand Field Theory*, McGraw-Hill, New York, 1962.

Cotton, F. A., *Chemical Applications of Group Theory*, Interscience, New York, 1964.

Figgis, B., *Introduction to Ligand Field Theory*, Wiley, New York, 1966.

Griffith, J. S., *The Theory of the Transition Metal Ions*, Cambridge Univ. Press London, 1961.

Jørgensen, C. K., *Orbitals in Atoms and Molecules*, Academic Press, New York, 1962.

PROBLEMS

1. Draw an energy level diagram for the d levels in a tetrahedral crystal field (like Fig. 9.2.1).
2. Consider the square planar complex of transition metal ions, $\mathbf{D}_{4h}$. Take the x and y axes along the metal–ligand bonds. Draw an energy level diagram for the d levels in a square planar crystal field (like Fig. 9.2.1). Compare octahedral, tetrahedral, and square planar energy level diagrams.
3. Make a table of the crystal field stabilization energies (CFSE) of d-orbital configurations in a tetrahedral field. For $(d)^n$ ($n = 4, 5, 6, 7$) how many configurations need be considered in tetrahedral symmetry? Why?
4. The experimental spectrum of $Ni(NH_3)_6^{2+}$ consists of bands at 10,750, 17,500, and 28,200 cm^{-1}. Assign these bands. Find Δ. Calculate the spectrum from Δ and the free-ion energy separation $E(^3F-^3P) = 15{,}800$ cm^{-1} for Ni^{2+}.

Appendixes

A
MATHEMATICAL TOOLS

An UNDERSTANDING of the elements of differential and integral calculus is assumed throughout the text. In this appendix a few mathematical tools of importance in quantum chemistry are presented.

A.1 OPERATORS, FUNCTIONS, AND COORDINATE SYSTEMS

$$f = f(x) \qquad (A.1.1)$$

In words, equation (A.1.1) reads, "f is a function of x." Implicit in (A.1.1) are the following notions:

1. f is defined on a certain *interval*, $a \leq x \leq b$, often written (a, b).
2. x *independently* takes on all values in this interval.
3. There exists a prescribed rule such that for each value of x in this interval there is a *definite value* of f.

The dependent variable f is a function of the independent variable x. There is no limit to the possible number of independent variables of which f is a function, $f = f(x_1, x_2, x_3, x_4, \ldots, x_N)$.

COORDINATES

The points x, y, and z, on the three Cartesian axes are examples of independent variables. Another coordinate system of three independent variables is used throughout the text: the *spherical polar* coordinates r, θ, and ϕ (see Fig. A.1.1).

Relations between Cartesian and spherical polar coordinates are

$$\begin{aligned} x &= r \sin \theta \cos \phi \\ y &= r \sin \theta \sin \phi \\ z &= r \cos \theta \end{aligned} \qquad (A.1.2)$$

Intervals are

$$\begin{aligned} 0 &\leq r \leq \infty & -\infty &\leq x \leq \infty \\ 0 &\leq \theta \leq \pi & -\infty &\leq y \leq \infty \\ 0 &\leq \phi \leq 2\pi & -\infty &\leq z \leq \infty \end{aligned} \qquad (A.1.3)$$

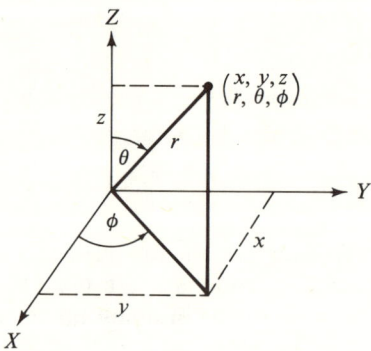

FIG. A.1.1 Relation between the Cartesian and spherical polar coordinates.

The volume elements for integration are

$$dV = dx\, dy\, dz \qquad dV = r^2 \sin\theta\, dr\, d\theta\, d\phi \qquad (A.1.4)$$

An important differential *operator* is

$$\nabla^2 = \frac{\partial^2}{\partial x^2} + \frac{\partial^2}{\partial y^2} + \frac{\partial^2}{\partial z^2}$$

$$\nabla^2 = \frac{1}{r^2}\frac{\partial}{\partial r}\left(r^2 \frac{\partial}{\partial r}\right) + \frac{1}{r^2 \sin\theta}\frac{\partial}{\partial \theta}\left(\sin\theta \frac{\partial}{\partial \theta}\right) + \frac{1}{r^2 \sin^2\theta}\frac{\partial^2}{\partial \phi^2} \qquad (A.1.5)$$

COMPLEX VARIABLES

In general the independent variable may be *complex*; $\sqrt{-1} = i$ is called the *imaginary number*, and the number c will be complex if $c = a + ib$, where a and b are real numbers. Similarly the complex variable is

$$z = x + iy \qquad (A.1.6)$$

where x and y are real variables. Transforming to spherical polar coordinates the complex variable is

$$z = r \exp i\phi \qquad (A.1.7)$$

The *complex conjugate* of a complex number, or variable, is obtained by placing a minus sign in front of each i. The complex conjugate of a complex number, variable, or function, is denoted by the star, *.

$$z^* = x - iy = r \exp(-i\phi) \qquad (A.1.8)$$

The *magnitude*, or absolute value, of z is $|z|$,

$$|z| = (z^*z)^{1/2} = (x^2 + y^2)^{1/2} = r \qquad (A.1.9)$$

A.1 Operators, Functions, and Coordinate Systems

A function of a complex variable is $f(z)$; the complex conjugate of the function is $f^*(z)$, and the magnitude of the function is $|f(z)| = [f^*(z)f(z)]^{1/2}$.

WELL-BEHAVED FUNCTION

A function f is said to be *well behaved* on the interval (a, b) if it satisfies three conditions:

1. f is single-valued, that is, for each value of the independent variable (variables) there is one and only one value of the function f. The converse need not hold: A given value of f may be associated with many values of x.
2. f is continuous (i.e. f is free of discontinuities and infinities).
3. f is square-integrable (i.e., $\int_a^b f^*f\, dV = N$, where N is a finite number called the norm of f).

ORTHONORMALIZED SETS OF FUNCTIONS

Consider the notation

$$\int_a^b f^*f\, dV = \langle f, f \rangle = N.$$

If the norm is unity, $N = 1$, the function is said to be *normalized* (or normalized to unity).

EXAMPLES $1/\sqrt{2\pi}\,\exp(i\phi)$ is normalized to unity on the interval $0 \leq \phi \leq 2\pi$

$$\int_0^{2\pi} 1/\sqrt{2\pi}\,\exp(-i\phi)\,1/\sqrt{2\pi}\,\exp(i\phi)\,d\phi = 2\pi/2\pi$$

But $\exp(ix)$ is not well behaved on the interval $0 \leq x \leq \infty$, that is, possesses no norm,

$$\int_0^\infty \exp(-ix)\exp(ix)\,dx = \infty.$$

Two functions $f_1(x)$ and $f_2(x)$ are said to be *orthogonal* on the interval (a, b) when

$$\int_a^b f_1^* f_2\, dx = \langle f_1, f_2 \rangle = 0 \qquad (A.1.10)$$

A *complete set* of functions on $(a, b), f_1, f_2, f_3, \ldots$, is a set of functions such that the arbitrary well-behaved function $g(x)$ may be expressed as a linear combination of the f_i. That is,

$$g(x) = c_1 f_1 + c_2 f_2 + c_3 f_3 + \cdots$$

$$g(x) = \sum_{i=1}^\infty c_i f_i \qquad (A.1.11)$$

(A.1.11) is the *expansion theorem*.

An orthonormal set of functions has the combined properties of orthogonality and normalization, so $\langle f_i, f_j \rangle = \delta_{ij}$. The Kronecker delta δ_{ij} has the values $\delta_{ij} = 0$, if $i \neq j$, and $\delta_{ij} = 1$, if $i = j$.

EXAMPLE $\sqrt{2}/\sqrt{\pi} \sin mx$, $m = 1, 2, 3, \ldots, \infty$ is an example of a complete set of orthonormal functions on the interval $(0, \pi)$. This set is not complete on the interval $(-\pi, \pi)$.

Assuming $f_1, f_2, f_3, \ldots$ form an orthonormal set; we obtain from (A.1.11)

$$\langle f_j, g \rangle = \sum_i c_i \langle f_j, f_i \rangle = \sum_i c_i \delta_{ij} = c_j \quad (A.1.12)$$

A set of functions $f_1, f_2, f_3, \ldots$ is said to be *linearly independent* if no member of the set can be expressed as a linear combination of the remainder. That is, the set f_i is linearly independent if the relation

$$c_1 f_1 + c_2 f_2 + c_3 f_3 + \cdots = 0 \quad (A.1.13)$$

is satisfied only for $c_1 = c_2 = c_3 = \cdots = 0$.

EXAMPLE The set $1, x, x^2, x^3, x^4, \ldots$ is linearly independent.

Any orthogonal set of functions is linearly independent. To prove this use process (A.1.12) to show that each coefficient in (A.1.13) is zero by orthogonality.

OPERATORS

An operator is a rule or recipe for converting one function into another function. Let $\hat{O}$ be the operator and f the function it "works on."

$$\hat{O} f = f' \quad (A.1.14)$$

f' is the resulting function (f' may be identical to f). A *linear operator* has the following properties, where c is a number:

$$\hat{O} c f = c \hat{O} f$$
$$\hat{O}(f_1 + f_2) = \hat{O} f_1 + \hat{O} f_2 \quad (A.1.15)$$

EXAMPLES d/dx is a linear operator; ∇^2 is a linear operator; $\sqrt{}$ is *not* a linear operator; $1/x$ is a linear operator.

A.1 Operators, Functions, and Coordinate Systems

If
$$\hat{O}f = kf \qquad (A.1.16)$$
where k is a number, then f is said to be an *eigenfunction* to $\hat{O}$ with *eigenvalue* k. The equation $\hat{O}f = kf$ is called an eigenvalue equation.

EXAMPLE   Let $\hat{O} = d^2/dx^2$ and $f = \exp(imx)$, then
$$\frac{d^2}{dx^2} \exp(imx) = -m^2 \exp(imx)$$
The eigenvalue is $-m^2$.

Hermitian operators have the following property:
$$\langle f, \hat{O}g \rangle = \langle \hat{O}f, g \rangle \qquad (A.1.17)$$
which is to say,
$$\int_a^b f^* \hat{O}g \, dV = \int_a^b g \hat{O}^* f^* \, dV$$
$\hat{O}$ is said to be a Hermitian operator if it satisfies (A.1.17) on the interval (a, b).

EXAMPLE d^2/dx^2 is Hermitian on the interval (a, b) if we require that $f(a) = f(b) = g(a) = g(b) = 0$ for all the functions considered.
Proof: Integrate by parts
$$\int_a^b f^* \frac{d^2}{dx^2} g \, dx = f^* \frac{dg}{dx}\bigg|_a^b - \int_a^b \frac{dg}{dx}\frac{df^*}{dx} dx$$
the integrated term vanishes; integrating the second term by parts
$$= -g\frac{df^*}{dx}\bigg|_a^b + \int_a^b g \frac{d^2 f^*}{dx^2} dx$$
Since the integrated term vanishes again, the result is proven.

Many of the properties of Hermitian operators are given in Chapter 2 in the discussion of the Sturm-Liouville equation. For example, Hermitian operators have purely *real* (rather than complex) eigenvalues, and the eigenfunctions of a Hermitian operator form a *complete, orthonormal set*.

Finally, we note that if a function f is *simultaneously* an eigenfunction to *two* operators then the operators will *commute*. Proof: Let the two operators be $\hat{O}$ and $\hat{P}$ with eigenvalues k and m, respectively; then $\hat{O}f = kf$ and $\hat{P}f = mf$.
$$\hat{P}\hat{O}f = \hat{P}kf = kmf$$
$$\hat{O}\hat{P}f = \hat{O}mf = mkf$$

or
$$\hat{P}\hat{O}f = \hat{O}\hat{P}f$$
that is,
$$\hat{P}\hat{O} = \hat{O}\hat{P}$$

When $\hat{O}\hat{P} = \hat{P}\hat{O}$ the operators are said to commute.

EXAMPLES d/dx and y commute, but d/dx and x do not commute [i.e., d/dx commutes with $\hat{O}(y)$ but does not commute with $\hat{O}(x)$].

A.2 VECTORS, MATRICES, DETERMINANTS, AND SIMULTANEOUS EQUATIONS

VECTORS

A vector is a quantity possessing both magnitude and direction. *Unit vectors*, **i**, **j**, and **k**, are vectors of unit length directed along the Cartesian axes x, y, and z, respectively. A vector may be expressed in terms of its *components* along the x, y, and z axes using the unit vectors. For example, in Fig. A.2.1 the vector **a** is directed along the x axis, $\mathbf{a} = \mathbf{i}a_x$, where a_x is the length of the vector **a**. Similarly, $\mathbf{b} = \mathbf{j}b_y$ and $\mathbf{c} = \mathbf{k}c_z$. The resultant vector, $\mathbf{r} = \mathbf{a} + \mathbf{b} + \mathbf{c}$ is

$$\mathbf{r} = \mathbf{i}a_x + \mathbf{j}b_y + \mathbf{k}c_z \tag{A.2.1}$$

The *length* of **r** is $r = |\mathbf{r}| = (a_x^2 + b_y^2 + c_z^2)^{1/2} = (\mathbf{r} \cdot \mathbf{r})^{1/2}$. The product $\mathbf{r} \cdot \mathbf{r}$ is called the dot (or scalar) product. The result is a scalar (a number) rather than a vector.

$$|\mathbf{r} \cdot \mathbf{a}| = ra \cos \theta \tag{A.2.2}$$

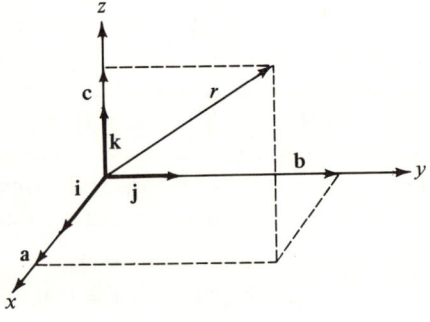

FIG. A.2.1

where θ is the angle between the vectors **r** and **a**. It follows that the dot product between perpendicular vectors is zero since $\theta = \pi/2$, that is, $\mathbf{i} \cdot \mathbf{j} = 0$, etc. The dot product is distributive and commutative $\mathbf{r} \cdot (\mathbf{s} + \mathbf{t}) = \mathbf{r} \cdot \mathbf{s} + \mathbf{r} \cdot \mathbf{t}$ and $\mathbf{r} \cdot \mathbf{s} = \mathbf{s} \cdot \mathbf{r}$.

Vectors may also be multiplied together by taking the cross (or vector) product, $\mathbf{a} \times \mathbf{b}$. The result is a vector with direction perpendicular to the plane determined by **a** and **b**. Let $\mathbf{a} \times \mathbf{b} = \mathbf{c}$. By definition, **c** has the components

$$\mathbf{c} = \mathbf{a} \times \mathbf{b} = \mathbf{i}(a_y b_z - a_z b_y) + \mathbf{j}(a_z b_x - a_x b_z) + \mathbf{k}(a_x b_y - a_y b_x) \quad (A.2.3)$$

From Fig. A.2.1 it is clear that **c** has only a component in the z direction, $\mathbf{c} = \mathbf{k} c_z = \mathbf{k} a_x b_y$. Vector multiplication by cross product is distributive but not commutative; it is anticommutative, $\mathbf{a} \times \mathbf{b} = -\mathbf{b} \times \mathbf{a}$. The magnitude of the cross product is $|\mathbf{c}| = ab \sin \theta$; again θ is the angle between **a** and **b**, thus the cross product between parallel vectors vanishes.

DETERMINANTS

The cross product may be written as a *determinant*.

$$\mathbf{a} \times \mathbf{b} = \begin{vmatrix} \mathbf{i} & \mathbf{j} & \mathbf{k} \\ a_x & a_y & a_z \\ b_x & b_y & b_z \end{vmatrix} \quad (A.2.4)$$

A determinant is a two-dimensional array with an equal number of rows and columns. A determinant has a certain *value* which is found by "multiplying it out" according to a certain rule (expansion in cofactors). Let det **D** mean determinant of the *array* **D**. Let the array **D** have *elements* D_{ij}.

$$\det \mathbf{D} = \begin{vmatrix} D_{11} & D_{12} & D_{13} & D_{14} & \cdots & D_{1N} \\ D_{21} & D_{22} & D_{23} & D_{24} & \cdots & D_{2N} \\ \vdots & \vdots & \vdots & \vdots & & \vdots \\ D_{N1} & D_{N2} & D_{N3} & D_{N4} & \cdots & D_{NN} \end{vmatrix} \quad (A.2.5)$$

The *cofactor* of the element D_{ij} is $(-1)^{i+j}$ times the determinant which is formed by striking out the ith row and the jth column of the original determinant (A.2.5). The cofactor of D_{ij} is denoted D^{ij}. With the cofactor defined it is now possible to give a general definition of the value of a determinant.

$$\det \mathbf{D} = \sum_{i=1}^{N} D_{ij} D^{ij} = \sum_{i=1}^{N} D_{ji} D^{ji} \quad (A.2.6)$$

EXAMPLE   Apply (A.2.6) to (A.2.4).

$$\mathbf{a} \times \mathbf{b} = \mathbf{i}(-1)^{1+1}\begin{vmatrix} a_y & a_z \\ b_y & b_z \end{vmatrix} + \mathbf{j}(-1)^{1+2}\begin{vmatrix} a_x & a_z \\ b_x & b_z \end{vmatrix} + \mathbf{k}(-1)^{1+3}\begin{vmatrix} a_x & a_y \\ b_x & b_y \end{vmatrix}$$

$$= \mathbf{i}(a_y b_z - a_z b_y) - \mathbf{j}(a_x b_z - a_z b_x) + \mathbf{k}(a_x b_y - a_y b_x) = \text{eq. (A.2.3)}$$

The *value of a determinant is zero* if

1. All the elements of a row (column) are zero.
2. All the elements of a row (column) are identical with, or multiples of, the corresponding elements of another row (column).

The value of a determinant is *unchanged* by taking linear combinations of the corresponding elements of rows (columns).

The value of a determinant *changes sign* if two rows (columns) are interchanged.

It follows from the definition of the zero-value determinant that

$$\sum_{i=1}^{N} D_{ij} D^{ik} = 0 \quad \text{if} \quad j \neq k \quad \text{and} \quad D_{ij} = D_{ik} \quad \text{for all} \quad i \quad \text{(A.2.7)}$$

because this is the value of a determinant with identical *j*th and *k*th columns and such a determinant has the value zero.

MATRICES

A *matrix* is a two-dimensional array. Unlike determinants a matrix possesses no "value." A matrix may be square or rectangular. The simplest matrix is the row or column, $\mathbf{v}^+$ or $\mathbf{v}$.

$$\mathbf{v} = \begin{pmatrix} v_1 \\ v_2 \\ v_3 \\ \vdots \\ v_N \end{pmatrix} \qquad \mathbf{v}^+ = (v_1 v_2 v_3 \ldots v_N)$$

Such row and column matrices are called *N-dimensional vectors*.

Two matrices are added by addition of their elements.

$$\mathbf{A} + \mathbf{B} = \mathbf{C}$$

$$\begin{pmatrix} A_{11} & A_{12} \ldots \\ A_{21} & A_{22} \ldots \end{pmatrix} + \begin{pmatrix} B_{11} & B_{12} \ldots \\ B_{21} & B_{22} \ldots \end{pmatrix} = \begin{pmatrix} C_{11} & C_{12} \ldots \\ C_{21} & C_{22} \ldots \end{pmatrix}$$

$C_{11} = A_{11} + B_{11} \qquad C_{21} = A_{21} + B_{21} \quad$ etc.

Two matrices may be multiplied to give a product matrix.

$$\mathbf{AB} = \mathbf{C} \qquad \text{(A.2.8)}$$

The elements of the product matrix are $C_{ij} = \sum_{k=1}^{N} A_{ik} B_{kj}$. For multiplication the matrices must be conformable, that is, if $\mathbf{A}$ is $M \times N$ then $\mathbf{B}$ must be $N \times L$ (i.e., $\mathbf{A}$ has M rows and N columns).

A.2 Vectors, Matrices, Determinants, and Simultaneous Equations

The *unit matrix* **1** has unity for elements along the diagonal and all other elements are zero, for example, the 3 × 3 unit matrix is

$$\mathbf{1} = \begin{pmatrix} 1 & 0 & 0 \\ 0 & 1 & 0 \\ 0 & 0 & 1 \end{pmatrix}.$$

The unit matrix has the property $\mathbf{1A} = \mathbf{A}$. If **A** is a square matrix and det $\mathbf{A} \neq 0$, **A** is said to be a *nonsingular matrix*. If **A** is nonsingular it possesses an *inverse matrix* $\mathbf{A}^{-1}$, such that

$$\mathbf{A}^{-1}\mathbf{A} = \mathbf{A}\mathbf{A}^{-1} = \mathbf{1} \qquad (A.2.9)$$

Two matrices which satisfy $\mathbf{AB} = \mathbf{BA}$ are said to *commute*.

SIMULTANEOUS EQUATIONS (SYSTEMS OF LINEAR EQUATIONS)

$$\begin{aligned} A_{11}c_1 + A_{12}c_2 + \cdots + A_{1N}c_N &= 0 \\ A_{21}c_1 + A_{22}c_2 + \cdots + A_{2N}c_N &= 0 \\ &\vdots \\ A_{N1}c_1 + A_{N2}c_2 + \cdots + A_{NN}c_N &= 0 \end{aligned} \qquad (A.2.10)$$

A system of equations like (A.2.10) is more succinctly written in matrix form.

$$\mathbf{Ac} = 0 \qquad (A.2.11)$$

In (A.2.11) the matrix **A** is $N \times N$ with elements A_{ij} and **c** is the N-dimensional column vector with elements c_i. If **A** is nonsingular then $\mathbf{A}^{-1}$ exists and equation (A.2.11) is solved immediately.

$$\begin{aligned} \mathbf{A}^{-1}\mathbf{A}\mathbf{c} &= 0 \\ \mathbf{c} &= 0 \end{aligned} \qquad (A.2.12)$$

This is a trivial solution to the problem. In a more interesting case **A** is a singular matrix (det $\mathbf{A} = 0$), in which case $\mathbf{A}^{-1}$ does not exist. Before discussing the solution of (A.2.11) for singular **A**, note that a very interesting equation can be reduced to the form (A.2.11).

$$\mathbf{Hc} = \lambda \mathbf{c} \quad \text{or} \quad (\mathbf{H} - \lambda \mathbf{1})\mathbf{c} = 0 \qquad (A.2.13)$$

(A.2.13) is called an *eigenvalue equation*. The effect of the matrix **H** is merely to multiply the *eigenvector* **c** by a constant scalar factor λ, the *eigenvalue*. From (A.2.12) it follows that nontrivial solutions to the eigenvalue equation exist if $(\mathbf{H} - \lambda \mathbf{1})$ is a singular matrix, that is, if

$$\det(\mathbf{H} - \lambda \mathbf{1}) = 0 \qquad (A.2.14)$$

This is the fundamental equation of the problem. In most cases of interest λ is unknown and we wish to find those particular values of λ for which

(A.2.14) is satisfied. Expanding $\det(\mathbf{H} - \lambda \mathbf{1})$ according to (A.2.6) yields a polynomial in λ called the *secular equation*.

$$\lambda^N + a_1 \lambda^{N-1} + a_2 \lambda^{N-2} + \cdots + a_N = 0 \qquad (A.2.15)$$

The secular equation, being of degree N, has N roots (some may be the same, i.e., *degenerate*) $\lambda_1, \lambda_2, \lambda_3, \lambda_4, \ldots, \lambda_N$, all of which satisfy (A.2.14).

To find the eigenvector $\mathbf{c}$, we note first of all that there is an eigenvector for each λ_n. Then $\mathbf{Ac} = 0$ has for its kth row

$$A_{k1} c_1 + A_{k2} c_2 + \cdots + A_{kN} c_N = 0$$

which is satisfied by $c_1 = A^{j1}$, $c_2 = A^{j2}$, etc., or $c_i = A^{ji}$, according to equation (A.2.7). j can have the value $j = 1$ to N. Furthermore the c_i are undetermined to the extent of a multiplicative constant. The ratio c_i/c_k is determined.

$$c_i/c_k = A^{ji}/A^{jk} \qquad (A.2.16)$$

After the ratios of the coefficients have been determined their individual values can be found from a normalization condition.

In sum, the eigenvalue equation $(\mathbf{H} - \lambda \mathbf{1})\mathbf{c} = 0$, is solved by first finding the λ_n from the secular equation (A.2.15). There is one *set* of coefficients c_i for each λ_n. The c_i are determined by (A.2.16) with $\mathbf{A} = (\mathbf{H} - \lambda_n \mathbf{1})$.

MATRIX TRANSFORMATION

The matrix multiplication

$$\mathbf{Av} = \mathbf{v}' \qquad (A.2.17)$$

is a transformation of the vector $\mathbf{v}$ into a new column vector $\mathbf{v}'$. $\mathbf{A}$ is an $N \times N$ matrix; $\mathbf{v}$ and $\mathbf{v}'$ are N-dimensional column vectors. A simple example of such a matrix transformation is the rotation of an ordinary vector in three-dimensional space. Let

$$\mathbf{r}_1 = \begin{pmatrix} x_1 \\ y_1 \\ z_1 \end{pmatrix}$$

be the vector before rotation about the z axis. Let

$$\mathbf{r}_2 = \begin{pmatrix} x_2 \\ y_2 \\ z_2 \end{pmatrix}$$

be the vector after rotation about the z axis by an angle ϕ. Then $\mathbf{r}_1$ and $\mathbf{r}_2$ are related by the matrix transformation

$$\begin{pmatrix} \cos\phi & \sin\phi & 0 \\ -\sin\phi & \cos\phi & 0 \\ 0 & 0 & 1 \end{pmatrix} \begin{pmatrix} x_1 \\ y_1 \\ z_1 \end{pmatrix} = \begin{pmatrix} x_2 \\ y_2 \\ z_2 \end{pmatrix} \qquad (A.2.18)$$

which is the same as

$$x_2 = \cos\phi\, x_1 + \sin\phi\, y_1$$
$$y_2 = -\sin\phi\, x_1 + \cos\phi\, y_1 \qquad \text{(A.2.19)}$$
$$z_2 = z_1$$

Note that **A** is a matrix *operator* because it transforms one vector into another vector.

REFERENCES

Anderson, J. M., *Mathematics for Quantum Chemistry*, Benjamin, New York, 1966.

Margenau, H., and Murphy, G. M., *The Mathematics of Physics and Chemistry*, vol. I, 2nd ed., D. van Nostrand, Princeton, New Jersey, 1956.

B
EVALUATION OF THE COULOMB INTEGRAL J_{1S1S}

$$J_{1s1s} = \langle |1s(1)|^2, (1/r_{12})|1s(2)|^2 \rangle$$

$$= \frac{Z^6}{\pi^2} \iint \frac{\exp[-2Z(r_1 + r_2)]}{r_{12}} \, dV_1 \, dV_2 \qquad \text{(B.1)}$$

In order to proceed with the integration we require an expression for $1/r_{12}$ in terms of the coordinates of electrons 1 and 2. A well-known expansion[1] is

$$\frac{1}{r_{12}} = \sum_{l=0}^{\infty} \sum_{m=-l}^{m=+l} \left[\frac{2}{2l+1}\right] \left(\frac{r_<^l}{r_>^{l+1}}\right) \Theta_{lm}(\theta_1) \Theta_{lm}(\theta_2) \exp[im(\phi_1 - \phi_2)] \quad \text{(B.2)}$$

where $r_>$ and $r_<$ are the greater and the lesser of r_1 and r_2, respectively, and

$$\Theta_{lm}(\theta_i) = \left[\frac{2l+1}{2} \frac{(l-|m|)!}{(l+|m|)!}\right]^{1/2} P_l^{|m|}(\cos \theta_i)$$

are the normalized and orthogonal angular functions described in Chapter 4.

$$dV_1 = r_1^2 \sin \theta_1 \, dr_1 \, d\theta_1 \, d\phi_1 \qquad dV_2 = r_2^2 \sin \theta_2 \, dr_2 \, d\theta_2 \, d\phi_2$$

Upon substituting (B.2) into (B.1) and integrating over the angular coordinates we make use of the orthogonality properties of the Θ_{lm} and the fact that a 1s function has angular dependence $\Theta_{00}(\theta)$. This limits the apparent infinite expansion to the single integral $l = m = 0$.

$$J_{1s1s} = \frac{Z^6}{\pi^2} \iint \frac{\exp[-2Z(r_1 + r_2)]}{r_>} \, dV_1 \, dV_2$$

$$J_{1s1s} = \frac{Z^6 (4\pi)^2}{\pi^2} \iint \frac{\exp[-2Z(r_1 + r_2)]}{r_>} r_1^2 r_2^2 \, dr_1 \, dr_2$$

$(4\pi)^2$ is the result of the integration over the angles. The integration over r_1 and r_2 is broken into two parts depending on which is the greater, as follows:

---

[1] H. Eyring, J. Walter, and G. E. Kimball, *Quantum Chemistry*, appendix V, Wiley, New York, 1944.

$$J_{1s1s} = 16Z^6 \int_0^\infty \left[\frac{1}{r_1} \int_0^{r_1} \exp[-2Z(r_1+r_2)]r_2^2 \, dr_2 \right.$$
$$\left. + \int_{r_1}^\infty \exp[-2Z(r_1+r_2)]r_2 \, dr_2 \right] r_1^2 \, dr_1$$

These are now standard integrals, which are evaluated to give

$$J_{1s1s} = \frac{5}{8}Z$$

C THE HARTREE-FOCK EQUATION FOR THE HELIUM ATOM AND OTHER CLOSED-SHELL SYSTEMS

THE HARTREE-FOCK equation for the helium atom, $(1s)^2$, follows directly from the application of the variational principle to the trial wavefunction. $\tilde{\Psi}$ is the trial wavefunction.

$$\tilde{\Psi} = \hat{A}\widetilde{1s}\alpha(1)\widetilde{1s}\beta(2) \qquad \langle \tilde{\Psi}, \tilde{\Psi}\rangle = 1$$
$$W = \langle \hat{A}\widetilde{1s}\alpha(1)\widetilde{1s}\beta(2), \hat{H}\hat{A}\widetilde{1s}\alpha(1)\widetilde{1s}\beta(2)\rangle \qquad (C.1)$$
$$\hat{H} = \hat{h}(1) + \hat{h}(2) + 1/r_{12}$$
$$W = \langle \widetilde{1s}\alpha(1), \hat{h}(1)\widetilde{1s}\alpha(1)\rangle + \langle \widetilde{1s}\beta(2), \hat{h}(2)\widetilde{1s}\beta(2)\rangle$$
$$+ \langle \widetilde{1s}\alpha(1)\widetilde{1s}\beta(2), (1/r_{12})\widetilde{1s}\alpha(1)\widetilde{1s}\beta(2)\rangle \qquad (C.2)$$

The $\widetilde{1s}$ trial radial wavefunction is purely real[1]; furthermore the spatial parts of the spin-orbitals $\widetilde{1s}\alpha$ and $\widetilde{1s}\beta$ are taken to be identical. The variational principle allows us to minimize W by variation of $\widetilde{1s}$ subject to the normalization condition (a constraint) $\langle \widetilde{1s}, \widetilde{1s}\rangle = 1$. If δ indicates an arbitrary variation, then $\delta W = 0$ and $\langle \delta\widetilde{1s}, \widetilde{1s}\rangle = 0$ is the normalization constraint. Altogether, to the extent of an undetermined multiplier $-\lambda$, we have for the variation of a given spin-orbital $\widetilde{1s}(i)$,

$$\langle \delta\widetilde{1s}(i), \hat{h}\widetilde{1s}(i)\rangle + \langle \delta\widetilde{1s}(i)\widetilde{1s}(j), (1/_{ij})\widetilde{1s}(i)\widetilde{1s}(j)\rangle$$
$$-\lambda\langle \delta\widetilde{1s}(i), \widetilde{1s}(i)\rangle = 0 \quad (C.3)$$

or

$$\langle \delta\widetilde{1s}(i)\{\hat{h}\widetilde{1s}(i) + \langle \widetilde{1s}(j), (1/r_{ij})\widetilde{1s}(j)\rangle_j\widetilde{1s}(i) - \lambda\widetilde{1s}(i)\}\rangle_i = 0$$

If (C.3) is to hold for any arbitrary variation $\delta 1s(i)$, then the quantity in braces must vanish.

$$\hat{h}(i)\widetilde{1s}(i) + \langle \widetilde{1s}(j), (1/r_{ij})\widetilde{1s}(j)\rangle_j\widetilde{1s}(i) = \lambda\widetilde{1s}(i) \qquad (C.4)$$

This is the equation for the Hartree-Fock orbital, 1s, that is, the Hartree-Fock equation. The subscript j on the braket indicates integration is over the coordinates of electron j; $\hat{V}_{\text{aver}}(i) = \langle 1s(j), (1/r_{ij})1s(j)\rangle_j$ is a function

---

[1] Therefore the complex conjugate terms in (C.3) have been omitted.

of coordinates of electron i. Defining $\hat{H}_{\text{eff}}(\text{helium}) = \hat{h}(i) + \hat{V}_{\text{aver}}(i)$ we have

$$\hat{H}_{\text{eff}}(i)1s(i) = E_{1s}1s(i) \tag{C.5}$$

where E_{1s} (i.e., $-\lambda$) is the 1s *orbital energy*. The result is quite simple for helium, there being no exchange integrals present in the variational energy. The general Hartree-Fock equation for the N-electron closed-shell system[2] is, in terms of the N occupied spin-orbitals φ_i

$$\hat{h}(i)\varphi_i(i) + \sum_{j=1}^{N} \left\langle \varphi_j(j), \left(\frac{1-\hat{P}_{ij}}{r_{ij}}\right)\varphi_j(j) \right\rangle_j \varphi_i(i) = E_i \varphi_i(i) \tag{C.6}$$

and

$$\hat{V}_{\text{aver}}(i) = \sum_{j=1}^{N} \left\langle \varphi_j(j), \left(\frac{1-\hat{P}_{ij}}{r_{ij}}\right)\varphi_j(j) \right\rangle_j$$

In general $\hat{V}_{\text{aver}}$ contains the exchange terms via the permutation operator $\hat{P}_{ij}$ which exchanges *electron* indices i and j.

Because $\hat{H}_{\text{eff}}$ *contains* 1s, and also *determines* 1s, the only solution to equation (C.5) is by iteration, that is, assume a form for 1s, solve for a new 1s, and then use the new 1s in $\hat{H}_{\text{eff}}$ as input for another cycle; continue the process until the output 1s is as close as desired to the input 1s, then *self-consistency* has been attained. The result is a self-consistent field wavefunction (the field $\hat{V}_{\text{aver}}$ is consistent with the orbital it generates). If the variation leading to self-consistency has been *complete* the resultant orbital is the Hartree-Fock orbital. The coupled set of N equations, of which (C.6) is one, are also solved iteratively.

---

[2] J. C. Slater, *Quantum Theory of Atomic Structure*, vol. 2, chap. 1, McGraw-Hill, New York, 1960.

D

THE BORN-OPPENHEIMER APPROXIMATION

THE HAMILTONIAN for the complete molecular system of electrons and nuclei is

$$\hat{H}_T = \sum_i -\tfrac{1}{2}\nabla_i^2 + \sum_K -\tfrac{1}{2}\nabla_K^2 - \sum_K \sum_i Z_K/q_{iK} + \sum_{i>j} 1/q_{ij}$$
$$+ \sum_{K>L} Z_K Z_L/Q_{KL} + \text{coupling terms}[1]$$

The q are electronic coordinates and the Q are nuclear coordinates. What was heretofore called the electronic Hamiltonian is

$$\hat{H}_e = \sum_i -\tfrac{1}{2}\nabla_i^2 - \sum_K \sum_i Z_K/q_{iK} + \sum_{i>j} 1/q_{ij} + \sum_{K>L} Z_K Z_L/Q_{KL}$$
$$\hat{H}_T - \hat{H}_e = \hat{H}_n = \sum_K -\tfrac{1}{2}\nabla_K^2$$

In the Born-Oppenheimer approximation (BOA) we assume that the electronic and nuclear motions are *separable*, in the sense,

$$\Phi(q, Q) = \Psi_Q(q)\chi(Q)$$

where $\Psi_Q(q)$ is the eigenfunction to $\hat{H}_e$ for fixed choice of the Q_K.

$$\hat{H}_e \Psi_Q(q) = E(Q)\Psi_Q(q) \tag{D.1}$$

Of course, as we saw in Chapter 5, $\Psi_Q(q)$ does display a dependence on the Q_K. For the case of the diatomic molecule, $E(Q) = E(R)$.

$\chi(Q)$ is the nuclear wavefunction. In order to obtain the nuclear Schrödinger equation, we assume that the nuclei move so slowly relative to electronic velocities that the fixed Q_K electronic eigenvalues $E(Q)$ provide the potential in which the nuclei move.

$$\left[\sum_K -\tfrac{1}{2}\nabla_K^2 + E(Q)\right]\chi(Q) = E\chi(Q) \tag{D.2}$$

Equation (D.2) is the Schrödinger equation we solved for vibrational

---

[1] J. O. Hirschfelder and W. J. Meath, "The Nature of Intermolecular Forces," in *Advan. Chem. Phys.* **12**, 1 (1967). These coupling terms are sufficiently small to be ignorable for some purposes, but are responsible for breakdown of the Born-Oppenheimer approximation.

motion of the diatomic molecule [with the additional assumption that $E(R)$ is proportional to $(R - R_e)^2$; the harmonic oscillator approximation]. Each electronic state of the molecule has a different eigenvalue, $E(Q)$ at each Q, therefore different nuclear wavefunctions.

Equation (D.1) is the working electronic Schrödinger equation of molecular physics, $\Psi_Q(q)$ is the electronic wavefunction, and $E(Q)$ is the molecular energy; all within the framework of the Born-Oppenheimer approximation. There is strong spectroscopic evidence that the BOA is quite good for the ground states of diatomic and polyatomic molecules, but not so good for the excited states of large polyatomic molecules. Breakdown of the BOA occurs when electronic eigenvalues, $E_1(Q)$ and $E_2(Q)$, are degenerate or nearly degenerate for certain values of Q. There is an interesting physical analogy to BOA breakdown which may clarify what happens. Consider a tub of fluid with movable walls. The movable walls are analogous to the nuclei which confine the electronic "fluid" to well-defined regions of space. If the walls oscillate very slowly, the fluid will remain essentially undisturbed, but rapid oscillations will "excite" the fluid into standing wave patterns other than the undisturbed "ground state." Alternatively, the fluid may be in a state of excitation which is so nearly degenerate with another state that even the slow oscillation of the walls will bring about transitions between these states. In either case, the approximation of separable motion for fluid and walls breaks down.

E THE HELLMANN-FEYNMAN THEOREM

CONSIDER A SYSTEM described by the electronic Hamiltonian $\hat{H}_e$ which depends on a parameter Q. Let $E(Q)$ and Ψ be the energy and wavefunction of the system, then

$$E(Q) = \langle \Psi, \hat{H}_e \Psi \rangle$$

$$\frac{dE(Q)}{dQ} = \int \frac{\partial \Psi^*}{\partial Q} \hat{H}_e \Psi \, dV + \int \Psi^* \frac{\partial \hat{H}_e}{\partial Q} \Psi \, dV + \int \Psi^* \hat{H}_e \frac{\partial \Psi}{\partial Q} \, dV \quad (E.1)$$

Using the Hermitian property of $\hat{H}_e$, the last term of (E.1) becomes $\int (\partial \Psi / \partial Q) \hat{H}_e^* \Psi^* \, dV$. Furthermore, because $\hat{H}_e \Psi = E(Q) \Psi$ and $\hat{H}_e^* \Psi^* = E(Q) \Psi^*$ we obtain for (E.1)

$$\frac{dE(Q)}{dQ} = \left\langle \Psi, \frac{\partial \hat{H}_e}{\partial Q} \Psi \right\rangle + E(Q) \left(\left\langle \Psi, \frac{\partial \Psi}{\partial Q} \right\rangle + \left\langle \frac{\partial \Psi}{\partial Q}, \Psi \right\rangle \right)$$

$$= \left\langle \Psi, \frac{\partial \hat{H}_e}{\partial Q} \Psi \right\rangle + E(Q) \frac{\partial}{\partial Q} \langle \Psi, \Psi \rangle \quad (E.2)$$

or finally,

$$\frac{dE(Q)}{dQ} = \left\langle \Psi, \frac{\partial \hat{H}_e}{\partial Q} \Psi \right\rangle \quad (E.3)$$

(E.3) is the general form of the Hellmann-Feynman theorem.[1] In particular, when dealing with the diatomic molecule we take $Q = R$, then $\partial \hat{H}_e / \partial R$ is constant or a one-electron operator and the expectation value in (E.3) is evaluated from the charge density function ρ. Explicitly, from $\hat{H}_e = \hat{H} + Z_a Z_b / R$, equation (5.3.1) follows.

ELECTROSTATIC THEOREM

Consider the diatomic molecule ($Q = R$), express $\hat{H}_e$ in Cartesian coordinates with nucleus a as origin. $\hat{H}_e$ is given by equation (E.4).

$$\hat{H}_e = \sum_i -\tfrac{1}{2} \nabla_i^2 - \sum_i (Z_b/r_{ib} + Z_a/r_{ia}) + \sum_{i>j} 1/r_{ij} + Z_a Z_b / R \quad (E.4)$$

From (E.3) [and equation (5.3.1)]

[1] R. P. Feynman, *Phys. Rev.* **56**, 340 (1939).

$$\frac{dE}{dR} = \int \rho(i)\left(\frac{\partial r_{ib}}{\partial R}\right)\frac{Z_b}{r_{ib}^2}\,dV - \frac{Z_a Z_b}{R^2} \tag{E.5}$$

or, taking nucleus b as the origin of coordinates,

$$\frac{dE}{dR} = \int \rho(i)\left(\frac{\partial r_{ia}}{\partial R}\right)\frac{Z_a}{r_{ia}^2}\,dV - \frac{Z_a Z_b}{R^2} \tag{E.6}$$

The symmetric combination of (E.5) and (E.6) in Force $= -dE/dR$ gives equation (5.3.2) because

$$\frac{\partial r_{ia}}{\partial R} = \text{component of } \nabla \mathbf{r}_{ia} \text{ in } R \text{ direction}$$

or

$$\frac{\partial r_{ia}}{\partial R} = \frac{\mathbf{R}\cdot\nabla \mathbf{r}_{ia}}{R} = \frac{\mathbf{R}\cdot\mathbf{r}_{ia}}{Rr_{ia}} = \cos\theta_a$$

F

ELECTRO-NEGATIVITIES AND COVALENT RADII

THE CONCEPT of electronegativity is often used in modern chemistry courses. Qualitatively, electronegativity is the ability of an atom to attract electronic charge density towards itself in bond formation. This ability depends on the state of the atom-in-the-molecule, the so-called "valence state" of the atom, for example, sp, sp^2, sp^3. Electronegativity *differences* between atoms are thus related to the *polarity* of their bond. Every chemist develops an intuitive feeling for the electropositive character of metals and the electronegative character of nonmetals, but a decision must be made on how to express this intuitive feeling *quantitatively* in terms of *observables*.

If A and B are atoms, the process

$$A + B \rightarrow A^+ + B^-$$

requires an amount of energy $(I_A - E_B)$, where I_A is the ionization potential of the orbital on A, and E_B is the electron affinity of the orbital on B, involved in the electron transfer. Similarly the process

$$A + B \rightarrow A^- + B^+$$

requires an energy $(I_B - E_A)$. If $(I_A - E_B) > (I_B - E_A)$, then $(I_A + E_A) > (I_B + E_B)$, and it requires *more* energy to form A^+B^- than A^-B^+. This is an indication that A^-B^+ is the favored polarity of the A—B bond, that is, A is more electronegative than B. Since the *degree* of polarity in the bond, from homopolar to ionic, should be proportional to the *difference* between $(I_A + E_A)$ and $(I_B + E_B)$, Mulliken defined[1] $\frac{1}{2}(I + E)$ to be the electronegativity of the valence state orbital having ionization potential I and electron affinity E. (I and E are expressed in electron volts.)

Unfortunately, neither I nor E is very often an observable (the valence state is not necessarily a spectroscopic state of the atom). I and E must then be determined indirectly from the parametrization of observed atomic spectra.[2] Electronegativities, $\frac{1}{2}(I + E)$, have been given for the valence states of many important atoms.[2-4] Table F.1 presents the electro-

[1] R. S. Mulliken, *J. Chem. Phys.* **2**, 782 (1934).
[2] G. Pilcher and H. A. Skinner, *J. Inorg. Nucl. Chem.* **24**, 937 (1962).
[3] J. Hinze and H. H. Jaffé, *J. Amer. Chem. Soc.* **84**, 540 (1962).
[4] H. O. Pritchard and H. A. Skinner, *Chem. Rev.* **55**, 745 (1955); *Trans. Faraday Soc.* **49**, 1254 (1953).

TABLE F.1 $\frac{1}{2}(I+E)\text{eV}^a$

Neutral atom	Atomic orbitals or hybrid orbitals[b]				
	p	(sp^3)	$(sp^2) = t$	$(sp) = d$	s
Be	2.93 (sx)	3.72	3.98	4.67	6.35 (sx)
B	4.22 (dxy)	5.73	6.26	6.76 (dxy)	10.33 (sxy)
	4.27 (sxy)			7.29 (ddx)	
	4.64 (ddx)				
C	5.92 ($tttx$)	8.18	8.95	10.46	14.84 ($sxyz$)
	5.96 ($ddxy$)				
	6.09 ($sxyz$)				
N	7.84 (d^2dxy)	11.49	12.23	14.41 (ddx^2y)	20.75 (sx^2yz)
	7.85 (t^2ttx)			15.88 (d^2dxy)	
	7.93 (ddx^2y)				
	8.06 (sx^2yz)				
O	10.86 (sx^2y^2z)	15.25	16.65	18.83	26.79 (sx^2y^2z)

[a] After G. Pilcher and H. A. Skinner, *J. Inorg. Nucl. Chem.* **24**, 937 (1962). Used by permission.
[b] Valence state configurations are given in parentheses: x, y, and z are $2p_x$, $2p_y$, and $2p_z$, respectively; d and t are the hybrids (sp) and (sp^2).
[c] To convert to the Pauling electronegativity scale, Pilcher and Skinner suggest division by 3.27, that is, $\frac{1}{2}(I+E)/3.27$.

negativities of the valence orbitals of the first-row atoms. Note that the electronegativity increases with increasing s character of the valence orbital.

Pauling[5] suggested a more empirical definition of the electronegativity, which has the advantage of completeness over most of the periodic table. These electronegativities do not distinguish between the valence orbitals of the atom, but assign the atom a single number based on the following argument. Let AB be a diatomic molecule having dissociation energy $D_e(\text{AB})$, the dissociation energies of the homopolar diatomics A_2 and B_2 are $D_e(A_2)$ and $D_e(B_2)$, respectively. Assume that the *hypothetical* bond energy of the *hypothetical* purely covalent (nonpolar) A—B bond is the geometric mean of $D_e(A_2)$ and $D_e(B_2)$. The difference between the true dissociation energy $D_e(\text{AB})$ and the geometric mean is taken to be a measure of the polarity of the A—B bond.

[5] The original electronegativity scale, first put forth in 1932. See L. Pauling, *Nature of the Chemical Bond*, 3rd ed., Cornell Univ. Press, Ithaca, New York, 1960.

$$\Delta_{AB} = D_e(AB) - [D_e(A_2)D_e(B_2)]^{1/2}$$

If Δ_{AB} (kcal/mole) is a measure of bond polarity then it may be related to the electronegativity difference between atoms A and B. The following relation was found to give a consistent set of electronegativities for most atoms:

$$x_A - x_B = 0.208\sqrt{\Delta_{AB}}$$

where x_i is the electronegativity. Since Δ determines the electronegativity *differences*, the Pauling scale is not determined until one atomic electronegativity is given an arbitrary value. The arbitrary choice is to take fluorine (the most electronegative element) to have x_F equal to about 4.

There are several other semiempirical models which are used to assign electronegativity values to the elements, in addition to the Pauling scale. However, because other electronegativities are convertible to the Pauling scale by multiplicative and additive factors, they serve a useful purpose only where substantial disagreement or inconsistency exists. Table F.2 presents a selection of atomic electronegativities on the Pauling scale.

Electronegativities are often disparaged because of their semiempirical origins and their overuse. Admittedly, electronegativities are a crutch, not to be used in place of analysis and careful thought. On the other hand, we find that the Mulliken electronegativities are carefully defined in terms of certain "observable" properties of the valence shell hybrid orbitals which participate in the bonding. Furthermore, there are a large number of physical and chemical properties of molecules which show positive correlation with electronegativity differences (bond polarities). Infrared, ultraviolet, and nuclear magnetic resonance spectra; reaction rates, equilibrium constants, and dipole moments, often can be correlated with electronegativity differences within a series of structurally related molecules. Because no precise theoretical treatments of the complex molecules involved are possible at the present time, the electronegativities serve to quantify vague electrostatic arguments. Electronegativities should not be belittled under such circumstances, but rather accepted as empirical atomic parameters which organize large amounts of experimental data. As an example of the purely empirical uses of electronegativities, consider the covalent radii. The constancy of bond lengths between atoms is well known, for example, the C—C single bond length is about 1.54 A. Empirically, if $r_A = \frac{1}{2}R_e(A_2)$ and $r_B = \frac{1}{2}R_e(B_2)$, then $R_e(AB) \cong r_A + r_B$. This is the additivity rule; r_A and r_B are called *covalent radii* of atoms A and B, respectively. Deviations from the additivity rule are noted for polar bonds. Schomaker and Stevenson suggested the empirical relation

$$R_e(AB) = r_A + r_B - 0.09(x_A - x_B) \quad \text{in angstroms}$$

which works for many single bond lengths. Single-bond covalent radii are given in Table F.2.

TABLE F.2 Electronegativities[a] and Covalent Single-Bond Radii

	x	r (Å)		x	r (Å)
H	2.1	0.36	Ge	2.02	1.22
He	—	—	As	2.20	1.21
Li	0.97	1.23	Se	2.48	1.17
Be*(sp)	1.43	0.89	Br	2.74	1.14
B*(sp^2)	1.91	0.80	Kr	—	—
C*(sp^3)	2.50	0.77	Rb	0.89	—
(sp^2)	2.74	0.74	Sr	0.99	1.91
(sp)	3.20	0.69	Y	1.11	1.62
N* p	2.46	—	Zr	1.22	1.45
(sp^3)	3.51	0.74	Nb	1.23	1.34
(sp^2)	3.74	—	Mo	1.30	1.29
(sp)	4.40	—	Tc	1.36	—
O* p	3.32	0.74	Ru	1.42	—
			Rh	1.45	1.25
(sp^3)	4.66	—	Pd	1.35	1.28
F	4.10	0.72	Ag	1.42	1.53
Ne	—	—	Cd	1.46	1.41
Na	1.01	1.56	In	1.49	1.44
Mg	1.23	1.36	Sn	1.72	1.44
Al	1.47	1.25	Sb	1.82	1.40
Si	1.74	1.17	Te	2.01	1.31
P	2.06	1.10	I	2.21	1.28
S	2.44	1.04	Xe	—	—
Cl	2.83	0.99	Cs	0.86	2.35
Ar	—	—	Ba	0.97	1.98
K	0.91	2.03	La	1.08	1.69
Ca	1.04	1.75	Hf	1.23	1.44
Sc	1.20	1.44	Ta	1.33	1.34
Ti	1.32	1.32	W	1.40	1.30
V	1.45	1.22	Re	1.46	1.28
Cr	1.56	1.17	Os	1.52	1.26
Mn	1.60	1.17	Ir	1.55	—
Fe	1.64	1.17	Pt	1.44	1.26
Co	1.70	1.16	Au	1.42	1.49
Ni	1.75	1.15	Hg	1.44	1.48
Cu	1.75	1.17	Tl	1.44	1.47
Zn	1.66	1.25	Pb	1.55	1.46
Ga	1.82	1.25	Bi	1.67	1.45

[a] Selectively chosen electronegativity values expressed on the Pauling scale. From E. Little and M. Jones, *J. Chem. Educ.* **37**, 231 (1960). Used by permission.
* Pilcher and Skinner, loc. cit.; Mulliken electronegativities converted to the Pauling scale.

G MOLECULAR SYMMETRY

THE IMPORTANT point groups are classified in the Schoenflies notation as follows:

A. *Point Groups with One Axis of Rotation*
  1. C_n ($n = 1, 2, 3, \ldots$). Symmetry elements are $C_n, C_n^2, C_n^3, \ldots, C_n^n = E$.

EXAMPLE   H–O–O–H C_2

  2. C_{nv} ($n = 1, 2, 3, \ldots$). Symmetry elements are C_n and $n\sigma_v$ (n mirror planes containing the C_n axis).

EXAMPLE   N—O $C_{\infty v}$ or any unsymmetrical linear molecule.

  3. C_{nh} ($n = 1, 2, 3, \ldots$). Symmetry elements are C_n and σ_h, a horizontal mirror plane (which is to say, perpendicular to C_n). A center of symmetry and therefore the inversion operation is present for n even. For odd n, $S_n = C_n \sigma_h$ is present (an improper rotation axis).

EXAMPLE   (H)(Cl)C=C(Cl)(H) C_{2h}

B. *Point Groups with One Principal n-fold Axis and n Equivalent Twofold Axes*
  1. D_n ($n = 2, 3, \ldots$). Symmetry elements are C_n and n twofold axes, nC_2', perpendicular to C_n.
  2. D_{nh} ($n = 2, 3, \ldots$). Further symmetry elements are a horizontal mirror plane σ_h and n vertical mirror planes $n\sigma_v$. For even n a center of symmetry is also present.

EXAMPLE H_2, the homonuclear diatomic molecule, or any linear symmetric molecule, $D_{\infty h}$.

3. D_{nd} ($n = 2, 3, \ldots$). Further symmetry elements are n vertical mirror planes $n\sigma_d$ which bisect the angle between C_2' axes.

EXAMPLE "Staggered" C_2H_6, D_{3d}. Also allene

$$H_2=C=C=C\begin{matrix}H\\ \\H\end{matrix}, \quad D_{2d}.$$

C. Point Groups Without a Unique Axis of High Symmetry

In this category are two chemically important point groups of the cubic system: T_d, the regular tetrahedral group and O_h, the regular octahedral group.

T_d The point group of CH_4. Symmetry elements are $4C_3$, $6\sigma_d$, and $3C_2$ (with $3S_4$ colinear).

O_h The point group of SF_6. This is the symmetry of a face-centered cube. The symmetry elements are $3C_4$ (with $3S_4$ and $3C_2$ colinear), $4C_3$ (with $4S_6$ colinear), $6C_2$, $3\sigma_h$, $6\sigma_d$, and a center of symmetry.

H
THE PROBABILITY OF ABSORPTION AND EMISSION OF RADIATION

IN SECTION 6.8 the connection between the Einstein coefficient for absorption and induced emission, B_{ij}, and the experimental molar extinction coefficient is given by equation (6.8.4). In this appendix we will connect B_{ij} to the quantum mechanical properties of the system. The approach is that of time-dependent perturbation theory.

The oscillating electric field $E(t)$ associated with the radiation of frequency ν subjects the system of charged particles to a time-varying perturbation. If e_i is the charge and x_i the coordinate of the ith particle, then the perturbation energy in an electric field $E_x(t)$ parallel to the x axis is

$$\hat{V}(t) = E_x(t) \sum_i e_i x_i \tag{H.1}$$

$$\sum_i e_i x_i = \mu_x \tag{H.2}$$

where μ_x is the component of the system's dipole moment in the x direction. The time dependence of the perturbation is expressible as

$$E_x(t) = E_x^0 [\exp(2\pi i \nu t) + \exp(-2\pi i \nu t)] \tag{H.3}$$

TIME-DEPENDENT PERTURBATION THEORY

Let $\hat{H}_0 \psi_n = E_n \psi_n$ be the time-independent Schrödinger equation for the system. At time $t = 0$ the system is in the zeroth state, that is, ψ_0 is the stationary state wavefunction at time $t = 0$, whereupon the system is subjected to a small perturbation $\hat{V}(t)$. At some later time t we can write the wavefunction in the general form of an expansion in the stationary states of the system (Chapter 2).

$$\Psi(0) = \psi_0 \exp(-iE_0 t/\hbar) \tag{H.4}$$

$$\Psi(t) = \sum_n a_n(t) \psi_n \exp(-iE_n t/\hbar) \tag{H.5}$$

The *probability* that a transition from ψ_0 to another state ψ_m has occurred in the time interval $t = 0$ to t is proportional to $|a_m(t)|^2$. To find $a_m(t)$ we use the time-dependent Schrödinger equation, (2.4.8), with the total Hamiltonian $\hat{H} = \hat{H}_0 + \hat{V}$.

$$i\hbar \frac{\partial \Psi(t)}{\partial t} = (\hat{H}_0 + \hat{V})\Psi(t) \tag{H.6}$$

Substituting (H.5) into (H.6) we obtain

$$i\hbar \sum_n \frac{da_n(t)}{dt} \psi_n \exp\left(-\frac{iE_n t}{\hbar}\right) = \sum_n a_n(t) \hat{V} \psi_n \exp\left(-\frac{iE_n t}{\hbar}\right)$$

which upon multiplication by ψ_m^* and integration over the space coordinates yields

$$\frac{da_m(t)}{dt} = \frac{1}{i\hbar} \sum_n a_n(t) \langle \psi_m, \hat{V} \psi_n \rangle \exp\left[\frac{i(E_m - E_n)t}{\hbar}\right] \quad (H.7)$$

At time $t = 0$, $a_m(0) = 0$ and all other $a_n(0)$ are also zero except $a_0(0)$, which is unity [equation (H.4)]. Thus for times sufficiently small we need only consider the term of (H.7) having $n = 0$.

$$\frac{da_m(t)}{dt} = \frac{1}{i\hbar} \langle \psi_m, \hat{V} \psi_0 \rangle \exp\left[\frac{i(E_m - E_0)t}{\hbar}\right] \quad (H.8)$$

Substituting (H.1), (H.2), and (H.3) into (H.8) we obtain

$$\frac{da_m(t)}{dt} = \frac{1}{i\hbar} \langle \psi_m, \mu_x \psi_0 \rangle E_x^0$$

$$\times \left\{ \exp\left[\frac{i(E_m - E_0 + h\nu)t}{\hbar}\right] + \exp\left[\frac{i(E_m - E_0 - h\nu)t}{\hbar}\right] \right\}$$

Integrating the last equation between $t = 0$ and t

$$a_m(t) = \frac{1}{i\hbar} \langle \psi_m, \mu_x \psi_0 \rangle E_x^0 \left\{ \frac{1 - \exp\left[i(E_m - E_0 + h\nu)t/\hbar\right]}{E_m - E_0 + h\nu} \right.$$

$$\left. + \frac{1 - \exp\left[i(E_m - E_0 - h\nu)t/\hbar\right]}{E_m - E_0 - h\nu} \right\} \quad (H.9)$$

The numerators in (H.9) are both small and of the form $b[1 - \exp(iat)]$ with the square magnitudes $b^2(2 - 2\cos at)$, that is, magnitude 0 to $2b$. Thus, for radiation of arbitrary frequency, a_m is small unless one of the denominators is quite small. This condition is satisfied if $E_m - E_0 - h\nu \approx 0$, or $E_m - E_0 = h\nu$, which is the frequency condition of quantum mechanics for absorption or emission of radiation (Chapter 2).

Neglecting the immaterial first term in (H.9) and integrating out the ν dependence (the integrand is small except for frequencies satisfying the frequency condition $E_m - E_0 \approx h\nu$) we obtain

$$|a_m(t)|^2 = \frac{4\pi^2}{h^2} |\langle \psi_m, \mu_x \psi_0 \rangle|^2 (E_x^0)^2 t \quad (H.10)$$

The probability of transition from state ψ_0 to state ψ_m in time t is proportional to t, the coefficient of t being the *transition probability*. Since the radiation density is $\rho(v) = 6(E_x^0)^2/4\pi$ it follows by comparison with equation (6.8.1) that, in general, the Einstein coefficient for transition between the ith and jth states is

$$B_{ij} = \frac{8\pi^2}{3h^2} |\langle \psi_i, \mu_x \psi_j \rangle|^2$$

In three dimensions equation (6.8.5) is the quantum mechanical expression for B_{ij}.

INDEX

AlH, 175
Alternant hydrocarbons, 309f.
Angular momentum, 14
   intrinsic (spin), 64ff.
   operators, 107, 110
   orbital (atom), 107
   orbital (diatomic), 160
Anharmonicity, 207
Antibonding, 140, 143
Antisymmetry, 66, 69f., 96, 100
Argon, 126
Arrhenius, S., 5
$AsCl_3$, 282
AsF_3, 282
AsH_3, 282
Asymmetric top, 202
Atomic units, 87
Atoms, 81ff.
   atomic theory (Dalton), 3
   atomic physics, 18ff.
   states and spectra, 109
   table of properties, 31, 104
Average values, 62
Avogadro, A., 3

B_2, 163, 207
Back donation, 347
Balmer, J., 27
Be, 105
Be_2, 163
Beer-Lambert Law, 235
BeF_2, BeF_4^{2-}, 7
BeH, 175
BeH_2, 260ff.
Benzene (see C_6H_6)
Berzelius, J. J., 4
BH, 175
BH_3, 270ff.

Blackbody radiation, 22
BN, 179
BO, 181
Bohr, N., 28
Bohr hypothesis, 29
Bohr magneton, 64, 240
Bohr model of atom, 28f., 64
Bohr radius, 31, 87
Bond, 3f.
   banana, 184
   electron pair, 185
   energy, 286
   order, 149, 308
Bonding MO, 140, 153
Born-Oppenheimer approximation, 136, 259, 377
Boundary conditions, 41, 57
Br_2, 169
Butadiene (see C_4H_6)

C_2, 164, 207
Cannizzaro, S., 4
Carbon (spectrum), 118
Cathode ray, 18
Center of mass, 16, 81
Central field approximation, 94
Centrifugal stretching, 210
CH, 175
CH_4, 272f., 286
C_2H_4, 71f., 298, 302, 315
C_2H_6, 286
C_4H_6, 71f., 303f., 315f.
C_6H_6, 5, 217, 238, 310f.
Character, 224
Character tables, 227
   C_{2v}, 222
   D_{3h}, 228
   O_h, 328

Charge density, of atoms, 101
  of CH_4, 274
  of diatomic molecule, 163ff., 169
  of H_2^+, 142
  Hartree-Fock, 126
  of H_2O, 279
  invariance, 161
  of π, σ orbitals, 298
Charge transfer spectra, 346
CH_3CHO, 245
Chemical shift, 247ff.
Cl_2, 169, 187
Class, 220
Classical mechanics, 13
CN, CN^-, 181
CO, CO^+, 181, 187
CO_2, 265f.
Coefficients, 49, 155
Commuting operators, 44, 61, 110, 227
Complementarity, 34
Complete set, 44
Complex, 325f.
Configuration, 68, 71, 99
Configuration interaction, 129
Conjugation, 71, 297f.
Conservation laws, 12
Constants of the motion, 12
Coordination, 6, 325
Coordination number, 325
Correlation, of atoms, 127
  of energy, 127
  of molecules, 186
Coulomb integral, of atoms, 97
  Hartree-Fock, 123
  of helium $1s1s$, 97, 369
  Hückel method, 302
  molecular, 157
  valence bond, 190
Covalent bond, 4, 6, 174f.
Covalent radii, 387f.
Crookes, W., 18
Crystal field theory, 327ff.

Dalton, J., 3, 215
Davisson-Germer experiment, 33
de Broglie relation, 32f.
Degeneracy, 75, 87
Delocalization, 71, 305
Determinant, 360
Determinantal wavefunction, 70, 96, 147

Diamagnetism, 65, 248, 251
Diatomic molecule, 135
Dipole moment, 177, 280f., 310
Dirac, P.A.M., 64
Direct product, 230
d orbitals, 86f., 325f.

Earnshaw's theorem, 7, 145
Eigenfunction, 40, 359
Eigenvalue, 40, 59
Einstein, A., 25f.
Einstein coefficient, 235, 397
Electromagnetic spectrum, 23
Electron(s), charge/mass ratio, 19f.
  correlation (see Correlation)
  paramagnetic resonance, 238f.
  Pi (π), 71, 297ff.
  Sigma (σ), 298ff., 319
  spin, 64
Electronegativity, 385
Electrostatic theorem, 144, 381
Energy, 14f.
Ethylene (see C_2H_4)
Exchange integral, 123, 157, 190
Exclusion principle (see Pauli principle)
Expectation value (see Average values)
Extinction coefficient, 236

F_2, 7, 166f., 169f., 185, 207
Fermi contact interaction, 245
Fluctuation potential, 128
Force constant, 303f.
Formaldehyde (see H_2CO)
Franck-Condon principle, 212
Frankland, F., 4
Free electron orbitals, 71ff.
Free valence, 308
Functions, 355f.

Gamma ray, 23, 198
Gay-Lussac, J., 3
Geminal, 130
g factor, 339ff.
Goudsmit, S. A., 64
Group theory, 219ff.

H, 28, 81ff., 245
H_2, 146ff., 158ff.
H_2^+, 135ff.
Hamiltonian, 39f.

Index **401**

Hamilton's equations, 16
Harmonic oscillator, 203, 231
Hartree-Fock method, 122ff., 156ff., 186ff.
Hartree-Fock-Roothaan method, 156ff., 186ff., 262ff.
HCl, 175
H_2CO, 287ff.
He_2, 149
He_2^+, 149
Heisenberg uncertainty principle, 35, 60
Heitler, W., 8, 188
Helium, 95ff., 102, 244, 373
Hellmann-Feynman theorem, 144, 381
Helmholtz, L., 321
Hermite polynomials, 204
Hermitian operators, 42f., 62f., 359
HF, 175, 192
H_2O, 3, 216ff., 233, 237, 276ff., 282
Hoffmann, R., 317
Homopolar bond, 3, 177
Hooke's law, 203
H_2S, 282
Hückel method, 301ff., 319
Hund's rule, 114
Hybridization, 183, 264, 271, 275
Hyperfine coupling, 246

I_2, 169
Infrared, 23, 198
Ionic bonding, 6, 177ff., 190
Ionization potential, 30, 119f.
Irreducible representations, 224

jj coupling, 114

K_2, 168
Kekulé, A., 4, 215, 310
KF, 177
Kinetic energy, 14
Kossel, W., 6

Ladder operators, 116
Laguerre functions, 86
Langmuir, I., 6, 185
LeBel, J.-A., 5, 215
Legendre equation, 83
Lewis, G. N., 6, 185
Lewis' electron pair theory, 6, 185, 264, 271, 280

Li, 100, 117
Li_2, 170, 207, 212
LiF, 177
Ligand field theory, 325ff., 350
LiH, 175, 177
Localized orbitals, 73, 181, 264ff., 286
London, F., 8, 188
Lyman series, 31

Magnetic moment, 64, 240
Magnetic resonance, 238ff.
Magnetogyric ratio, 240
Matrix, 360
Matter waves, 32, 71
Maxwell, J. C., 21
MgH, 175
Microwaves, 23
Millikan, R. A., 19
Molecular orbital, 71, 137ff.
Moment of inertia, 199
Momentum, 13f., 33, 39, 107f.
Mössbauer spectroscopy, 198
Multiplet, 111
Multiplicity, 111

N_2, 165, 170ff., 180, 182ff., 187, 207
Na_2, 168
NaF, 7, 187
NaH, 175
Newton, I., 3, 13, 21
NF_3, 282
NH, 175
NH_3, 280ff.
NO, 181
NO_2, 282
Normal coordinates, normal modes, 232ff.
Normalization, 42
Nuclear magnetic resonance (NMR), 238ff.
Nuclear spin 238ff.

O_2, 167, 207, 211, 282
Operator, 38ff., 40f., 358
Orbital, 59
Orgel diagram, 340
Orthogonal, 43
Oscillator strength, 237
Overlap, 138ff.

Paramagnetism, 65, 239

Particle in a box, 56ff.
Pauli, W., 64
Pauli principle, 66ff.
Pauling, L., 269, 325
PCl_3, 282
Periodic system of the elements, 102
Perturbation method, 49
PF_3, 282
PH, 175
PH_3, 282
Phase, 92
Photodissociation, 214
Photoelectric effect, 25
Photon, 25
Pi (π) electrons, 71, 297ff., (see also Electrons, Pi)
Planck's constant, 24
Planck's law, 24
Point group, 215ff., 391f.
Polarization, 21
Population analysis, 287ff.
Potential energy, 15
Probability, 55ff., 101f., (see also Charge density)
Projection operator, 225f.

Quantum theory, 22ff.

Radio waves, 23
Rb_2, 168
Reduced mass, 81
Resonance, 241, 267ff.
Rigid rotator, 199
Rotation, 199ff.
Russell-Saunders coupling, 110ff.
Rutherford, E., 28
Rydberg constant, 28
Rydberg-Ritz equation, 27
Rydberg series, 292

S_2, 168
Schrödinger equation, 36ff.
Screening of nucleus, 97, 119f.
Secular equation, 155, 364
Selection rules, 118, 201, 207, 212, 235ff.
Self-consistent field, 124, 158, 374
SH, 175
Shielding coefficient, 247

Sigma (σ), 298ff., 319, (see also Electrons, Sigma)
SiH, 175
Slater determinant, 70, 96
Slater orbitals, 120
SO_2, 282
sp^n hybridization (see Hybridization)
Spectrochemical series, 334
Spherical harmonics, 86, 93, 330
Spin (see Electron, spin or Nuclear spin)
Spin-spin coupling, 247ff.
Strong field complex, 339f.
Sturm-Liouville theory, 41ff.
Symmetry, 151, 215ff., 391

Tanabe-Sugano diagrams, 340
Tetrahedral bonding, 3, 5, 272ff.
Thiele, J., 5
Transition elements, 103, 325ff.
Transition moment, 236

Uhlenbeck, G. E., 64
Ultraviolet, 23, 198
Uncertainty principle, 35

Valence, 4ff.
Valence bond theory, 8, 189ff.
Variational principle, 46ff.
Vectors, 360f.
Vibration, 203f., 231ff.

Walsh diagram, 283f.
Wave equation, 36
Wavefunction, 36
Wavelength, 22, 33
Wave particle duality, 33
Well-behaved function, 357
Werner, A., 6
Wheland-Mann formula, 312
Wolfsberg, M., 321
Wollaston, W. H., 3
Woodward, R. B., 317

x-rays, 23, 33, 198

Zeeman, 241, 244
Zero-point energy, 206